CHIEF ENGINEER
OF THE UNIVERSE

Albert
Einstein

ONE HUNDRED AUTHORS
FOR EINSTEIN

History of Knowledge is a new series from Wiley-VCH.
Internationally acclaimed experts bring new perspectives to the history of knowledge and introduce readers to hitherto unknown worlds of research and its conflictual history. The series is published in cooperation with the Max Planck Institute for the History of Science.

The three volumes *Albert Einstein – Chief Engineer of the Universe:*
Einstein's Life and Work in Context
One Hundred Authors for Einstein
Documents of a Life's Pathway
have been published to accompany the exhibition of the same title: *Albert Einstein – Chief Engineer of the Universe*, which was conceived by the Max Planck Institute for the History of Science on the occasion of The Einstein Year 2005.

Editor	Jürgen Renn
Editorial Team	Sabine Bertram, Lindy Divarci, Tanja Starkowski Wolf-Dieter Mechler, Christoph Lehner (German edition)
Translators	Dieter Brill, Robert Culverhouse, Lindy Divarci, Nancy Joyce, Susan Richter, Ann Robertson
Image Editors	Hartmut Amon, Edith Hirte, Tanja Starkowski
Design/Production	Regelindis Westphal Grafik-Design, Berlin Antonia Becht, Berno Buff, Anja Gersmann, Norbert Lauterbach
Image Editing	Satzinform, Berlin
Print/Binding	NEUNPLUS1 – Verlag + Service GmbH, Berlin

Fritz Thyssen Stiftung
FÜR WISSENSCHAFTSFÖRDERUNG

This publication was made possible due to the kind support of the
Fritz Thyssen Foundation for the Advancement of Science, Cologne

Booktrade edition
ISBN-10:3-527-40574-7
ISBN-13:978-3-527-40574-9

The essay volume accompanying the exhibition *Albert Einstein - Chief Engineer of the Universe* is published in German under the title *Albert Einstein – Ingenieur des Universums. Hundert Autoren für Einstein*
Book trade edition ISBN- 3-527-40579-8

Further titles:
Jürgen Renn (Ed.): *Albert Einstein – Ingenieur des Universums. Einsteins Leben und Werk im Kontext*,
Berlin: WILEY-VCH, 2005.
Book trade edition ISBN-3-527-40573-9
Jürgen Renn (Ed.): *Albert Einstein – Ingenieur des Universums. Einsteins Leben und Werk im Kontext*
together with *Dokumente eines Lebensweges*, two-volume-set. Berlin: WILEY-VCH, 2005.
Book trade edition ISBN-3-527-40569-0
Jürgen Renn (Ed.): *Albert Einstein - Chief Engineer of the Universe. Einstein's Life and Work in Context*
together with *Documents of a Life's Pathway*, two-volume-set. Berlin: WILEY-VCH, 2005.
Book trade edition ISBN-3-527-40571-2

Jürgen Renn (Ed.)

CHIEF ENGINEER
OF THE UNIVERSE

Albert
Einstein

ONE HUNDRED AUTHORS
FOR EINSTEIN

WILEY-
VCH

WILEY-VCH Verlag GmbH & Co. KGaA

Albert Einstein
Chief Engineer of the Universe
Exhibition in the Kronprinzenpalais, Berlin
from 16 May to 30 September 2005
www.einsteinausstellung.de

Organizers

MAX-PLANCK-GESELLSCHAFT

Max-Planck-Society
for the Advancement of Science

MAX PLANCK INSTITUTE
FOR THE HISTORY OF SCIENCE

within the framework of the Einstein Year 2005

A joint initiative of the Federal Government,
science, industry and culture

Design and Implementation

IGLHAUT PARTNER
+

www.iglhaut-partner.de

Sponsors

KULTURSTIFTUNG
DES
BUNDES

 Federal Ministry
of Education
and Research

Stiftung
Deutsche Klassenlotterie Berlin

An Exhibition without Walls –
interactive and online –
with the kind support of the
Heinz Nixdorf Foundation

SIEMENS

 Fritz Thyssen Stiftung
FÜR WISSENSCHAFTSFÖRDERUNG

KTF
THE KLAUS TSCHIRA
FOUNDATION gGMBH

ROBERT BOSCH STIFTUNG

as well as the Wilhelm and Else Heraeus
Foundation, and the Central European
University, Budapest

Exhibition Associates

Deutsches Museum

 HEBREW UNIVERSITY JERUSALEM

UNIVERSITÀ DEGLI STUDI
DI PAVIA

Media Associates

DW-TV
DEUTSCHE WELLE

RUNDFUNK BERLIN-BRANDENBURG

3sat

CONTENTS

Jürgen Renn

Preface

This book is dedicated to the change in our world-view that was brought about by the work of Albert Einstein. It covers essays and aperçus by more than 100 authors, who deal with the meaning of this change, with its history, and with its consequences for the present.

Toward the end of the Weimar Republic a thin, polemic pamphlet was published with the title *100 Authors against Einstein*. It was a kind of harmless harbinger of all the havoc that was to proceed from a Germany that forced Einstein to emigrate, silenced the voices of reason and, through war and holocaust, led the world to the abyss. When this text appeared, the theory of relativity already long enjoyed an acceptance in physicists' circles that placed it above such propagandistic attacks. Nevertheless, in certain sectors of science and outside of science, many found it difficult to accept that a physical theory could change the everyday understanding of such fundamental concepts as space and time – and, indeed, that science was at all capable of influencing our lives so deeply. When this incomprehension allied with anti-Semitism and resentment, their proponents often became willing fellow travelers and collaborators of the Nazis.

The theory of relativity and Einstein's other accomplishments do not require written vindication today either, not least because they have long since conquered our daily routine in other ways, such as through technical achievements like satellite navigation, which would not work without the knowledge supplied by the theory of relativity. Nevertheless, the consequences of the Einsteinian revolution for our understanding of the world have still not been sounded out in full and are hardly common property, even though they affect so many aspects of how we understand our own culture.

It is the joint concern of all the authors who participated in this volume to make a contribution to this understanding. They would like to convey Einstein's influence and work beyond all scientific barriers, and wish to make comprehensible their importance for both our daily life and our worldview.

Often it was simple, everyday questions that served as the point of departure for Einstein's scientific breakthroughs, such as the question of what moves the needle of a compass. The questions Einstein struggled with often enjoyed a long historical tradition, like the question as to whether there is really such a thing as an atom.

In every case they were comprehensible questions, which bring home not only the mysteries of nature, but also the charm of its investigations. A number of these questions still occupy science today. For instance, how does gravity relate to the other forces of nature? There are other questions it must put forward over and again, if science wants to continue to do justice to the moral standards Einstein set: what contribution can science make to the solution of the greatest problems of humanity?

Einstein also placed a high value on the understandability and accessibility of knowledge. His science does not belong in the ivory tower: it affects us all.

This book is dedicated to the same tradition, ultimately rooted in the European Enlightenment. It starts with Federal Chancellor Gerhard Schröder's speech, which opened the Einstein Year 2005 in Germany and placed it in this tradition. Next is an essay by Yehuda Elkana, who reinterprets the tradition of the Enlightenment in Einstein's sense and emphasizes the importance of a reflective understanding of science.

The book closes with a number of written comments by renowned scientists, and also by artists and science policy-makers, expressing their view of Einstein's legacy.

At the core of this volume, however, is a multitude of essays, approaching Einstein's life and work from very different perspectives. Their variety conveys to the reader the pleasure of wandering between Einstein's different worlds, just as he did a century ago.

Their order corresponds to the three main parts of the exhibition "Albert Einstein: Engineer of the Universe", on the occasion of which this book was produced.

The first section concerns the relationship between worldview and knowledge acquisition, that is, the question of how worldviews emerge from insights, and how insights can cause worldviews to collapse.

The essays in this section concern the great forerunners of Einstein, from Giordano Bruno, to Isaac Newton, to Karl Schwarzschild, as well as the prerequisites of the Einsteinian Revolution in developments of mathematics and experimental physics.

The essays in the second section are dedicated to Einstein's life, private subjects like his love for music, and the central scientific aspects of his life's work, such as the concept of inertia and gravitational lensing. The spectrum of topics covered is broad, providing an impression of the vitality of current Einstein research.

The third section deals with topics of contemporary science. This concerns astrophysics and cosmology, just as quantum mechanics and particle physics, but also the political legacy of Einstein's life and work.

The essays address quite controversial issues, which are intended to inspire reflection and, occasionally, contradiction.

The authors include outstanding scientists from all over the world, who have managed to impart an overview of the contemporary state of research, even for readers without any relevant scientific training.

For Einstein himself, such broad knowledge, which he also obtained by reading popular science books, may have been the decisive prerequisite for his breakthrough in 1905.

Today such broad survey knowledge is important for anyone who wants to value the central role of science in our society at its true worth. This volume is an invitation to partake actively in the knowledge it provides us, and not least of all to ensure that the voice of reason is never again silenced.

Such a project could only be realized with the help of partners. A special thanks goes to all the authors who contributed to this book as well as to Dr. Wolf-Dieter Mechler, Dr. Christoph Lehner, Lindy Divarci, Sabine Bertram, and Tanja Starkowski for the editorial work. I would also like to thank all sponsors and partners of the exhibition *Albert Einstein – Chief Engineer of the Universe*, in whose framework this book appears. The scholarly work on this publication was made possible by the generous support of the Fritz Thyssen Foundation for the Advancement of Science, Cologne.

Address at the Launch of the Einstein Year
on 19 January 2005

Honored guests, ladies and gentlemen,
It was almost as if nothing had changed since his lifetime. Last year, when a television network called upon the Germans to elect the greatest German of all time, Albert Einstein was lucky to make it to tenth place. In contrast, when Time magazine chose its man of the twentieth century, the choice was – of course – Einstein. The reasons given at the time: "He was the embodiment of pure intellect [...] the genius among geniuses, who discovered, merely by thinking about it, that the universe was not as it seemed." Einstein is, in fact, considered

Federal Chancellor Gerhard Schröder, 1 February 2005

one of the most important physicists of all time, and is perhaps the most famous scientist of the twentieth century. He revolutionized science with his ideas, which Max Planck characterized as "speculative natural science," and with them changed the world. Young people all over the world have idolized him for his moral integrity and non-conformism – and he has become a genuine cult figure.

Exactly 100 years ago Einstein's groundbreaking findings and revolutionary theories laid the keystone for his later mythical stature. In just a few months he published the foundations of quantum theory, the special theory of relativity, and what is probably the most famous formula in the histo-

ry of science, $E = mc^2$, defining the relationship between mass and energy. Thanks to the many Einstein Year publications, we know that Einstein was an above-average pupil. His outstanding talent for mathematics and natural sciences became apparent at an early age. We are well aware that the legend tells a different story. But the myth that comforted so many schoolchildren and encouraged so many parents to be more understanding has been discredited by historians: Einstein never flunked.

Yet his proverbial genius did not fall from heaven, either. Einstein was curious and inquisitive his whole life long. Even in his youth he acquired comprehensive basic knowledge of the natural sciences by reading everything he could get his hands on. It was, by the way, popular science books that kindled his enthusiasm for the subject. He was also distinguished by an almost childlike craving for discovery and a passionate desire to get to the root of everything. Einstein was not the kind of person to be satisfied with surface appearances. He was a true workaholic when it came to achieving his goals. Others may yet explain Einstein's genius conclusively. But so much is certain: it was a mixture of talent, intuition, attitude, and hard work.

The brilliant scientist apparently liked to work in seclusion, sometimes even as a recluse. He was nevertheless aware of the social interaction he required for his scientific achievements. He needed inspiration from colleagues and friends, he sought exchange in intellectual debate, he referred to previous works and the findings of other scientists, and he was grateful for the possibilities opened up to him by his school and university education. Incidentally, the liberal climate in Switzerland, and later the great freedom at the Kaiser Wilhelm Society were certainly advantageous as well. Again, this is something that should not be forgotten, now of all times. At the height of his fame Einstein repaid the world for his tremendous gifts by imparting knowledge in a comprehensible form. He thus gave us many works, ranging from the book *Über die Spezielle und die Allgemeine Relativitätstheorie* in 1917, which is probably completely over most of our heads, to the lecture on *What Workers Must Know About the Theory of Relativity*. Most of those present here – myself included – probably would have understood that one.

I dare say that the meaning and intelligibility of science has been the subject of discourse and debate throughout history. And today, too, I believe that making science accessible is as relevant as ever, that the potential derived from so doing is tremendous. The first demand on science is to make the fruits of its research, which is supported by public funds, of use to society. This means much more than just transforming the findings of theoretical research into new procedures and products. That's certainly part of it, but by no means all. Another expectation we place on science concerns the responsibility of researchers and scientists to impart their knowledge, just as Einstein did before them, in a thoroughly understandable manner so that others can share in it, and so build bridges between the world of research and schools, today's seats of learning. Such bridges can encourage children and teenagers to develop more enthusiasm about the fascinating field of science.

But demands are also placed on society as a whole, requiring above all a policy that conceives of science and education as a single unit. This is the very core of our innovation offensive, of our research and education policy, which encompasses day care and full-day schools, as well as the funding of top universities. This is also the flip side of our "Agenda 2010". We have to free up resources for innovation by redirecting the subsidies of the past into investment in the future. Our economic future lies in the talents and gifts of our people. One of the great tasks of society as a whole, and

of politics in particular, is to foster these skills through education and training, but also to stretch them by tapping their creative potential to the full.

The media, too, should fulfill their programming obligations far better then they have to date, by reporting on the findings of scientific research in yet

more popular shows. I know that a few positive starts have been made, but these must be followed up on and consolidated. For instance, providing clear and intelligible reports on nanotechnology, biotechnology, medical technology and particle physics would be well worth the effort. Only by working together will we be able to generate in our country a climate of openness, learning and innovation for as many people as possible, which

"E" for Einstein: Sculptures over two meters high advertise the exhibition *Albert Einstein: Engineer of the Universe* in public spaces. In front of the Federal Chancellery, the first "E" was erected on 1 February 2005, bearing information on Einstein's life and activity in Berlin

we expressly want to do, because it is imperative for our future. Any country as poor in raw materials as Germany must cultivate such a climate in this radically changing world.

If we are to hold our own in the 21st century, not only nationally, but in the international arena as well, we must develop a new culture of science, a culture in the tradition of Goethe, Schiller, Humboldt and indeed Einstein, which is however

Albert Einstein as a young man, 4 November 1898

adapted to today's demands on a modern information society. This, of course, relates directly to the perception of science and the value we place on it in this society, and the degree to which science is taken for granted in everyday life.

Let me illustrate what I am getting at with a sports analogy, sport being another important part of our culture. Many of the older members of this audience can probably list all the players who helped win the World Cup in Berne in 1954. But could we also name eleven German scientists who have won the Nobel Prize since World War II? I think anyone who is not a physicist, medical researcher or chemist would find it difficult to do so, and that is something we must set out to change.

When we talk about what science is, reference is often made to the irreconcilable nature of freedom and responsibility. Although this is a popular dichotomy, I don't believe it is entirely accurate. Freedom and responsibility are neither opposites, nor do they simply complement each other. Freedom at its best includes responsibility. The Basic Law of our country guarantees the freedom of science, research and teaching. The example of Einstein makes clear that his epochal achievements were only made possible by this very freedom of thought and research, as well as freedom from state regulation and economic dependence, which is also highly relevant today.

This is why it is important to remember, especially at a time when debate is closely focussed on economic utility, that only theoretical research conducted in freedom and independence, research without constraints and yet not boundless, is able to provide new knowledge – the most important raw material for future prosperity. By this we mean the kind of research performed at the Max Planck Society and the other research organizations with strong international reputations, which have earned a prestige we can use just as well domestically. I believe this country can also be proud of its scientific achievements. Yet a risk remains. We cannot eliminate the double-edged nature of scientific research, and we should not even attempt to do so. The results of research can also be directed against its original purpose. We must remain aware of this danger, as did Einstein, who once said: "What good is a formula if it does not keep people from killing each other?"

I have spoken of a new culture of science. We want to use the Einstein Year to help this culture achieve its breakthrough. I have the impression that our chances of doing so are fair. For a whole year we have talked about the necessity of structural reform in our country – not without reason – and we sometimes had difficulty explaining why we are pushing on with it. But the answer is clear: we need it not only to change and repair the structures, but also in order to mobilize resources for science, for education, for training and for the ensuing developments. Research in Germany is not

this. All surveys show that an overwhelming number of Germans are not only open to technical progress, but that many are true technophiles. This is certainly one reason for us to stop and think before railing against an allegedly widespread technophobia. As we know, many problems in this world of ours, from hygiene, medical care,

Einstein and the Chancellery

only conducted at the highest level, but also still enjoys a leading position in the world. We can be proud of this and should let the scientists know we are. But that is not enough. We must say it much more often and make others, outside circles like these, conscious of it as well.

And to this I would like to add: A new culture of science means that we have to scrutinize our perspective on science. For me this means stressing even more strongly the enormous potential of scientific and technological progress, without neglecting the risks. It is important to place the right emphases. On occasion I doubt that we do

the environment to a life in dignity for every individual, cannot be permanently solved without further scientific development and without technological progress.

Of course, it must be our concern to transform scientific findings into innovative products and methods more quickly than has so far been the case. But at the same time theoretical research needs the assurance that it can be pursued without any pressure for direct applicability, and without the obligation to constantly prove its utility. When Einstein wrote down the basic theoretical equation for the laser in 1916, he did not have the slightest inkling of the innovations that would follow from it five decades later. He wanted to understand nature, and in particular the emission of light. This and nothing more was his immediate concern. That so infinitely much has become of it

with France and became involved in the campaign for a new, democratic beginning in Germany. To the end Einstein, subjected time and again to the most vicious anti-Semitic baiting, fought against the gathering strength of National Socialism and for the defense of democracy. This serves not only as a reminder to us, but also as a call for action. We must always pay attention to this part of Albert Einstein's thought, not only during the Einstein Year, but also as a task set by this commemoration for the future. It can only be hoped that in an anniversary year like this many intellectuals and scientists will also work to promote this form of involvement in society. The voice of those who time and again advance science through their achievements, in whatever discipline, is a voice that also is

Albert Einstein in his study in Princeton (photo: Eric Schaal)

may be traced back to him, but did not directly motivate him in his research. It is not only as a great scientist that Albert Einstein deserves to be remembered and emulated. For him responsibility was not something that touched his professional life alone. Einstein felt just as much responsibility as an intellectual in society and for society.
The Einstein Year should also serve to honor and preserve his political legacy. Einstein was a passionate democrat with a highly developed sense of social justice. He was an unshakeable pacifist, who time and again raised his voice against nationalism and racism. After the outbreak of World War I, when large parts of the German intelligentsia joined in the jingoist patriotism taking hold around them, Einstein published with Georg Friedrich Nicolai and Wilhelm Foerster an anti-war manifesto entitled "Appeal to the Europeans." After the war he soon pushed for reconciliation

needed in social and political discourse – nationally and internationally. This, too, we can learn from the life of Albert Einstein. I would like those who work in science, and do so with their entire heart, to participate actively in political debates more than they have in the past. I also hope they will increasingly intervene – in the truest sense of the word – in the social discourse, and not only to strengthen their direct area of expertise or to demand resources, but also to discuss the state of our society.
I have already addressed Albert Einstein the scientist, the humanist and the moralist. But there is another facet of this altogether extraordinary person that I believe is of special importance. In my estimation he is also a symbolic figure because he was a global citizen. As a scientist, internationality and the cosmopolitanism associated with this were a matter of course. Today, more than in

his day, science knows no political borders and thus cannot recognize any such frontiers. Science manifests itself as a process extending across the world, in which cooperation and competition generally complement each other fruitfully. The World Wide Web, with its present great utility and even greater potential, originated not least from scientists' need to communicate with each other freely across borders. But Einstein's idea – or better, vision – of internationality extended far beyond networked collaboration among researchers. His ideal was a global domestic policy, run by a world government on the basis of binding regulations and binding laws. As we know, his dream of a world government has not been fulfilled and – we must bear in mind – certainly will not be fulfilled any time soon. But with the United Nations and international law, the community of states is equipped with the tools to resolve international conflicts, overcome nationalism and to guarantee freedom and stability in the world, because with these institutions conflicts can be resolved diplomatically, politically, and thus peacefully.

I believe Einstein would agree when I say that this is why we may not weaken the United Nations through the proposed reforms, but must rather strengthen it yet further, so that it can live up to its unique role in an ever more interlinked world. I think Einstein would welcome a project such as is currently being pursued by the Secretary-General of the United Nations.

Of course, for all his genius and impressive achievements, Einstein was also a human with contradictions and weaknesses. And why should he not have had such contradictions and weaknesses, too? We will certainly hear more about these as the year continues. But in closing, I wish to call attention once more to my priority in this anniversary year. In the next months we should all seize the opportunity to let what I called the "new culture of science" take root in our country. This year should be the occasion for undertaking the serious and enduring attempt – to adapt the words of the master himself – to ensure that by the end of the year even more people like Einstein, and that in any case a few more understand him.

Yehuda Elkana

Einstein's Legacy

Opening lecture
for Germany's
Einstein Year, on
19 January 2005,
at the *Deutsches
Historisches
Museum*, Berlin

Germany has chosen to dedicate this year, 2005, to Albert Einstein on the 100th anniversary of his 'annus mirabilis'. It is dedicated to the man Einstein, a German and a Jew who had to leave Germany because of the Nazis, never to return – a sheer accident that he did not perish in the Holocaust; it is also dedicated to his scientific œuvre, and to his humanistic, political and science-political legacy. It is a courageous and noble decision in which *Wissenschaft*, *Kultur* and *Wirtschaft* participate. It is courageous because Einstein was a very independent critical spirit, who claimed not to belong to any nation or culture, although he was very consciously a Jew. Thus, this is a major opportunity and not less so also a major challenge.

Out of the myriad of themes one could choose for discussion, I have decided to choose one central theme – that of *Befreiung* – and to follow in a brief survey the implications of this attitude in many walks of life, from science to politics.

Einstein was a *Freigeist*, and his self-appointed, conscious task was to be a liberator – a *Befreier*. In this he continued a great German cultural tradition established by Kant, Goethe, and simultaneously with Einstein, by Ernst Cassirer.

Einstein was a *Befreier* from all conventions, constraints, limitations – from everything that might be in the way of a free rein of the imagination (*Fantasie*).

Einstein's all-important five papers, all written in the period of a few months in 1905, while he was a clerk in the patent office in Bern, and thus not part of a university, were the first clear demonstration of using his unfettered imagination.

For him no established Truth looked sacrosanct; he started by challenging the very foundation of successful modern science, namely Newtonian Mechanics. And already then he showed that creative thinking could proceed liberated from any support, be it experimental or even mathematical: it was a pure conceptual flight of the imagination. A few years later, after he had been

Princeton,
9 February 1950,
a drawing by
Josef Scharl

invited to Berlin, the First World War erupted, and with it came a popular support for the war which bordered on mass hysteria – a 'madness' as Einstein described it – supported fully by the leaders of the academic and cultural elite. While 93 leading academics signed a war-supporting

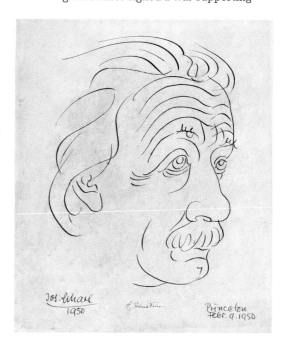

appeal *An die Kulturwelt*, Einstein again showed his independence from any constraints or social pressures, by being one of only four who signed an *Aufruf an die Europäer* deeply disapproving of the war.

As against the entire scientific establishment, Einstein thought and taught that there was no such thing as a scientific method, thus liberating scientific work from a strongly constricting pedagogical principle, which then, like very often today, cut the wings of imagination of many a budding creative scientist, very often crushing the inherent curiosity and potential love for science. For many a young person today such a constraint results in a turning away from science and technology altogether.

Einstein was not an anarchist, and he did not think that in science, or for that matter in politics, 'anything goes'. Imagination must be given free rein, but in due course the resulting theoretical edifice must be subjected to the control of the senses and the experimental result. That was an integral part of his realism, his belief that out there a real world existed independent of, and uninfluenced by, human intervention or even knowledge.

Einstein freed science and philosophy from the ruling positivism of the 19th and early 20th centuries. Positivism was a deep cultural commitment to facts and to the primacy of facts over theory, and to the belief that facts need not be interpreted, that they are independent of any context. Yet the issue is very relevant today and for all of us: we are living in a world where facts, political facts, are not heeded. Think only what such an attitude means when we are dealing with peace and war and the lives of millions of people. Einstein's understanding of himself was that he had aspired all his life – and succeeded – to liberate himself from what he called 'the merely personal'. He contemplated the physical world at large – as well as the social world – uninfluenced by previous theory, by any dogma or by self-interest, with absolute, fearless courage and serenity. Almost a hundred years later, after two world wars, after Hitler, Coventry, Dresden, Hiroshima, Gulag, we cannot afford this olympic distance, irrespective of whether we believe in the immediate efficacy of our actions. Max Brod, who had met Einstein in Prague, published a biography of Kepler modeled on Einstein. It bordered on a caricature of the cold scientist who obsessively cares only for his theories. If we go beyond Einstein in our demands on ourselves and our age, we still follow in Einstein's footsteps when we look courageously in the face of the historical mirror and, free of conventions, we make normative claims.

I would not have emphasized this need to go beyond Einstein, while learning from him, had it not been so relevant for our times: we live at a time when those with strong right-wing social and political attitudes, are full of energy for action, while the center-liberal academic and intellectual circles have almost abdicated. This is strongly the case in America, but it is beginning to be felt in Europe too. In or-

..Wichtig ist, dass man nicht aufhört zu fragen."

Albert Einstein

On the red carpet in the Schlüterhof of the German Historical Museum: At the opening ceremony of the Einstein Year 2005 on 19 January, the guests are inspired with a quote by Einstein

der to overcome this apathy, or feeling of help-
lessness, it is not enough to think through rationally what should take place, while personally
continuing our routine daily lives; we must feel
it through and act on the normative demand of
'what follows'. There is a need for the value-free
scholar to yield to the actively 'caring scholar'.
This is of paramount social and political importance.

The Federal Government of Germany called for
a culture of innovation, and for the creation of a
much more creative and efficient higher educa-

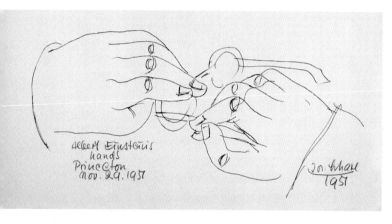

*Albert Einstein's
hands, Princeton,
29 November 1951,
a drawing by
Josef Scharl*

tion system in Germany, and even for a new social
contract between *Wissenschaft, Wirtschaft* and
Gewerkschaften to create 'partners for innovation'. This call is activist in its very formulation,
and not a placid reliance on the forces of the
market to do the job. It is certainly what Einstein
would have endorsed in general and in detail.
However, here too, in the spirit of this legacy, we
must go beyond what Einstein could or would
have thought about.

The quest for innovation must be liberated from
being couched in the merely actual; it needs planning on a much longer time-scale than the usual
horizon of industry and/or politics. Globalization,
the acute problems of poverty, socially spreading
diseases like HIV/AIDS, – which all thrive on
acute social and economic inequality and poverty
– need long-term rethinking way beyond the intel-

lectual scope that the two-hundred year long tradition of Enlightenment thinking has presented us
with. Einstein had the right intuitions, but not the
conceptual tools to show us the way to rethink
our heritage. This rethinking has to face a world
where none of our convenient dichotomies hold:
the precise separation between Church and State;
the sharp distinction between nature and culture;
a clear distinction between the local, and a strong
quest for the universal neglecting the local; mis-
reading the local Western universals for the gen-
uinely global. All this is gone and we must cope
with the problems as we try to repair the ship
of our conceptual tool-kit while floating in mid-
ocean (following the brilliant metaphor of Otto
Neurath). And this can be achieved only – and
this Einstein knew in depth – if our knowledge
of the world is based on reflection and is contex-
tualized. When broken down this means:

The quest for innovation must be liberated from
the constraining, and, in the final account, shortsighted separation between basic and applied research. Einstein's own work amply demonstrates
the mutual interdependence of basic research
and applied research. Industry used to know this
when it fared economically better. Now, under
economic constraints, it forgets its own glorious
achievements which mostly followed from not
separating basic from applied research. The area
of study, which aptly catches these historical developments and what follows from them, could
be called 'political epistemology of research'.

Not instead of being better funded, but in addition
to it, the universities have to rethink the meaning
and process of doctoral studies even in the natural
sciences, not to speak of the social sciences and
the humanities. What Einstein teaches us is that
doing science cannot be separated from reflection
upon science, by the same scientist and while
doing science; it is not enough that philosophers
of science be responsible for epistemology, while
scientists stop being engaged in epistemology, or,

at best engage in it after their retirement, when they can no longer influence their own creation of new knowledge. Let us remember that creating new knowledge, and at the same time continuously contextualizing it, was part and parcel of a rich

much attention given to the reintroduction of reflection/epistemology into the training of doctoral students, and little attention paid to rebuilding the reflective disciplines. Indeed History and Philosophy of science were latecomers to Germany. Even today, German universities are abolishing Chairs in History of Science to their, and the country's, own peril.

If Europe and Germany will not take upon themselves this part of Einstein's legacy, it will boomerang back to science, universities, and indeed on innovation.

Parallel to the need by the new partnership to rethink the public understanding of science, energy must continuously be spent on expanding the 'open access' to knowledge movement, which is a necessary prerequisite to be able to act globally, and to counteract widespread poverty in the world by empowering the poor with usable know-ledge, and giving them the knowledge-based tools for 'aspiring' and finding their 'voice'.

Much has been said recently – but often channeled in the wrong direction – about 'elite education'. The bad name of 'elite' stems from the historical concept of hereditary elites, enjoying unjustified social status and financial privileges. In Einstein's spirit, an elite is constituted by individuals who know how to strive for ever higher, self-imposed standards of quality and who achieve beyond what their background would have pushed them to achieve. Through its overemphasis on democratic account-

Yehuda Elkana and the subject of his lecture

European and German tradition before Nazi times. All great thinkers, in all branches of knowledge, tended to reflect publicly about their own work. Yet the Nazi regime eliminated all that. Some of this tradition migrated to, and flourished for a while, in America. After the war, Europe, but mainly Germany, consciously rebuilt first of all the positive areas of knowledge. There was not

ability in the name of transparency, the present social system stands in the way of the emergence of such a self-appointed elite. Not that accountability and transparency are not needed, but elites must be free to exercise judgment – it is an essential part of the task of an elite – and this task is by definition non-democratic. Scepticism against authority is a prerequisite for having elite universities. That is what is meant by the repeated em-

environment". ("*Das Misstrauen gegen jede Art Autorität [...] eine skeptische Einstellung gegen die Überzeugungen, welche in der jeweiligen sozialen Umwelt lebendig waren.*")
This attitude is important in the liberation of science from any specific method (as referred to above), but also in his politics, which to many seemed naïve. It was anything but naïve. I would characterize it as dialectical pragmatism.

The walls on all sides of the Schlüterhof glow with Einstein's handwriting and the famous formula $E = mc^2$

phasis that universities – elite universities – must be meritocratic.
These were aspects of Einstein's role as 'liberator'. Actually, all exemplify that liberation from authority – any authority – is an important part of Einstein's legacy. Already in 1901 he had said in a letter, "German worship for authority (*Autoritätsdusel*) [...] is the greatest enemy of truth".
Later, when writing his intellectual autobiography for the Schilpp volume in 1946, he described his characteristics as: "Suspicion against every kind of authority [... a skeptical attitude towards the convictions which were alive in any specific social

Einstein, in 1939, wrote to President Roosevelt warning him that Germany might be working on the development of an atomic bomb and therefore America should engage in research on it. Then, after Hiroshima, he repeatedly urged nuclear disarmament – this was neither unreasonable, nor naïve; it focused on the essential at each point of time. The same is true when he simultaneously supported the establishment of a Zionist state, and warned against emerging strong nationalistic tendencies among the Zionists. Both points were focusing on the absolutely essential.

If you permit me one personal remark: when I, as a Holocaust survivor, enjoy the warm reception by German democracy today, I am following the spirit of Einstein. I love Israel and feel a deep loyalty towards it, and hope for its continued existence, and at the same time I warn against strong nationalist tendencies which may endanger the democratic character of the state. This attitude is in the same spirit. And when I publicly called for 'The need to forget" against the political manipulation of the Holocaust in Israel (by right-wing and left-wing governments equally), and at the same time I oppose tendencies by some in Germany who wish to 'close the chapter' of the Holocaust, I do not think that I am being inconsistent. Rather, I concentrate on the real issue in each context.

Summarizing a quick tour, I have tried to derive from Einstein's life and thoughts, guidance for a love of knowledge and science, for democratic internationalism, for a science policy which encourages long-term innovation, for social and political engagement rooted in enlightened social partnership between the main pillars of society, and for a free-ranging imagination which – accompanied by reflection, and relying on an all-persuasive critical spirit – will foster love of science, technology and innovation among people.

Very abbreviated version of the lecture manuscript

Jürgen Renn
Ulf von Rauchhaupt

In the Laboratory of Knowledge

Ideas and concepts are, even in the sciences, subject to historical processes. They thereby change, react with each other, and occasionally enter conflicts and combustions, in which the elements of knowledge are newly composed. The mixture of ideas from classical physics encountered at the end of the 19th century three such combustions. Cooled by philosophically and historically grounded critique, the foundations of modern physics emerged from them in the year 1905 through Einstein's revolutionary papers.

These groundbreaking publications were not revelations of a lone mastermind. In fact he culled from a catalytic concoction of ideas that had brewed together over the course of centuries. Einstein apparently has something to offer to almost everyone. In the hundredth anniversary of his "miracle year," it is tempting to distill from his myth a kind of magic potion, from which new innovative power could be gained. But how exactly did it come to the unparalleled innovative spurt, in which the third-class technical expert at the Patent Office in Bern turned classical physics on its head? There is no shortage of recipes for Einstein's creativity. Some declare that his brain was the main ingredient, some his supposedly childlike mind, some his relationship to women, and some even believe that Einstein's genius proves that fantasy is ultimately more important than knowledge. From the perspective of a history of knowledge, such home-made recipes are worth little. The conditions of the Einsteinian revolution cannot be grasped without taking a look at the entire intellectual laboratory from which the new ideas of 1905 emerged. A simple glance makes apparent that Einstein's innovation was not a mere piece of alchemy following a simple recipe, but rather the result of a transformation of knowledge, in which the preservation of knowledge handed down was at least as important as its changes. So what are the roots of the revolution of 1905? The concepts and ideas that it changed go back to

Democritus (around 460–370 B.C.) and his teacher Leucippus traced all phenomena back to the smallest indivisible particles, which move in an otherwise empty space. This was to solve a main question of pre-Socratic philosophy: how the conviction that only what is unchangeable can really exist could be reconciled with the changeability of phenomena. Atomism solved this question by instilling the quality of imperishability in the atoms and the quality of changeability in their arrangement. How this arrangement is to be determined remained an open question. That was one of the reasons why, well into the Modern age, atoms had such a hard time holding their ground against Aristotelianism.

Plato (427–348 B.C.), in contrast to his pupil Aristotle, was of the opinion that concrete phenomena are not the key to achieving the knowledge of the world. Phenomena are but the image of ideal principles, and these principles would lead to the goal of knowledge. For him, the principles behind the things of nature were fundamentally of mathematical nature. Another element of Plato's thought, which was to exert influence on science for centuries, was the critique of the view that what we see – or imagine in the categories trained by everyday experience – is what is actually true.

Aristotle (384–322 B.C.) was the son of a physician; as a natural scientist, he proceeded from concrete phenomena. He categorized these in a painstakingly elaborated system of nature structured by means of philosophical concepts, which remained the model for scientific knowledge well into the modern age. His theory of nature unifies a systematization of human everyday experience with a synthesis of various pre-Socratic attempts to reveal the unity behind the variety of phenomena. In contrast to such thinkers as Democritus, however, this uniform original basis of everything material is not of a discrete nature, but rather a continuum that can be divided at will.

Pages 24–25
Joseph Wright of
Derby (1734–1797)
"An Experiment
on a Bird in the
Air Pump," 1768

Nicolaus Copernicus (1473–1543) was trained as a doctor and church jurist; his main profession was as an administrative official of a Polish Bishopric. His idea that the earth, along with the other planets, could revolve around the sun (rather than being the center of all heavenly motions) had the actual goal of simplifying and improving the prediction of the planetary positions. Above all it was to fulfill the Aristotelian demand for the circularity and uniformity of heavenly motions, which had been lost through wily adjustments made to the predominant geocentric model that were necessitated by actual observations.

Galileo Galilei (1564–1642), like Descartes, sought a unified view of natural phenomena. In his attempt to develop a model to counter that of Aristotle he made use of atomistic conceptions. He developed a new mechanics, whose empirical foundation was the practical knowledge of the engineers of his day. Galileo was presumably the first person to direct the newly developed telescope toward the sky. There he discovered the craters of the Moon as well as the four large moons of Jupiter, which he interpreted as evidence for Copernicus' heliocentric model of the world. The physical justification of the heliocentric model demanded an explanation for why we usually do not sense the motion of the Earth around itself and around the Sun. From Galileo's considerations emerged what became the relativity principle of classical mechanics: If an experimental arrangement moves homogeneously and in a straight line relative to a stationary laboratory, then the processes within it will take place exactly as in the stationary laboratory. Newton later granted this idea the status of a basic principle.

the beginnings of science in antiquity. They are concepts which come from that intuitive everyday knowledge with which we orient ourselves practically in our daily life – such as those which our language uses to designate the changes of bodies and the interactions between them. The first of these concepts originate from the systems of thought of the Greek philosophers before Socrates. Their systems were set up to compete with the myths of the gods, from which they broke by universalizing processes of nature: They singled out one process of nature, from which they derived everything else. Thales of Miletus, for instance, traced everything back to water and its transformations; his compatriot Anaximenes, to the compression and dilution of air. The arbitrariness of explanations of this kind was striking at even an early stage. Parmenides of Elea attempted to elude it by replacing the universalization of natural processes with a reflection on the language with which we speak about reality. Thus Parmenides arrived at the opinion that "being" was plainly indestructible, unique and of highest simplicity. The great theories of nature in Antiquity, atomism, Platonism and Aristotelianism, picked up on the pre-Socratic approaches at explanation and developed them further. But they, too, essentially remained dependent on a reflection of intuitive everyday knowledge as their empirical foundation. Thus atomism evolved from the shift of an intuitive model of thought into the microworld: Democritus, its main representative, conceived of atoms as small, indestructible units of being, which moved in an absolutely empty space. With this view atomism appeared to defy the critique of Parmenides. Yet this could also be conceived of as a epistemological critique. That was the case for Plato, who declared that the only legitimate point of departure for obtaining knowledge was a world of ideas, from which he inferred the existence of abstract concepts in mathematics. Aristotle, in contrast, put together a core inven-

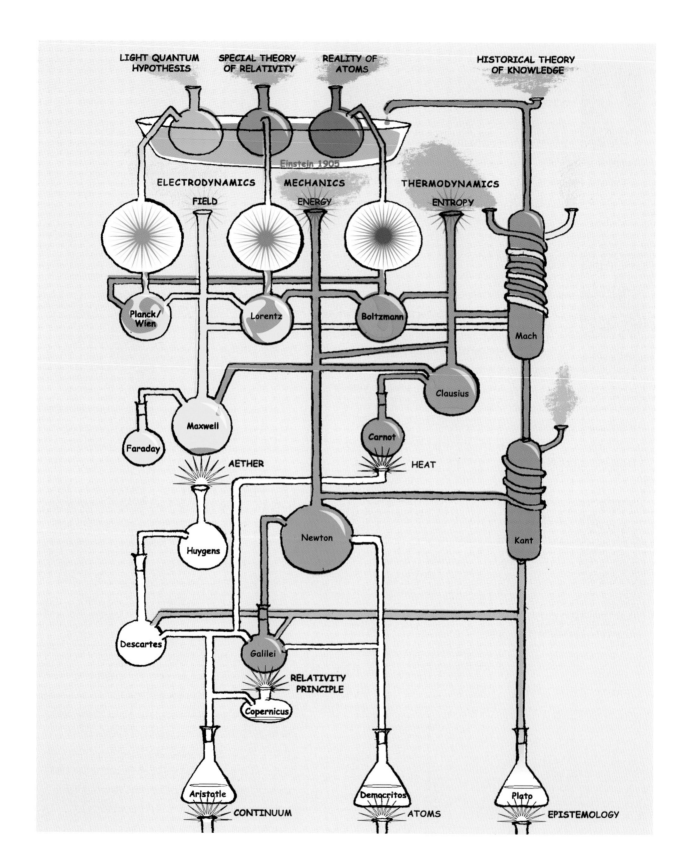

René Descartes (1596–1650), like all preclassical natural scientists, took from the inventory of ideas of the Ancient natural theories. In spite of this, he opposed Aristotle's natural philosophy. Like Newton later, he attempted to overcome the Aristotelian separation between earthly and heavenly phenomena and to understand them on the same conceptual foundation. In contrast to Newton, he assumed that matter was continuously dispersed.

Isaac Newton (1643–1727) is considered the founder of mathematical physics. In his influential main work he succeeded in tracing back the force that makes an apple fall from a tree and the force that links the Moon to the Earth – two phenomena regarded as completely different until that time – to the very same mechanical principles. Newton assumed the existence of an absolute space, against the background of which all motion could be described, as well as universal time, flowing synchronously everywhere in the universe. Further, he believed that changes in forces had immediate effects in the entire universe, rather than propagating from their source with finite speed.

Immanuel Kant (1724–1804) impressed by Newton's achievements, performed scientific research himself. In his main philosophical work, which concerns the question of whether certain knowledge is possible, he defined Newton's concepts of an absolute space and a universal time as prerequisites for our being able to perceive at all. This removed time and space from the sphere of that which could be investigated empirically and made it more difficult for later generations to overcome the Newtonian idea of time and space. On the other hand, Kant's critique of the idea that things are "in themselves" the way they appear opened up a way to call into question even the most fundamental concepts of our everyday intuition.

tory from the rich store of concepts of intuitive everyday knowledge, whose universal validity he attempted to ensure through a comprehensive theoretical system. These are concepts like "matter"

Dialogue on the origin of gravity

Dialogue on the origin of cosmology

and "form," whose relationship is located at the center of his conception of physical reality. Aristotle had modeled his concept on the experience of natural or artificial processes of creation. He divided the physical world into different concentric spheres and differentiated between the heavenly aether and four earthly elements. These he portrayed as continua which fill up everything, such that there is no empty space anywhere.

At the beginning of the Modern Age, Copernicus proposed an interpretation of heavenly bodies which brought new order to the knowledge handed down from earlier periods. This was not a bolt from the blue, however, but occurred on the basis of a long tradition of observing and calculating astronomy. His relocating the center of the heavenly motions from the Earth to the Sun was soon to have effects for terrestrial physics as well. For the Copernican system put into question not only the Aristotelian view of the cosmos as an "onion" of concentric spheres, but was also a challenge to the conventional understanding of motions. In particular, the differentiation between true and apparent movements became an explosive question, whose study ultimately led to the introduction of the modern concept of the relativity of motion. However, modern science was not only the result of the long-term processes of accumulating and transforming knowledge. It was also a consequence of the rise of new elites, who were able to employ the potential of this knowledge to master practical challenges. For them, the Aristotelian worldview handed down from Antiquity represented a framework that was gradually transformed by the newly acquired knowledge into a new worldview, the worldview of classical mechanics. But for this, too, intuitive everyday knowledge remained the decisive foundation. Both the modern atomism of Galileo and Newton and the ideas of a space-filling continuum championed by Descartes and Huygens were linked directly to conceptions of Antiquity. In the 19th century this ultimately became the all-pervading aether as the carrier of light and electromagnetic phenomena.

Hence, classical mechanics appeared to be the foundation of all other natural sciences, if not of all sciences. This is especially true for its concepts of space and time, which were declared by philosophers like Kant to be the steadfast foundation of human thinking. Over the course of the 19th

Christiaan Huygens (1629–1695), the son of a Dutch poet who was friends with René Descartes, was influenced strongly by Cartesian ideas. For instance, Huygens conceived of light as continuous wave – in complete contrast to Newton, for whom light was composed of particles. Huygens' understanding of light as a wave proved extraordinarily fruitful for optics, but required the assumption of an "aether," a ubiquitous medium whose vibrations could be regarded as these waves.

Nicolas Léonard Sadie Carnot (1796–1832) was originally a military engineer. Proceeding from the problem of how to improve the efficiency of the newly invented steam engine, he created a theory of all possible heat engines. Carnot held heat to be a kind of substance. He was able to show that two reservoirs of material of different temperature are always required to convert heat into mechanical work and that the maximum efficiency possible is dependent only on the temperature difference.

Michael Faraday (1791–1867) began his career as an assistant in a chemical laboratory and had very little preparatory training in Newton's mathematical physics. Thus he was all the more unencumbered to develop from the experiments he performed with magnets and electricity a conception of those forces which were irreconcilable with the Newtonian concept of force. According to Faraday these forces do not always operate along the lines connecting the power source. Moreover, changes in these forces were propagated only with a finite velocity. Faraday's concept of electric and magnetic fields was later to be brought into mathematical form by Maxwell.

Rudolf Clausius (1822–1888) was one of the fathers of the modern theory of heat, also known as thermodynamics. He generalized the transformability of heat into mechanical work – in a steam engine, for instance – to a conception in which work and heat are merely different forms of energy. He formulated the conservation of energy in such processes as the first law of thermodynamics. However, if only the first law were true, it would still be conceivable that an object simply cools off and drives an engine with the temperature difference. As the second law of thermodynamics Clausius put forward Sadie Carnot's insight that this is impossible, because a temperature gradient is always required to perform mechanical work. In order to formulate this in mechanical terms, he introduced a new physical quantity, "entropy." Thus the second law reads: The entropy of an isolated quantity of matter can never decrease in a closed system.

James Clerk Maxwell (1831–1879) did for the theory of electric and magnetic phenomena more or less that what Isaac Newton had done for mechanics: The Scotsman combined the existing observations and concepts (above all those by Michael Faraday) to construct a comprehensive mathematical theory, of classical electrodynamics. This is based on four basic equations, which also yield waves with a given velocity of propagation. Maxwell conceived of these electromagnetic waves as vibrations in the aether, just as sound waves are vibrations in air. By recognizing that light, too, must be of this nature, he made optics, an independent field until that time, part of electrodynamics.

Today we know **Ernst Mach** (1838–1916) primarily from the unit "Mach" which expresses flight velocities as multiples of the speed of sound. Yet the Austrian physicist was also an influential philosopher of science. Mach also occupied himself quite thoroughly with the history of physics. The insight that the basic concepts used to describe nature have grown over history made him a critical epistemologist, who refused to allow any concepts that did not refer to directly observable phenomena. Therefore he also disputed the existence of atoms. Einstein's critical scrutiny of concepts like "atom" and "aether" can be traced back in great part to Mach's influence.

century, the mathematical concepts of space and time penetrated everyday life ever further, through economic and technical developments such as the standardization of measurement systems. At the same time, classical physics increasingly used cognitive resources from the area of practical knowledge, which was ever more expanded by technologies no longer purely mechanical – such as the steam engine. Researching engineers like Sadie Carnot played an important role as well, examining such technologies primarily from the perspective of benefiting knowledge. Inversely, science became increasingly useful. Over the course of the 19th century it grew to become a large-scale business, where there was no longer room for arguments about world systems, but where knowledge was produced within precisely delineated disciplines in accordance with each of these disciplines' methods. The concepts of mechanics remained the common point of reference, and initially became even more important as a result of attempts to explain even non-mechanic interactions like electricity, heat and light through mechanical models. Following Michael Faraday's considerations of lines of electric and magnetic force and in the tradition of Huygens' optics, James Clerk Maxwell attempted to explain electromagnetic phenomena, including light, using a mechanical aether. Mechanical models were also taken as the foundation to explain heat phenomena, as in as the idea that heat is nothing other than the motion of atoms.

With the growth of knowledge, however, it became more and more difficult to formulate consistent mechanical theories. Increasingly, aether and atoms suffered from a "Figaro problem": at the same time they had to fulfill ever more numerous functions in different fields – from optics to thermodynamics, which ultimately proved irreconcilable in the framework of classical physics. At the same time, Maxwell's electrodynamics and Rudolf Clausius' thermodynamics were emancipated to become independent sub-fields of physics, whose central concepts "field" and "entropy" no longer required mechanical grounds. Philosophers and historians of science like Ernst Mach took this occasion to scrutinize critically the basic concepts of mechanics.

In this manner a completely new situation emerged toward the end of the 19th century. Classical physics decomposed into three sub-areas: electromagnetism, mechanics and thermodynamics. Fault zones formed between these conceptual continents, in which borderline problems accumulated and the probability of a scientific earthquake grew.

These borderline problems constituted central research objects for Wilhelm Wien, Max Planck, Hendrik Antoon Lorentz and Ludwig Boltzmann, the masters of classical physics. All of them investigated these borderline problems, each primarily from the vantage points of his respective continent of knowledge. They lacked the comprehensive perspective from which the explosiveness of these problems could have attracted their attention.

This comprehensive perspective was taken up by Einstein. For him, the works of the masters of classical physics on its borderline problems were the point of departure for his scientific revolution of 1905. Einstein's largely autodidactic studies and his discussions with the friends of the Bohème of his college days in Zurich functioned as the intellectual reactor in which the insights of Wien, Planck, Lorentz and Boltzmann were then transformed into the foundations of a new physics. The substance of this revolution was not new, but the result of a centuries'-long accumulation and reorganization of knowledge. What was new, however, was the conceptual structure this knowledge now received through a kind of Copernican revolution. In place of the aether, new concepts of space and time took stage. In place of the wave conception of light, a dualism of waves and particles

Ludwig Boltzmann (1844–1906) was at loggerheads with some of his colleagues because he was convinced that matter must consist of atoms. For him, heat was nothing other than their chaotic motion, and the temperature of a body a measure of the statistical mean of the kinetic energy of its atoms. The Viennese physicist was also able to give an atomistic and thus a tangible meaning to the entropy introduced by Rudolf Clausius: According to Boltzmann, entropy is essentially the number of the various possibilities for an accumulation of atoms to arrange themselves and to move, without changing the macroscopic appearance (such as temperature or volume filled).

Hendrik Antoon Lorentz (1853–1928) was the sage of theoretical physics in his day. Proceeding from Maxwell's electrodynamics, the Dutchman assumed the existence of an all-pervasive aether at rest in Newtonian absolute space. According to this hypothesis, the aether actually should have become noticeable during movements of the measurement apparatus – however, this could not be proven experimentally: The speed of light always proved to be constant, in every experimental arrangement. In order to eliminate this problem, Lorentz created a theory that formally anticipated the statements of the special theory of relativity – such as the "Lorentz contraction" of bodies during their movement through the aether.

In 1894 **Wilhelm Wien** (1864–1928) applied the thermodynamic concepts "temperature" and "entropy" to electromagnetic radiation. Two years later he published a formula for the spectrum of the thermal radiation of a black body. Later it became apparent that his formula describes only the short-wave end of this spectrum. Nevertheless Wien's achievements constituted important preliminary work for Planck's radiation formula.

As a high school student, **Max Planck** (1858–1947) was dissuaded from studying physics, as there was believed to be nothing more to discover in this field. In his effort to systematize the physical knowledge of his time, he attempted an explanation of thermal radiation, that is, the electromagnetic radiation given off by a body of a certain temperature. This was a phenomenon in the boundary area of two classical theories which originally had been largely separate: As a radiation phenomenon it belonged to electrodynamics, but to the extent that it was brought about by heat, it belonged to thermodynamics. With methods culled from Ludwig Boltzmann's theory of gases, in 1900 Planck succeeded in deriving a formula for the spectrum of thermal radiation in a black cavity at a given temperature – what he called a "black body." To do this, however, he had to make the auxiliary assumption that energy can only be exchanged in finitely small portions, known as "quanta." Not until Einstein did it become apparent that this assumption could not be reconciled with James Clerk Maxwell's electrodynamics.

Even as a pupil and a student, **Albert Einstein** (1879–1955) studied the classics of 19th century physics. But he also read philosophical works that dealt with the question as to the possibility of certain knowledge about the material world. His epoch-making publications in 1905 were answers to contradictions that had broken out among the previous generation of scientists, between the classical theories of electrodynamics, mechanics and thermodynamics: The light quantum hypothesis gave the core assumption behind Max Planck's radiation formula an experimentally verifiable theoretical explanation.

The special theory of relativity re-interpreted Hendrik Antoon Lorentz' discussions and thus arrived at a completely new understanding of space and time. Einstein's interpretation of an old observation, known as "Brownian motion" of particles floating in a liquid, contributed to ending a bitter debate about whether or not there was such a thing as atoms and molecules.

emerged, which no longer could be grasped with the concepts of mechanics. And in place of classical atoms and their motions, a new kind of particles appeared, whose existence no longer remained purely hypothetical, but whose properties could only be described statistically.

The prerequisites for Einstein's revolutionary innovation of physics thus lay as much in the way he knew how to maintain the knowledge passed down to him as in his overview, which allowed him to recognize its limits. An important element of Einstein's overview was his critical awareness of the long-term development of knowledge. This awareness is presumably also the reason why, after the breakthrough in 1905, it did not escape his notice that the revolution of classical physics was anything but concluded. How worthwhile it was to scrutinize the handed-down knowledge became apparent in the general theory of relativity of 1915, which, in turn, drew essential impulses from Mach's historical-philosophical critique, and which solved the riddle of gravity in a surprising way. While Newton's concept of gravitational force could still be grasped with the help of our intuitive knowledge, Einstein now explained gravitation through the curvature of space and time. This explanation emancipated natural science even more from its conceptual dependency on intuitive everyday knowledge. On the other hand, the epistemic categories of space and time had again become objects of physical knowledge. The material world thus became less intuitive, but at the same time, more accessible to investigation. Over the rest of the 20th century, this process was to continue in the question as to the basic structures of matter; even today it has not concluded.

Henning Vierck

Comenius and Einstein as Educators

What is the point of putting Comenius, an educator at the beginning of the Modern Age, on the same level as Einstein? That means comparing apples and oranges. Would it not make more sense to compare Einstein with Galileo, a physicist of that epoch?

Scholarship in a man without virtue is "like a golden ring in the snout of a swine" (Spr.11,22) Johann Amos Comenius (1592–1670)

He who has acquired special knowledge must develop an active sense for what is beautiful and morally good. Otherwise he is more like a well-trained dog than a harmonically developed creature.
Albert Einstein

my pupils; I only attempt to provide the conditions in which they can learn." This quote is found in the Internet no less than 48,900 times. Since he rejects a prerequisite of education — instruction – Einstein is obviously not an educator. Every child knows that. Is there anyone who, at least as a child, did not admire the picture with his tongue stuck out? Einstein himself denies any authority. A comparison with Comenius, the classical scholar of education, appears impossible.

Johann Amos Comenius, in Czech "Komensky." Lithography from a contemporary portrait, 1845

True, the objection could come that Comenius also wrote physics, the *Naturall Philosophie Reformed by Divine Light.* Yes, but would Comenius have had a chance in a confrontation with Einstein on the subject? After all, his physics was not successful, that of Galileo certainly more so. Thus let us forget this comparison based on the history of physics and react instead to a different circumstance. Namely, it is possible to draw attention to the converse fact that Einstein occasionally expressed his views on issues of education. So why not compare the educator Einstein with the educator Comenius, a well-preserved old apple with new one that is still fresh? In this case Comenius would at least have a chance against Einstein.

Yet unfortunately this comparison too, is somewhat treacherous, for one of the most popular quotations attributed to Einstein is: "I never teach

Teacher and pupil (from: Comenius, *Orbis sensualium pictus,* 1658)

Anyone who knows the Orbis pictus by Comenius, certainly the most frequently published schoolbook in the world, has also seen the picture in its introduction of the teacher with his index finger raised. Is this gesture a reference to punishment with the rod? Is the threat of violence perhaps a method of Comenius' education? Einstein rejects such procedures. He says: "It appears to me worst when a school works mainly with the means of fear, coercion and artificial authority. Such treatment annihilates a healthy attitude to life, the sincerity and self confidence of the pupil."

Although it thus appears that Comenius and Einstein cannot be compared as educators, it may yet be possible to balance the experience of one's school days against that of the other. After all, Comenius utters thoughts similar to Einstein's: "Of many thousands, I too am one, a poor human being, whose delightful spring of life, the blossoming years of youth, were spoiled with scholastic humbug." Comenius thus also rejects a school system founded on authority.

Certainly, the methods of teaching in Comenius' day were different than those in Einstein's. One bewails the nearly absurd disputations of a religious school, the other the drilling of a state school that verges on brutality. Both are in agreement, however, in their rejection of authoritarian education. Yet do Einstein and Comenius even know the alternative to such education? It can be claimed that Einstein did. Having attended not only the Luitpold Gymnasium in Munich, a renowned preparatory grammar school, but also the liberal canton school in Aarau, he writes: "Through the comparison with six years of schooling at a German gymnasium run on the basis of authority, I became vividly aware of how superior the education about how to act freely and take personal responsibility is to education which is upheld by drill, external authority and ambition. True democracy is not an empty illusion."

Is it not disturbing that Einstein and Comenius – despite living centuries apart – made similar experiences in school? Aren't children still suffering such torments today? Has nothing at all changed in education? Indeed, Einstein suggests from comparative experience: True democracy is not an empty illusion.

Comenius did not experience democratic conditions, but did aspire in his invitation to the training of "seeing, speaking and acting oneself." He regarded the self-determination of man as the "method of paradise." His method of education was not the threat of violence, not the raised in-

Albert Einstein on the way to a lecture in Berlin

dex finger, but rather a voluntary exchange of knowledge between pupil and teacher. We may call it – certainly also in Einstein's sense – a method of democracy.

Comenius even inversed the relationship between teacher and pupil, in resistance to the structure of government of his age: "Since the teacher is only servant, not lord, only co-shaper, not re-shaper of nature, let him not press the pupil to anything with violence."

The teacher on the cover illustration of Orbis pictus does raise his index finger, but he does not do so in admonition, but rather to point out the light that joins pupil and teacher. Free exchange requires a "general tool," and according to Comenius this is the light. The pupil determines himself. He points to his own head, in which – through sensory organs, through "portals of things," the world and the teacher are perceived.

Perhaps Comenius, just like Einstein, did not instruct, but only attempted to create the conditions under which his pupils could learn. But how do

chure to me, but even more so that you give me such positive testimony as a docent and ponderer. The value of the theory of relativity for philosophy appears to me to be that it displayed the dubiousness of certain concepts, which were also recognized as tokens in philosophy. After all, concepts are empty when they cease to be linked

Albert Einstein's graduating class in Aarau, 1896

these circumstances look? What are democratic conditions in school? Einstein praises and reproves not only the instruction he was granted. He also taught, even when he did not have any official obligation to do so, as in Berlin – albeit only students, and for them it may be easier to create true conditions for learning. To one of the students that attended his lecture course, Hans Reichenbach (1891–1953), he later writes: "I am truly very glad that you want to dedicate your outstanding bro-

firmly to experience. They resemble social climbers who are ashamed of their family background and want to disavow it."

Concepts must remain closely connected to what is experienced. This rule certainly belongs to the method of democracy. Comenius says: "No one is allowed to repeat something that he does not understand, or to understand anything he cannot express. For he who does not express his intellect's perception is a statue, and he who babbles away something he does not understand is a parrot. We, however, educate the individual and want to do so without detours. This is possible everywhere where language keeps step with things and things with language."

Einstein and Comenius are educators, especially and primarily because they only wanted to create

This is what happens too, to the lively spirit which goes without serious employment: It entangles itself in vain, curious and quite perishable things and becomes the cause of its own decline."

Once again: So what does animation or accompaniment by a teacher consist in? What kind of grain is it that he gives his pupils to mill? Is it

Johann Amos Comenius: *Vorpforte der Schulunterweisung.* A schoolroom in the 17th century

the conditions under which it is possible to learn. Both take children and teenagers seriously as researchers who want to comprehend and impart something. The teacher, according to Einstein, may never "strangle the holy inquisitiveness of research [...], for besides animation, the main thing this delicate little plant requires is freedom." And Comenius says: "Human nature is free, loves self-determination and hates compulsion. Therefore it wants to be guided to where it strives, and not pulled, pushed, or coerced."

But what does animation or accompaniment by a teacher consist in? Comenius says: "And if one does not give a mill any grain during operation, any raw material for flour, it wears itself away, grinds down its stones to dust with a crashing racket, sustains damage or even breaks to pieces.

the teacher's own questions? Or is it questions that come from pupils? Certainly both, but never questions to which teachers or pupils can give prepared answers. A democratic school is not an institution of teaching, but rather a research institution. Mills that grind finished flour sustain damage and ultimately become the cause of their own decline.

Lectures held by Einstein deviated radically from what was usual at the time, because he, in his own words, had the "course finished neither in his head nor in his notebook." Einstein did not read from his notes, did not recite something prepared, did not instruct. According to his conviction it is an "optical illusion" to believe that "everything one has to say is self-evident."

Katja Bödeker

Time in the Embryonic Stage

The relativistic idea of time violates fundamental everyday conceptions. It seems absurd that time elapses more slowly from the perspective of a system in motion than from the perspective of a system at rest. Is it not a matter of course that dates and times possess an absolute validity, that is, one independent of the state of motion of the reference system?

The relativization of time thus presented Einstein's adversaries with a favorite target. Most of the contributions to the polemic *100 Autoren gegen*

common sense. For the latter of these, the conception of an absolute time appeared to be immediately given and unquestionable. The absolute time of common sense presents itself as a continuous flow, which moves uniformly, and in which all events in the world take place.

Common sense usually comes along quite simply and as a matter of course. Everyday conceptions about time appear familiar enough to everyone. Upon closer inspection, however, it appears that these conceptions are by no means directly given.

Jean Piaget in his study

Einstein published in 1931 take as their subject the motion of time in the special theory of relativity: The "relativization of simultaneity" is "nonsense," it is an "unexecutable thought" that "there should be many 'times' simultaneously," and the relativity of simultaneity as a theory can only be "comprehended with humor." But what is put forth against this "nonsense"? Among the courts appealed to are "universally valid time," "irrefutable laws of thought" or, more prosaically – simple

On the contrary, the concept of an absolute, homogeneously passing time, in which all events have an unambiguous place, already presupposes a complex structure of cognitive operations. The idea of an irreversible sequence of events is one element of the everyday concept of time. Wheat grows after it has been sown. Wheat is harvested after it has grown. And because the grain was sown before it grew, sowing also took place before harvesting. The concept of time also implies the idea of duration: When two events start at the same time and stop at the same time, they last the same amount of time; if one of the two start before the other, but both events end at the same time, then one of the two lasted longer than the other.

In day-to-day dealings with time these principles rarely come to light. They are a matter of course – even banalities, which we may always presuppose, but seldom reflect upon. All the greater is our

Merve: (nudging Yasemin in the shoulder) Born at the same time!

Katja Bödeker: And seven years ago? How old were you seven years ago?

Merve: I think she was three. I was zero!

Katja Bödeker: [...] (to Merve) Do you think you were zero seven years ago?

Merve: Yes.

Katja Bödeker: (to Yasemin) And how old were you seven years ago?

Manar, age 10:
Past, present and
future are localized
on different worlds

astonishment when we confront situations in which these "common sense rules" are violated. Merve and Yasemin are seven years old, as they reported at the beginning of the interview, and attend the second grade of a primary school in Berlin.

Katja Bödeker: You are seven now, right?

Yasemin: Yes.

Katja Bödeker: So how old were you one year ago?

Yasemin: Six.

Yasemin: Two.

Up until the second questions everything seems to come naturally. If Yasemin is seven years old now, then she was six one year ago. Merve's friendly nudge in the shoulder appears to show that Merve concludes that, as the girls are the same age now, this means that they were born at the same time, and further presupposes that both were also the same age one year ago. The surprise does not set in until to the third question

Afrtim, age 11: The left side presents the world without time, the right side with time

is answered. For Merve, the fact that both girls were "born at the same time" – which Merve has stated before – is not irreconcilable with the assertion that she and her friend were not the same age at a point in the past. The coordination between her own life-time and that of her friend – a precondition for a universal and homogeneous time – no longer takes place. So does Merve presuppose here that each of the two girls gets older in her own time? Does this mean that the idea of a single, homogeneously passing time for all events is not a matter of course?

Jean Piaget, known as the pioneer of child psychology, states in the foreword to his book about the development of the concept of time in the child that it was Albert Einstein who inspired his studies on time. During the *Davoser Hochschulkurse* in the spring of 1928, Piaget had the opportunity to discuss with Einstein questions on the epistemo-

logical status of the concept of time: Is the subjective intuition of time directly given? Is the intuition of time primary, or does it presuppose the intuition of speed? Piaget approached these questions by asking children about their understanding of time.

For Piaget, studying children's knowledge of time was not an end in itself. In order to understand a concept like that of time – according to Piaget's idea – the knowledge of its development is indispensable. On the other hand, he believed that studies on the development of the concept of time should not be directed exclusively to the historical genesis of conceptions of time in science and philosophy, and thus should not only deal with those stages in the development of the concept of time which were already products of theoretical work. In order to find the "origin of the understanding of time," it was necessary to look at the "larval stage," the embryonic stage of the concept of time. According to Piaget, studying the mental development of children makes an "intellectual embryology" of this kind possible.

Action is the "larva" of the concept of time for Piaget. Performing even the simplest tasks demonstrates the ordering of time, which also younger children must be aware of; before drinking, the bottle is opened. Before climbing up on the table, first the chair has to be scaled. In order to get to

main conflated with courses of action or with the paths covered: the path to the meadow is further than the path to the woods, therefore it takes longer.

However, practical temporal orderings are not sufficient to provide a concept of a homogeneous time that would be independent of courses of events or of spatial shifts and thus would allow coordinations of different actions. In the early stages of intellectual development, time disintegrates

Büşra, age 10: The future and the past

into a multitude of "local" times. Piaget thus asserts "that the unity of time by no means imposes itself in the first stages of its development." The many local times remain centered to changes of place or to individual actions. When time is not yet conceived of as independent of sequences of

the big meadow, first you have to leave the house, then cross the street and finally walk through the small woods to the clearing. Chronological outlines like these do allow for judgements about the sequence and the relative duration of parts of an action. Yet such practical temporal orderings reactions or events, the concept of "simultaneity" has no meaning either, because this presupposes a coordination of different motions. The understanding of age and its change, too – an idea about which even kindergarten children bear a wealth of experience – remains bound to life events: We have already met Merve. Kübranur is seven years old, just like Merve, as she reported at the beginning of the interview.

Katja Bödeker: So how old were you two one year ago?

Merve: I was six – then I had my birthday and I was seven. I didn't have my birthday and then I was still six.

Kübranur: I was six, too.

Katja Bödeker: And how old were you eight years ago, Kübranur?

Merve: I know, you were six, too. That's right, she was six, too.

Jessica's representation of a clock

Spatial representation of temporal sequences. Amin, age 9, painted rythms

Katja Bödeker: Eight years ago?

Merve: Yes.

Katja Bödeker: Why do you think that she was six, too?

Merve: Because she didn't have her birthday yet.

Kübranur: I was in nursery school then.

Even Merve's first statement is illuminating: Although she determines her age one year ago correctly, from the perspective of "common sense," she apparently advances not arithmetic considerations as the reason for her answer, but rather memories of her own birthday. Birthday

means a change in one's own age, and without a birthday there can be no change in age. Thus her classmate Kübranur could have been six even eight years ago, because she hadn't had her birthday yet.

A different concept of time becomes clear in the following statements:

Ceyda is ten years old and attends the fourth grade of a primary school in Berlin.

Stefanie Giese: Now, say I celebrate my birthday only every two years. Then I get old more slowly, don't I?

Ceyda: No, that is not what matters. That only has to do with whether you celebrate. If that were the case, then when you don't celebrate you stay nine. Although you turn ten!

Stefanie Giese: Do you get older whether you want to or not?

Ceyda: Yes, whether you want to or not.

Stefanie Giese: Does that mean that there is time

and I can't influence it at all? I can't make time longer or shorter?

Ceyda: No, I don't think so. Because an hour passes in an hour.

In this case, time is that uniform process independent of human will and effort with which we are familiar. If the concept of a uniform time hardly foists itself upon us – as Piaget asserts–, then what leads to its development?

Piaget gives a surprising answer: the concept of speed. In the local times linked with action, the different velocities of motions are not taken into consideration. Not until children compare and coordinate motions of different velocities with each other do they attain a concept of time that is independent of the completion of individual motions. Piaget's proposal is astonishing because, in the tradition of classical mechanics, we generally regard space and time as basic concepts, whereas we conceive of velocity as a quantity derived from space and time. This is generally the way the concept of velocity is introduced in physics classes: velocity = distance / time.

His observations of children lead Piaget to reverse this relationship. It is the concept of traversed space and the concept of speed which are primary the concept of an absolute, i. e. of an action-independent time however differentiates itself from the two basic concepts.

In this manner the relativistic concept of time may violate fundamental everyday intuitions: "But here the historical and genetic perspective can show how little we may trust intuition – which always remains related to a certain level of intellectual development [...]

As paradox as it appears, the relative duration and the local times of Einstein's theory of relativity are related to absolute time just as absolute time relates to *Eigenzeit* or the local times of childish intuitions."

The interview extracts are taken from the workshop "World Pictures," which took place from May to October 2004 at the Comenius Garden in Berlin Neukölln in preparation of the Einstein Exhibition. I would like to thank Stefanie Giese for her kind cooperation

Renate Wahsner

Absolute Space: Mach vs. Newton

According to a widespread perception, Isaac Newton's "bucket experiment" was intended to prove the existence of absolute space, which Ernst Mach was later able to refute, thus inspiring Albert Einstein to substantiate the general theory of relativity. While this view is not completely incorrect, it fails to recognize not only the true achievements of these three scholars, but also the character of physics.

Mach believed Newtonian mechanics to be correct in principle, but in certain respects not clearly

Isaac Newton, painting by John Vanderbank, 1725

physical" assumptions. For this reason he criticized the concepts of absolute space, absolute time, and the concept of mass in classical mechanics. Mach perceived absolute space as a metaphysical specter. He wanted to do away with it in favor of something that was in principle more accessible by experience. In his view, a reinterpretation of the law of inertia would make this possible. An approach to this reinterpretation was given, he believed, in the critique of the Newtonian bucket experiment.

Albert Einstein, portrait between 1915 and 1920

portrayed. According to him, this was a result of the fact that the foundations of mechanics could not be found entirely *a priori*, nor entirely through experience, which entailed an imprecise and unscientific treatment of these foundations and basic concepts. He saw the way out of this situation in excluding from mechanics, or from science in general, all assumptions that could not be checked by means of experience, above all the "meta-

Newton had written that in human matters, not inappropriately, we employ not *absolute* places and motions, but rather relative ones, whereas in natural philosophy, that is, in physical science, we must *abstract from the senses*. "It can of course be the case that no truly resting body exists to which one could relate places and motions."

After this fundamental observation he explained that absolute and relative motions differ from each other due to *centrifugal forces* from the axis of motion. He claimed that these forces do not exist for a relative rotary motion, but certainly do exist for an absolute rotary motion (and indeed,

Ernst Mach,
photogravure
by Charles Scolik,
1910

the size of these forces depends on the size of the absolute motion).

Newton then intends to illustrate (and surely not prove) this assertion of the mechanics he had elaborated and tested empirically by means of a mental experiment, as follows. First a bucket is hung up on a very long rope, and then spun around in circles until the rope becomes quite stiff from the rotation; and then the bucket is filled with water and held steady along with the water. (Thus a suitable mechanism was built to set the vessel in rotation.) When the bucket is now caused to rotate (in this case through the untwisting of the rope), the surface of the water will initially be flat (the water is at rest with respect to the bucket's surroundings, just as before the vessel started moving); when then the force gradually begins to affect the water (the bucket pulls it along), the water begins to rotate noticeably. Little by little it moves away from the middle and climbs up the bucket walls, taking on a hollow (concave) shape.

At the beginning, claims Newton, when the *relative* motion of the water was greatest in relation to the bucket (the water was at rest with respect to the surroundings of the bucket, while the bucket was rotating in relation to its surroundings), this motion did not give rise to any effort to remove itself from the axis. Not until the relative motion of the water diminished did the aforementioned effort become apparent, and then grew until the rotary motion of the water (with respect to the surroundings of the bucket) became greatest, namely at the point at which the water itself was at rest *relative* to the bucket (since it executes the same motion as the bucket). In summary:

– The centrifugal force in the water is not caused by a true interaction between the mass bucket with the mass water (which would have had to exist from the beginning), but rather by what is known as the law of inertia, the said centrifugal force.

– Because a motion can be a motion (or rest) only with reference to something, the very obvious centrifugal motion of the water is a motion with

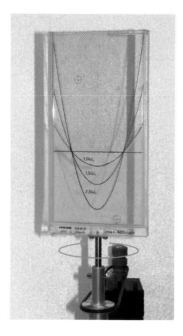

Centrifugal vessel (Newtonian bucket) When the vessel is at rest, the water level is flat. During rotation it takes on a parabolic form

"Instead of relating a body in motion K to Newton's absolute space, we will observe its relationship to the bodies of the universe directly, by relating its motion to the center of mass of all the matter in the universe."

Mach believed that this reinterpretation showed that the "specter of absolute space" could be replaced by the mass of all cosmic matter and thus provide a "cosmic foundation" for the law of inertia. Because – as the bucket experiment illustrates – this depends on the rotary motion of the water *relative* to the masses of the bucket-water system, it is easy to conclude or presume that the same effect of the centrifugal motion of water occurs when the situation is reversed, with the surrounding masses in rotation and the system of the bucket and water at rest.

The idea of providing a cosmic foundation for the law of inertia was later elucidated by Einstein in another form. He claimed that the inert mass of a body is determined by all other masses in the universe, intensifying this hypothesis to become "Mach's Principle" in 1918. However, this principle does not state what Mach precisely intended to demonstrate, but is formulated with a view to developing physics further, beyond the classical mechanics that had proved insufficient to explain this principle. Initially, Einstein held the general theory of relativity as a fulfillment of Mach's principle, but he changed his view on this point around 1920.

With regard to Newtonian mechanics, Einstein was not averse to the concept of absolute space. In fact, he emphasized that acceleration in classical mechanics can only be determined as acceleration *against space as a whole*. And thus he concluded: "The geometrical reality of spatial concepts were thus joined by a new function of space that determined inertia. When Newton declared space to be absolute, he presumably meant this real meaning of space, which for him entailed that his space had to be assigned a certain defi-

respect to the space surrounding the system of the bucket and the water. Newton called it absolute space. He designated the motion as a true motion with respect to this space.

Newton admits that it is quite difficult to recognize the true motions of the individual bodies and to differentiate them from the apparent motions, "because the parts of that stationary space in which the bodies truly move cannot be recognized with the senses." *Time, space, place* and *motion* are known to everyone, and normally, according to Newton, they are conceived of exclusively with respect to the senses, which brings with it certain prejudices. In order to eliminate this prejudice of the senses, he thus differentiated between absolute and relative, true and apparent, mathematical and everyday motions.

As mentioned, Mach saw this as a starting point to reinterpret the law of inertia, a law he did not dispute, but indeed held to be a fact recognized by Galileo. However, he believed that a great uncertainty existed in the law of inertia because it does not state with respect to which body the direction and speed of the body in motion is meant. In order to eliminate such uncertainty, Mach proposed:

nite condition of motion." Einstein's view was that Newton could have given his absolute space another name. For – in his words, "the only thing that is essential to regard the acceleration of the rotation as something real is that another, imper-

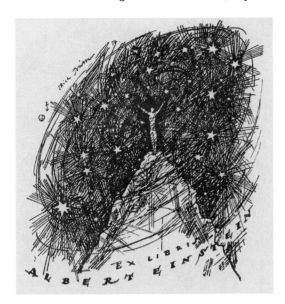

ceptible thing – besides the objects that can be observed – be regarded as real."

Mach was mistaken when he believed that his reinterpretation could make absolute space superfluous. For his reinterpretation is only successful when assertions are made about the motion of the center of mass with respect to absolute space, thus presupposing the spatial concept of Newtonian theory. Mach knew this implicitly, for he

claimed that his formulation of the law of inertia was equivalent to the Newtonian one and therefore also implied the same difficulties: "In the one case we cannot get hold of absolute space, in the other case only a limited number of masses can be accessed by our intellect, and the implied smma tion thus cannot be completed." Mach held these difficulties to be in principle, but believed that his formulation had made scientists aware that the law of inertia, just like other basic laws of mechanics, are based on experience that however had not – and indeed never could be – concluded.

A physical theory cannot be criticized with the statement that everything is relative "in reality." A physical theory must always set something as absolute. It may not be forgotten, however (and that is what happened in Mach's day) that this something is only set as absolute, and may not declare that the connection formulated within it exists in and of itself.

"Absolute" space, "absolute" time, and "absolute" motion are in fact abstractions, just as Mach writes. Yet no physical theory can do without them. Mach's modification of Newtonian mechanics, which attempts to replace absolute space by all the matter in the universe, does not liberate theory from this necessity. For the *total mass of matter* in the universe is just as much an absolutum as absolute space.

Bookplate for Albert Einstein from Erich Büttner, Berlin 1917

For further discussion of this theme, see the remarks and epilogue of the editors in: Ernst Mach, *Die Mechanik in ihrer Entwicklung – historisch-kritisch dargestellt*, edited by R. Wahsner and H.-H. v. Borzeszkowski, Berlin 1988

Falk Müller

Why Does a Light Mill Revolve?
A Historical Look

The memoirs of the generation of physicists before Einstein frequently mention a strange device that was a source of great excitement in their youth and during the period during which they were gathering their first experience with the physical sciences: the radiometer, also known as the "light mill." The radiometer was introduced to the Royal Society as a scientific instrument for the first time in a lecture held in the spring of 1875 by William Crookes, a chemist and London

pied Crookes in the early 1870s. The first involved measurement of the chemical element thallium discovered by him in 1862, which he was performing using a precision scale. The scale, enclosed in an evacuated glass container to achieve greater precision, recorded strange oscillations when a heated object was brought in the vicinity of the scales pan, such that Crookes thought he was on the trail of as yet unknown attractive and repulsive effects. With various types of scales and rotat-

Left:
Sir George
Gabriel Stokes

Right:
Sir William Crookes

businessman. In its classic form, the radiometer can still be seen today in display windows as a scientific toy: an evacuated glass bulb containing a rotor with a crossbar, to the ends of which vertical plates, blackened on one side and metallic on the other, are attached, and in which the rotor begins to revolve in sunlight or when a heated object approaches it. Soon after the first radiometer was introduced, it was replicated by a number of instrument makers, among them Heinrich Geissler, a glassblower in Bonn, from whom it received the name "light mill." Through such reproductions the instrument spread quickly as a novelty of physics, which inspired the fantasy and scientific curiosity of a number of scientists and laymen in the 1870s. The construction of the radiometer involved synthesizing the two areas of research that had occu-

ing pendulums constructed by his assistant Charles Gimingham, among the effects he investigated was whether light and warmth exert a direct influence on the force of gravity. As ambitious as these investigations were, Crookes attempted to explode the existing borders of the field of physics with his other project. Together with a number of colleagues and friends, he had begun with the scientific investigation of spiritualistic phenomena in the early 1870s and had come upon a new "psychic force." Using sensitive scales and torsion beams (which were also fused into glass vessels), Crookes sought an organic explanation for the inexplicable effect of spiritualistic media on material bodies, and for a time believed in a mineral source of these phenomena. In the following years Crookes hosted various

WORLDVIEW AND KNOWLEDGE ACQUISITION   49

Why Does a Light Mill Revolve? A Historical Look

mediums in his home, with whom he performed experiments to investigate phenomena such as levitation, glowing and other events. Even though Crookes was subjected to criticism for this research, some of it quite sharp, there was a generally benevolent interest in Crooke's spiritualist research in the upper echelons of Victorian society. A demonstration of his experiments in front of the Royal Society, however, was prevented by its secretary, the physicist and mathematician George Gabriel Stokes. Crookes was not allowed to show his experiments until they were linked with another explanation.

In his *Treatise of Electricity and magnetism*, published in 1873, the Cambridge physicist James Clerk Maxwell had predicted that electromagnetic waves – to which visible light also belongs – exert a mechanical effect on massive bodies. With suitable instruments it was also possible, for instance, to measure the "pressure" of light. This thesis suggested an explanation of the repulsion effects, and Crookes is believed to have developed suitable instruments with his pendulums and the radiometer. In this case, however, the radiometers would have had to revolve with the blackened surfaces facing forward, which was not generally the case. An argument that spoke against this hypothesis and for another explanatory model came from the British physicist Arthur Schuster. He and a number of other scientists believed that an explanation for these effects could be found in neither the field of spiritualistic speculations nor electro dynamic theories, but rather in a research area just starting to emerge: kinetic theory of gases. A variety of different models were possible in this theory, each of which assigned an important role to the interaction between the remainder of gases still contained in the glass bulb and the differentially heated light and dark surfaces of the plates. Schuster believed that the discussion with Crookes could be decided by a simple experiment in which the entire glass body was sus-

pended and allowed to revolve. If the force were generated within the glass bulb, that is, through the differentially heated surfaces of the rotor, then, in keeping with the conservation of momentum, the glass body would have to show a movement in the opposite direction of the rotor. If a radiation pressure were exerted on the surfaces of the rotor from outside the bulb, the glass body would have to revolve slowly in the same direction as the rotor due to the friction at the tip of

James Clerk Maxwell

the steel point of the rotor. That was the theory. In practice Schuster's experiments did show a movement in the opposite direction; however, Crookes was not able to reproduce these experiments with unambiguous results.

Crookes had another reason for his aversion to the gas theory: the gas molecules could not play any role at all, since in his opinion, the mercury vacuum pumps he used – and which had been improved repeatedly by his assistant – removed all gases from the glass bulb, thus generating a "perfect" vacuum. The way in which Crookes and his contemporaries realized and learned to master the multifarious problems connected with the generation and measurement of vacuum is another story. The history of the radiometer is hence also a history of vacuum technology at the end of the 19th century.

Through a number of improvements, and after many, many setbacks, in the early 1880s Crookes at long last believed that his instruments were up to the task that Einstein declared superfluous two and a half decades later: the measurement of the viscosity of physical aether.

layer that drives other molecules away, consequently providing negative pressure. Thus, while the effects that occur in this manner on the two sides of the rotor could compensate for each other at normal pressures, a different situation resulted when the more strongly heated, faster molecules

Left:
Electric radiometer by Puluj, 1880s

Center:
Schematic drawing of an electric radiometer

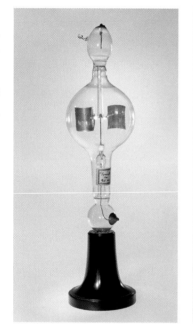
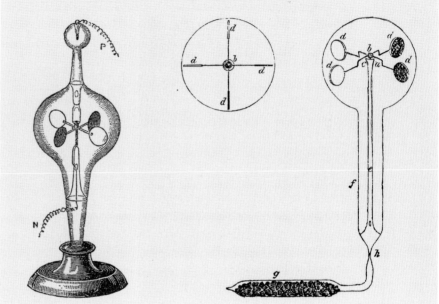

Drawing of a radiometer with carbon reservoir

An important step on the way to a better understanding of the radiometer movement was realizing how complex the apparently simple device actually was. In addition to the assumptions that the surfaces of the plates, and later also the gas molecules located in the glass bulb, played an important role, now the wall of the vessel obtained a function as well. The Irish physicist Johnstone Stoney summarized these new perspectives in several clear imagery models. The molecules that came in contact with the light surfaces, which were heated to a lesser degree, received a slight additional impetus, while those that struck the dark surfaces were sped up considerably more. According to Stoney, these "material rays," accelerate vertically with respect to the surface, but then continuing parallel and for the most part avoiding collisions with each other, constitute a

of the darker side propagated up to the glass wall as the vacuum improved, losing their energy there. Then more molecules would hit the dark side than the light side and these would cause the dark sides to drive the rotor's movement. For Stoney the radiometer was a power and heat machine analog to the steam engine, in which the surfaces of the plates served as the source of heat, the gas molecules as the medium of transportation, and the glass surfaces as a condenser or cooling element. The radiometers move because a different energy flow occurs between the respective surfaces of the plates and the glass surface. Osborne Reynolds and James Clerk Maxwell were not convinced by Stoney's explanation and searched for an alternative explanation. While the theoretical models available at the time allowed for the portrayal of the forces in action and the

WORLDVIEW AND KNOWLEDGE ACQUISITION **51**

Why Does a Light Mill Revolve? A Historical Look

processes that resulted, Reynolds and Maxwell developed a new hypothesis based on the systems of equations of kinetic theory of gases with which they were familiar. In their opinion the processes at the surfaces of the plates played no role at all, because they cancelled each other out. The inter-

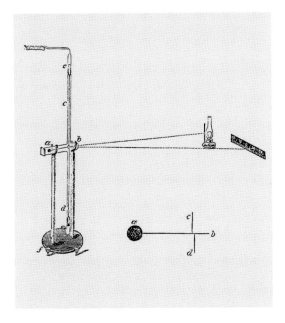

esting processes took place at the edges of the plates, for it was there that molecular "slide" processes took place between the colder and the warmer sides, exerting a tangential force on the

rotor. Researchers continued to concern themselves with radiometer effects even after the widespread acceptance of this explanation, among them Albert Einstein, who attempted to partially rehabilitate Stoney's theory. Ultimately it became apparent that the movement of the rotor resulted from an overlaying of various processes. Thus it is only possible to tell the story of the radiometer as a complex interplay in which the formulation of mathematical theory, the construction of theoretical models, instrumental and experimental practice each made original and independent contributions. Crookes was less interested in proving theories, most of which he did not understand. For him models and figures served as inspiration for the construction of new instruments and the execution of experiments – and in this case he presumably enjoyed his greatest influence in applying the means and methods of researching molecular processes to another field of research in the late 1870s: cathode rays. Through experimental synthesis he reached the conclusion that the cathode rays must consist of a directed movement of fast, negatively charged particles – a hypothesis that found a certain degree of confirmation shortly thereafter through the introduction of electrons to physics.

Drawing of a
torsion pendulum

Lidia Falomo
Carla Garbarino

The Leyden Jar

While electrical phenomena were being observed and theories as to their origin formulated, several instruments to demonstrate electricity were built during the first two decades of the 18th century. Otto von Guericke built the first electrostatic machine in ca. 1660. This was a gyrating iron rod

Eusebio Sguario, *Dell'elettricismo, o sia delle forze elettriche de' corpi svelate dalla fisica sperimentale*, Venice, 1746

topped by a sulfur sphere that became electrified when a hand stroked it. In subsequent years, these studies accelerated and, in short, the marvelous and exciting electrical phenomena conquered the whole of Europe. Experiments multiplied and everything was electrified – from everyday objects to the human body, from prepared tables to the young ladies who gave electric kisses (above). The young men sometimes became so electrified that "their bodies became surrounded by light, the way the saints with halos used to be

depicted." Not only old hands and amateurs but also some "electrifying" physicists (Stephen Gray and Charles Du Fay were two of the first) began to frequent salons and squares, popularizing electricity and entertaining.

Meanwhile, electrostatic machines and instruments for demonstrating electricity in bodies continued to be improved. In the former, the sulfur sphere was replaced by a glass one which was turned by a handle, and the hand was replaced by friction pads. Ways of storing the electricity generated were introduced. These were generally a metal bar hanging from a silk thread, one end of which was near the rotating sphere or connected with it by means of metal wires.

While early theories to explain the observed electrical phenomena were being formulated, something truly wonderful happened. In 1745, Ewald Jürgen von Kleist, the Dean of Camin and an amateur scientist, electrified a wire fixed to a hand-held container of alcohol and a little mercury. When the other hand touched the wire, he received a terrific shock. Some time later, the same thing happened to Petrus van Musschenbroek, a mathematics and physics professor at Leyden (according to another version it happened to the lawyer Andreas Cuneus while he was repeating some of Musschenbroek's experiments). Trying to electrify water via a wire in contact with an electrostatic machine, Musschenbroek held the bottle in one hand, contradicting all the then applicable rules, which stated that it should be on an isolating stand. He also touched the wire dipped in water with his other hand and got an huge shock, so much so that in a letter written after the experiment he wrote that he felt himself "struck in his arms, shoulders and breast, so that he lost his breath, and it was two days before he recovered from the effects of the blow and the terror" (Joseph Priestley, *The History and Present State of Electricity, with Original Experiments*, London, 1767, p. 83).

The news of this experiment began to spread to the rest of Europe, especially to France, where two letters from Leyden arrived in 1746. One was to the famous physicist Jean-Antoine Nollet. He wrote: "Since the gentlemen have not indicated precisely who first carried out this experiment, I have decided to call it the Leyden Experiment, as it has been known ever since." (Jean-Antoine Nollet, *Leçons de physique expérimentale par m. l'Abbé Nollet*, v. 6, Paris, 1780, p. 478). Given that "The more wonderful the electrical phenomena become, the stronger is our wish to discover its causes."

Nollet repeated the experiment several times and suggested an explanation based on his own conception of electricity. Electrical matter, he said, was an extremely fine, elastic fluid around and within all bodies, into and out of which it went through the pores in the material of which they were composed. The person carrying out the Leyden Experiment felt a strong shock because the electric fluid within him was struck from two opposite sides the moment the conductor sparked. Through the conducting rod connected to the Leyden jar (right) the electric fluid produced by stroking the glass sphere was pushed into the bottle. It had difficulty penetrating the thickness of the bottle and partly flowed back through the conductor. So it rushed more forcibly to the finger near it, as this was a more readily penetrable means. This in-coming current met the outgoing one from the finger itself to the conductor. Hence, there was a violent shock caused by the meeting of the two currents. The person holding the bottle got a counterstroke because of the rebound from this first reaction. The counterstroke was particularly violent, because the electrified glass gave extra "energy" to the electrical matter filtering thorough it.

While elaborating this complex theory, Nollet continued entertaining people. He made a chain of people holding hands feel the shock, when the first person was holding the jar and the last touched the conductor projecting from it. He once made two hundred people jump in this way. On another occasion, in the King's presence, he demonstrated the same phenomenon with 180 soldiers. He even gave as many as six hundred

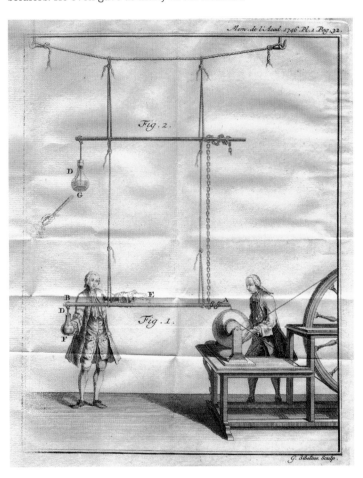

people holding hands the same experience. Other curious experiments included having the shock felt at a distance, even through flowing water, lighting up pictures and books, making animal forms made of conducting material move as if alive, like the electric fish and the electric spider, etc. News of these experiments spread not only throughout Europe, but also to America, where Benjamin Franklin worked with great theoretical and experimental skill. It is to him that we owe a particularly elegant explanation of the

Jean-Antoine Nollet, *Observations sur quelques nouveaux phénomènes d'électricité*, Histoire de l'Académie Royale des Sciences, année, Paris 1746

Leyden experiment, one that has particular importance in the history of electricity (see fig. below). He assumed there was a single electric fluid with two contrary electrical states: positive, corre-

restored in the bottle by inward communication or contact of the parts; but it must be done by a communication form'd without the bottle between the top and bottom, by some non-electric [in mod-

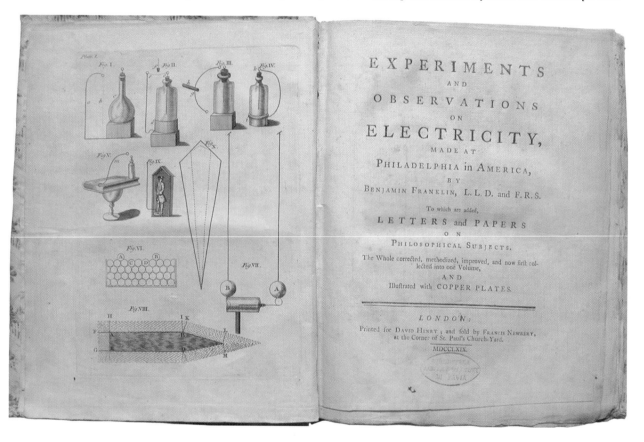

EXPERIMENTS
AND
OBSERVATIONS
ON
ELECTRICITY,
MADE AT
PHILADELPHIA in AMERICA,
BY
BENJAMIN FRANKLIN, L.L.D. and F.R.S.

To which are added,
LETTERS and PAPERS
ON
PHILOSOPHICAL SUBJECTS.

The Whole corrected, methodized, improved, and now first collected into one Volume,
AND
Illustrated with COPPER PLATES.

LONDON:
Printed for DAVID HENRY; and fold by FRANCIS NEWBERY,
at the Corner of St. Paul's Church-Yard.
MDCCLXIX.

Benjamin Franklin, Experiments and observations on electricity, made at Philadelphia in America, by Benjamin Franklin, London, 1769

sponding to "being full of electrical fire," and negative, corresponding to "being void of said fire." In the Leyden jar, both of these states were present, the former on the upper surface and the latter on the lower. Given that the electric fluid could not pass through the glass, balance was violently reestablished when the two parts were brought together externally by means of a conductor (which could be a person). This was how Franklin put it: "At the same time that the wire and top of the bottle, is electrized positively or plus, the bottom of the bottle is electrized negatively or minus, in exact proportion: i.e. whatever quantity of electrical fire is thrown in at the top, an equal quantity goes out of the bottom. [...] The equilibrium cannot be

ern terms "conductor"], touching or approaching both at the same time; in which case it is restored with a violence and quickness inexpressible; or touching each alternately, in which case the equilibrium is restored by degrees. [...] Here we have a bottle containing at the same time a plenum of electrical fire, and a vacuum of the same fire; and yet the equilibrium cannot be restored between them but by a communication without, though the plenum presses violently to expand, and the hungry vacuum seems to attract as violently in order to be filled." Benjamin Franklin, "Letter III from Mr Benj. Franklin in Philadelphia, to Peter Collison, F.R.S. London" [11 September 1747], in experiments and observations on electricity, made at Philadelphia in America, by Benjamin Franklin, London, 1769, pp. 13–14.

It was soon realized that water and other liquids used in the Leyden experiment functioned merely as conductors which could be replaced by other conductors, such as a metallic covering stuck to

The Leyden jar (or more precisely the combination of the hand holding the jar, the glass and the water, or the two metal armatures separated by glass) was the first condenser in electrical history. The first important modification was made by Franklin himself who started using a sheet of

Leyden jars, Pavia

the bottle or a hand. The Leyden jar in different shapes, cylindrical, pyriform, cuboid, rectangular, etc., spread in great numbers to all the physics cabinets of the time, where they were used for decades.

glass with metal coating instead of the bottle. Ever since that time, the condenser has been continually modified and used right up to the present day, sometimes in ways very different from the original apparatus.

Lucio Fregonese

Volta's Battery in Einstein and Infeld's
The Evolution of Physics

In their famous 1938 book *The Evolution of Physics*, Albert Einstein and Leopold Infeld give a decisive role to Alessandro Volta's electrical battery in the conceptual development of physics. Volta's new instrument, announced in 1800, is credited with having raised no less than "the first serious difficulty" against what our two authors call the "mechanical view" of nature. By the latter expression, they mean the philosophical position that all phenomena can be explained on the basis of the following two main assumptions.

Alessandro Volta

Matter is an aggregate of microscopic immutable particles which interact with mutual forces. For any pair of such particles, the force felt by one particle lies along the joining line with the other and is a function of their mutual distance only. How was this general representation of nature so decisively challenged by Volta's battery? Einstein and Infeld assert that it was mainly as a result of the discovery, by Hans Christian Ørsted in 1820, that the current of the battery causes magnetic effects. The figure (above right) shows how they present the case.

The loop is a circular-shaped wire lying on a vertical plane. The object at the centre is a magnetic needle. Initially, no current passes through the wire and the needle lies on the vertical plane on which the circle also lies. If the current is switched on, the needle turns and becomes perpendicular to the plane of the circle, as shown in the figure. Einstein and Infeld argue that this turning action results from forces which are *perpendicular* to the plane of the circle and that forces so oriented cannot be obtained if one adopts the mechanical view.

Volta's battery then enters the general plot somewhat indirectly, but it is described in detail, with analogies and differences compared to Volta's original arrangements. At a certain point a quotation from his writings is also reported. All these points are interesting and one is naturally stimulated to investigate them.

The following, our authors say, is "what is known as a voltaic battery":

"There are several glass tumblers, each containing water with a little sulphuric acid. In each glass two metal plates, one copper and the other zinc, are immersed in the solution. The copper plate of

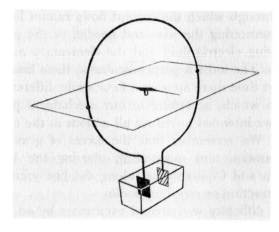

Magnetic effect generated by the current of a battery. Drawing from Einstein/Infeld

one glass is connected to the zinc of the next, so that only the zinc plate of the first and the copper plate of the last glass remain unconnected."

This is similar to Volta's "crown of cups" version of the battery (above). Z and A are plates of zinc and copper respectively.

A comparison with the above description reveals, however, two important differences. Einstein and Infeld assert that each cup has a couple of zinc and copper plates, but we see that Volta omitted

Alessandro Volta's battery, "crown of cups" version

the zinc plate Z in the first cup and the copper plate A in the last cup.

The other difference is in the arc-shaped connections between the copper plate A of one cup and the zinc plate Z of the next cup. Einstein and Infeld consider these joining arcs as simple conductors and do not specify further on them. On the contrary, Volta prescribed arcs made of two heterogeneous halves, copper on the left and zinc on the right, "soldered together in any part above that which is immersed in the liquor."

Both differences are understood if we compare the modern chemical interpretation of the battery,

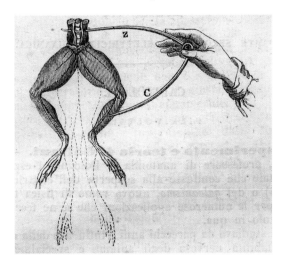

by which Einstein and Infeld were visibly influenced, with Volta's contact electricity interpretation.

In the chemical theory, each cup with its couple of immersed plates Z, A forms an "element." The engine, so to say, which in this element impels the current is the chemical reactions between the solution and the plates Z, A. The joining arcs serve simply to connect the various elements in series and it is inessential what metal they are made of. The addition of the zinc plate Z in the first cup and of the copper plate A in the last cup is convenient because two extra elements are thus included in the battery.

For Volta, things were quite different. The solution, he maintained, works simply as a conductor and the impelling engine is localised where the heterogeneous halves of the joining arcs are soldered together. This was an application of his theory of contact electricity, according to which the simple contact between heterogeneous conductors is enough to create an electrical imbalance between them. Not being in contact with a different metal, both the plate Z in the first cup and the plate A in the last cup do not add any impulsion and can be omitted. Volta noticed that the immersed plates experience chemical reactions, but interpreted this as a secondary effect of the contact-produced current.

Volta's theory of contact electricity went through various complex phases. All started with the work of Luigi Galvani on frogs. Among many other things, in 1791, Galvani announced that muscular contractions are obtained in a skinned frog leg (left) when the spinal nerve and the naked muscles are connected with a metal arc.

Galvani maintained that this phenomenon is caused by the discharge of a proper kind of "animal electricity" between the internal and the external parts of the muscles. This discharge takes place through the spinal nerve when it is connected to the external parts of the muscles with the metal arc. He noticed and reported that muscular contractions are generally stronger when the metal arc is composed of two heterogeneous halves instead of one metal only, but he did not derive major implications from this.

Volta adhered initially to Galvani's theory of animal electricity but soon became dissatisfied with it. He was struck in particular by the much larger efficacy of the bimetallic arc in comparison to the monometallic arc. With complex thinking and experimenting, he composed a very different picture and it is interesting that reasoning on the causal explanation of muscular contraction played an important part here.

Luigi Galvani's attempt to prove an "animal electricity" by means of frog-thigh contractions

As Volta saw things, Galvani regarded his postulated discharge of animal electricity between the inner and the outer parts of the frog legs as the sole and total cause of the observed contraction. Volta became convinced that this causal relation was wrong because the strength of the assumed

Luigi Galvani

direct cause was much smaller than the strength of the effect this cause was supposed to produce. He had in fact concluded from his studies that, if existing, animal electricity can be only extremely weak in strength, while the muscular twitching was generally vigorous. Using the language of modern science metaphorically, we could say that Volta became convinced that Galvani's animal electricity had too little *energy* to be the true direct cause of muscular contraction.

To solve this difficulty, Volta introduced a two-step causal chain. The *energy* of muscular contraction actually comes from a faculty proper to the muscles, which, like others, he called "irritability." If stimulated by some agent, the nerves transmit a signal which causes muscular contraction indirectly by activation of the irritability of the muscles. In the case of Galvani's twitching frog legs, the stimulating agent on the nerves is for Volta a very weak current impelled where the heterogeneous conductors forming the circuit nerve-arc-muscles come into mutual contact. This electrical stimulation of the nerves awakens muscular irritability and the observed contraction ensues accordingly.

In Volta's perception, his notion of contact electricity was not given the full legitimacy it deserved and this put him on the path towards the battery. In 1796, he managed at last to show experimentally that two different metals in mutual contact take up opposite charges.

But soon various species of fish capable of giving strong electrical shocks, especially the torpedo, were pointed out as direct evidence that animal electricity exists. Volta took this as a further move against contact electricity and decided that definitive proof was needed. This, he thought, was spectacularly achieved when he finally invented the battery.

The new tool was the result of attempts to reproduce the functioning of the electrical organ of the torpedo from the standpoint of contact electricity. Volta knew that this organ was composed of many little columns, made in their turn of numerous layers piled up together. Inspired by this structure, he piled up (right) numerous bimetallic couples Z–A with a layer of wet cardboard between one couple and the other.

The electrical shock caused by the instrument resembled that of the torpedo and Volta could now claim that the animal's organ was powered by contact electricity impelled where the various tissue layers of the little columns touch each other. The name "artificial electrical organ" he originally chose for the battery reveals beautifully this line of thought. Because of its shape, he also called it "column apparatus." It is worth noting that the situation is now opposite to that of the "crown of cups" case, in the sense that the column apparatus now ends with double electrodes Z–A which are meaningless in the chemical theory and contribute to the impulsion in the contact theory.

The current of the battery was a completely new phenomenon and one naturally asks how Volta conceived it.

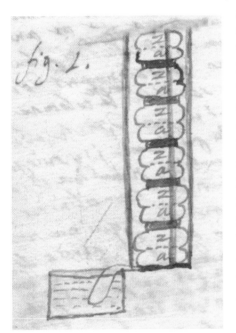

He presented different views, one of which appears in the only quotation of his that Einstein and Infeld chose for their book. In this quotation, taken from the paper announcing the battery, Volta assimilates the current of the battery to a series of successive discharges from a group of "Leyden jars" connected together. The Leyden jar was one of the most common 18th-century electrical devices, consisting basically of a glass jar with the internal and external surfaces covered with thin metal foils. The main property of this arrangement is its capacity to store large quantities of opposite charges on the two surfaces. If the metal foils covering the surfaces are connected with a conducting wire, a sudden discharge of the opposite charges follows through the wire.

Volta held that the current of the battery is like the discharge one would obtain from a group of Leyden jars "feebly charged, which act unceasingly, or so that their charge after each discharge re-establishes itself."

With this idea of a set of self-recharging Leyden jars, he could reduce the new unknown situation to familiar notions. This is a good example of the intricate interplay of conservativism and innovation in science. On this occasion, his attitude was conservative. More bold was his previous assumption that the *energy* of an effect cannot exceed that of its direct cause. However, he did not raise this to the level of a general principle and on other occasions he did not conform to the precept. To blame him for this would be unfair because he was facing unknown complexities of nature. Thanks to Einstein and Infeld, we have had occasion to form an idea of how he handled the intricacies he met and to meditate on how the new instrument he obtained as a by-product contributed to changing the fundamental interpretation of the physical world.

Alessandro Volta's battery in the form of a column

Fabio Bevilacqua
Stefano Bordoni

Electromagnetic Induction:
Symmetries and Interpretations

"It is well known that Maxwell's electrodynamics – as usually understood at present – when applied to moving bodies, lead to asymetries that do not seem to attach to the phenomena."

The first words of Einstein's 1905 paper *On the Electrodynamics of Moving Bodies* are devoted to theoretical asymmetries in the Maxwell-Lorentz explanation of electromagnetic induction. Thus electromagnetic induction and, particularly, the relative motion between magnets and conducting coils, is placed at the root of Einstein's theory of relativity.

The simplest way to discover electromagnetic induction is the relative motion between a magnet and a conducting coil: the motion produces an electric current in the coil. The phenomenon is symmetric from the kinematical point of view: the induced current is the same in both cases – resting magnet and moving coil, or resting coil and moving magnet.

A similar current arises when we replace the magnet with another coil carrying an electric current. The magnetic effect of electric currents had already been discovered by the Danish scientist H. C. Ørsted in 1820. The motion of this current-carrying coil produces an induced current in the second coil: the first is named "primary circuit" and the latter "secondary circuit." We can realize the same effect even without motion: if we change the electric current in the primary circuit, an electric current appears in the secondary circuit. To the change in the electric current, a corresponding change in the magnetic effect follows. We notice that Ørsted discovered the magnetic effect of moving electricity and Faraday, conversely, discovered the electric effect of moving magnets. On these two effects are based two now widespread devices: the electromagnetic engine and the electromagnetic generator. In an electromagnetic engine, the current flowing through a coil, placed between the poles of a magnet, makes it rotate. In an electromagnetic generator, the

rotation of the coil between the poles of a magnet makes an electric current flow through it. Starting from 1831, M. Faraday analysed all different occurrences of the electromagnetic induction and pointed out the already quoted symmetry: the phenomenon depended only on the relative motion between coil and magnet. He explained the induced current as a result of the interaction between the conducting coil and the *lines of force* spread by the magnet through the surrounding space. These lines correspond to the lines drawn by iron filings thrown on a piece of cardboard placed upon a magnet. The amount of the induced current depends on the number of lines of force crossed, during the motion, in a standard range of time. Faraday refused the then widespread explanation, due to French scholars, of electromagnetic phenomena in terms of forces acting at a distance between charged bodies or currents. On the contrary, he considered electromagnetic induction as the result of a contiguous action propagating through space, aether or matter.

J. C. Maxwell often stated his agreement with Faraday's conceptual model of contiguous action. The core of his theory is the concept of *field*, a specific condition of space and matter when stressed by electric and magnetic actions. In Maxwell's theory, the interactions between electric and magnetic fields and also between the fields and their sources (electrified bodies and electric currents) take place in the ether, at finite speed. Following the conceptual model of the mechanics of fluids, he imagined a magnetic flux crossing the surface surrounded by the coil. An electric current is induced in the coil when the flux changes over time. This explanation is suitable for both kinematical arrangements: coil in motion or magnet in motion. Nevertheless, when he investigated the "causes" of that change of magnetic flux, a new asymmetry appeared. In his *Treatise* (1873), he presented an equation for the "electromotive intensity" induced in the conducting coil. It contains two different mathematical terms. The first term represents the magnetic force experienced by the inner electric fluid of the conducting coil in motion through a magnetic field and the second term represents the electric force induced in the coil by the changing magnetic field of the approaching magnet. Translating into Faraday's words, in the first case the coil "cuts"

charged particles of matter (electrons). These particles, either at rest or in motion, are the sources of electric and magnetic fields. On the other hand, these fields act on the electrons by means of a force split into two different parts: an electric force and a magnetic force.

Lorentz was actually concerned with the property of symmetry of electromagnetic phenomena. He

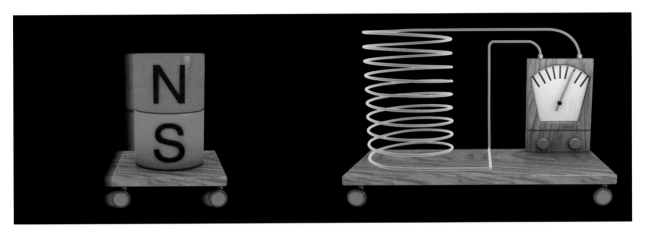

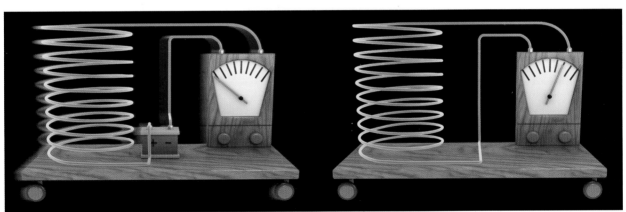

the lines of force, in the second case the lines of force "cut" the coil. Two different mathematical terms account for two experimentally symmetric arrangements.

From the theoretical point of view, the symmetry is lost. This is the query Einstein referred to in his introduction of the famous paper.

In the last years of the 19th century, the Dutch physicist A. H. Lorentz gave a new interpretation of Maxwell's theory. He introduced electrically

assumed their invariance when observed and described from different inertial reference frames. The presence of two different terms – electric and magnetic – is counterbalanced by the fact that electric and magnetic fields transform into each other, according to the different frames of reference.

This relative identity – electric or magnetic – of the fields is a consequence of the general approach to invariance of the physical laws of Poincaré and

The induction of a current through a moving magnet and a moving coil respectively

Electromagnetic Induction: Symmetries and Interpretations

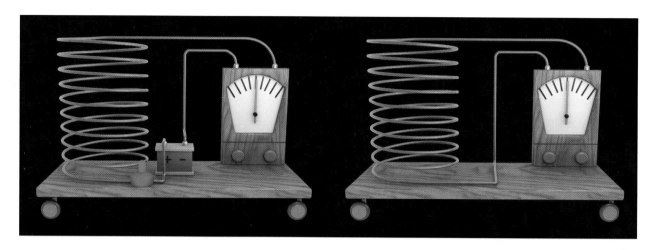

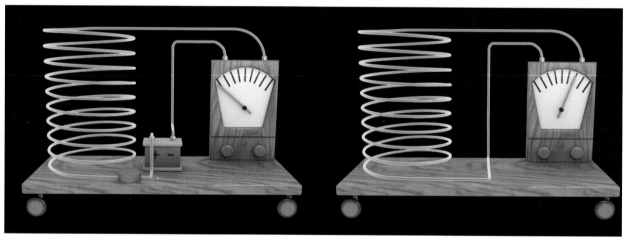

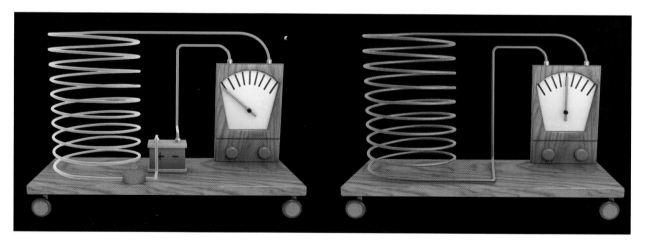

Einstein. In the 1905 paper, Einstein built his theory on two wide-ranging principles:
1. the invariance of physical laws, for both mechanics and electromagnetism (principle of relativity),
2. the invariance of electromagnetic waves speed in empty space, after having cleared away the aether.

In such a way, every asymmetry disappears in the interpretation of electromagnetic induction. The magnetic force on the moving coil and the electric force on the coil, arising from the moving magnet, are two different aspects of the same unified electromagnetic force.

Faraday's and Einstein's theories are surely symmetric theories. In Faraday's we find conductors and lines of force intersecting each other. In Einstein's, we find a deep inner symmetry between electricity and magnetism. Maxwell's and Lorentz's theories present both symmetric and asymmetric explanations. In Maxwell, we find the symmetric flux rule and the asymmetric explanation of the "causes" of induction. In Lorentz', we find symmetric transformation equations for E and B and the asymmetric account in terms of electric and magnetic forces.

Left side: Induction through a change in the current in a stationary coil
Above: Current is switched off
Center: Current is switched on
Below: Current flows uniformly

Shaul Katzir

Electricity and Heat:
The Connections Between Two Invisible Forces

As students of nature try to get to know its forces and laws, they often apply connections between different agents. These connections are divided into two types: causal (the generation of one force by the other) and analogical. While the former link the forces more tightly, the latter are helpful in understanding these forces and in formulating theories. Both kinds of connections between electricity and heat were exploited by students of these forces. The ideas about these connections, and thus the following brief history, were shaped by the changing concepts of the forces and by experimental evidence for their interaction.

At the end of the 18th century both electricity and heat were explained by theory of subtle or imponderable fluids. According to this doctrine, central attractive and repulsive forces between such fluids – which penetrate into matter – and ordinary matter accounted for cohesion and affinity, light, heat, electricity and magnetism. Electricity was explained by either one or two fluids. In the former case, electric fluid repels electric fluid but attracts matter, while matter devoid of electric fluid (minus electricity) repels matter. In the 1770s and 1780s Johan Carl Wilcke (1732-1796) used the analogy with the behavior of electric fluid to elucidate the concepts of latent heat (the heat needed to melt ice without changing its temperature) and specific heats (the amount of heat needed to raise standard mass of a given material by one degree). In parallel to the electric theory, Antoine Lavoisier (1743-1794) explained the effect of heat by the caloric (the fluid of heat), which holds the constitutes of matter apart against their own attractive force. The demise of the caloric theory, with the rise of the view of heat as a kind of motion in the second quarter of the 19th century and the rise of electromagnetism in the same period, ended this basic analogy between the two forces.

Still, similarities between heat and electricity did not totally disappear. The most important of these analogies was that between the flows of heat and electricity. Georg Ohm (1789-1854) based his famous 1827 theory of the electric conduction of metals on Joseph Fourier's (1768-1830) 1822 theory of heat flow. His success in deriving analogous equations to those of Fourier for electric current indicated to him an "intimate connection" between the two phenomena. Others shared Ohm's view. This view stimulated Gustav Wiedemann (1826-1899) and Rudolph Franz (1827-1902) to their discovery of an approximate constant relation between the conductivity of heat and electricity in 1853. In turn, their empirical law supported an assumption of the intimate connection between the two phenomena, whose exact mechanisms were a subject of speculation until well after the discovery of the electron. Using a combination of empirical and theoretical arguments, Ludwig Lorenz (1829-1891) claimed in 1872 that the Wiedemann-Franz constant depends linearly on the temperature, a claim that he verified experimentally by 1881. The electron theories of Einstein's time (notably that of Paul Drude (1863-1906)) explained Lorenz's relation on the assumption that both heat and electric currents in matter result from the motion of its electrons.

While the mechanical theory of heat broke the analogy between the forces of heat and electricity, it brought attention to the heat produced by electricity. In particular, it stimulated a quantitative-empirical study of the production of heat by the electric current by James Joule (1818-1889), who

Antoine Laurent Lavoisier

formulated a law for the heat evolved by current in 1841. This is the kind of heat generated by electric heaters. Following his subsequent work and that of others during that decade, the effect was viewed as a conversion of electric energy into heat. As such it found a natural place in the new thermodynamics formulated at the beginning of the 1850s, which viewed it as a good example of an irreversible effect. This process, however, seemed to teach little about heat, electricity or the specific connection between them. Potentially more telling were effects of heat on electricity – the generation of electricity by heat – and reversible effects. These effects suggested some ideas (unfortunately mostly unfruitful) about the nature of electricity and its relation with matter. Interestingly their discovery also involved questions about the relations between electricity and other forces, such as magnetism and what was then called the galvanic circuit (the electric circuit in today's terms).

The first effect of this kind was coined pyroelectricity. In 1756 Franz U. T. Aepinus (1724–1802) recognized that a known effect in which a heated crystal named tourmaline attracts and repeals ashes is of electric nature. He found that the crystal's two opposite ends show opposite electric behavior simultaneously, which he attributed to opposite minus and positive charges. The crystal is thus polar like a magnet. This observation supported the nascent and important analogy between electricity and magnetism. This analogy continued to appeal to later students of the field. Aepinus thought that the effect is due to heat. However, three years later John Canton (1718–1772) showed that electricity appears "only by the increase, or diminution of heat." A year later he found out that other minerals share the phenomenon. Later research showed that the electric properties of these depend on internal quality rather than on external shape. This indicated that the phenomenon is closely dependent on the

inner structure of matter. That was the idea of René-Just Haüy (1743–1822). Around the turn of the 18th century he formulated a theory of crystal structure, which among other things claimed that the appearance of pyroelectricity depends on the crystalline structure of the minerals. On the analogy with magnetism and his molecular view of crystals, he suggested that the molecules of pyroelectric crystals (like tourmaline) are polar. These polar molecules are responsible for the appearance of the macro effect. However, despite further research of the phenomenon, it did not receive a detailed theory. Neither heat nor electricity was the subject of the research that led to the first observation that heat can generate current. Thomas J. Seebeck (1770–1831) studied the relation between the so called galvanic circuits and magnetism, following Hans Christian Ørsted's (1777–1851) discovery of the magnetic effect of such a circuit in 1820. Ørsted, like most contemporaries, viewed "galvanism" as an electric current due to an electric tension in a wet or dry metallic cell (battery); the tension originated in the contact between two metals. Seebeck, however, thought that the magnetism of the galvanic circuit is not a direct effect of an electric current, that it is "independent of all external influences" including electrical. He also supposed that dissimilarities between the shapes of the two metallic junctions rather than their environments are the source of the effect. Experiments on different shapes of the junctions of two metals led him to the discovery that a difference in the temperature of two junctions alone produces, what he saw as a magnetic polarization in the circuit. However, Ørsted and his French colleagues immediately conceived that differences of temperature generate an electric current

Paul Drude

(although it was observed only by its magnetic influence), and coined it thermoelectricity. In Germany, however, during the 1820s and later, many viewed the effect as independent of electricity. Questions of electric conductivity in metals, the way their resistance produces heat and the connection between electricity and chemistry led Jean C. A. Peltier (1785–1845) to the discovery of an

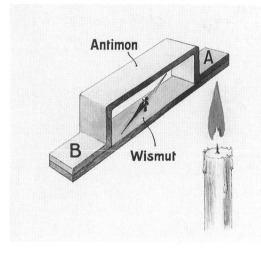

Generation of electricity through heat differential. Representation of an old exhibition board at the *Deutsches Museum*. Seebeck discovered that when the position at which two different kinds of metals touch is heated or cooled, electric current results. Experimental arrangement: two strips of metal, one bismuth and the other antimony, whose ends A and B are soldered together. When one soldered joint was heated, an electrical current resulted, deflecting a magnetic needle. Current also occurred (from the opposite direction) when a soldered joint was cooled

effect converse to Seebeck's. Peltier conjectured that heat was produced when the current breaks the electrochemical bonds within the metal. Since Humphry Davy (1778–1829) had shown that electricity sometimes destroys and sometimes builds chemical bonds, Peltier expected to observe cooling in the latter case. That he did not find, but he discovered that of the two junctions between the metals (bismuth and antimony) one became hotter and the other cooler. This change of temperature at the junction of two dissimilar metals is called the Peltier effect. Unlike the regular heating that is described by Joules's equation, this is a reversible effect that can be also used to cool a junction. In 1838 Heinrich Lenz (1804–1865) succeeded in freezing water by this effect.

Like pyroelectricity, thermoelectricity did not receive a general theory that connects Seebeck's and Peltier's effects and explains their origin. As an unexplained reversible phenomenon of heat,

thermoelectricity was one of the first subjects to be treated by the new thermodynamics. This was done by William Thomson (1824–1907), one of the founders of thermodynamics, in 1851. With the thermodynamic equations and the empirically informed assumption that thermoelectricity is a reversible phenomenon, Thomson predicted theoretically a third effect, which is often named after him: that in a homogenous conductor with a cooler and a hotter end, a current would produce or absorb heat, depending on its direction. In the following few years he carried out a few sensitive experiments, by which he eventually succeeded in verifying his theoretical prediction.

Thomson's thermodynamic theory accounted well for the central phenomena but could explain neither their causes nor specific features. For example, the theory showed that a current should exist but not its direction. Why does the heat flow with the (positive) electric current in copper and in a reverse direction in iron? All that Thomson could conclude is that the phenomenon is connected to a directionality in the matter, which he related to the crystalline structure of the metals. Independent experiments that showed marked differences in the thermoelectric properties depending on the direction of the current supported Thomson's view. In this sense thermoelectricity suggested ideas, but not firm propositions, about the connection between matter and electricity.

Production of electric tension, or electric motive force, by heat characterizes both pyro- and thermoelectricity. Moreover the first quantitative rules about the development of charge in pyroelectricity – established by Jean Gaugain (1811–1880) in 1856 – showed that the effect is linear, like thermoelectricity. This enabled Gaugain to suggest that pyroelectricity is caused by a hidden thermoelectric process. In 1886 Pierre Duhem (1861–1916) elaborated that hypothesis. However, most contemporaries rejected the hypothesis on its disagreements with various experiments.

Thomson suggested a different explanation based on its analogy to magnetism. He assumed that pyroelectricity originated in a permanent polarization inside the crystal, which is changed by heat. Jacques and Pierre Curie (1856–1941, 1859–1906) assumed that the electric polarity is of the crystal's molecules. This assumption led them to the discovery of a new phenomenon named piezoelectricity: the generation of electricity by a change of pressure in crystals. Following the Curies' work, physicists suggested that heat generates electricity through thermal motions of the molecules. However, no convincing detailed model of this mechanism was suggested.

The connections between heat and electricity were still expressed best in general thermodynamic equations. These equations did not explicate the inner connection between them and the way each agitates the other. This was also true for piezoelectricity. General thermodynamics supplies a description of a relationship between forces, effects and phenomena, not an explanation. To explain the connections between heat and

Ørsted demonstrates the deflection of the magnetic needle by the electrical current, 1820

electricity in matter one needed to know better the structure of matter that was still unknown around 1900. In retrospect we can also recognize the need to know the quantum mechanical laws. Even by themselves, these were found to be insufficient due to the complexity of matter and to the large number of particles and forces that interact in these effects.

Thomas Jung

Is Radiation Healthy or Does It Make Us Sick?

Radiation

While experimenting with cathode rays in 1895, Wilhelm C. Röntgen discovered a new kind of rays that could penetrate matter. Because their physical properties were largely unknown at the time, he called them "X-rays" (in the German-speaking world these rays are called "*Röntgen-Strahlen*"). The discovery of X-rays was followed a few weeks later by the first description of radioactive radiation by Antoine Henri Becquerel and Marie and Pierre Curie.

Röntgen's publications, like those by Becquerel and the Curies, caused a great sensation in science and society. The display of the first X-ray images during Röntgen's first lecture about his discovery made their benefit for medicine visible to everyone. Now diagnoses were possible without invasive operations. Very quickly it was recognized that X-rays had positive effects on many skin ailments and chronic illnesses like arthritis. As early as 1896 the first reports appeared about cancer therapy using X-rays. Primarily thanks to the Curies' work, soon radium was also introduced to the young science of radiation therapy as the first radionuclide.

In 1902, Ernst Rutherford summarized the early investigations on radioactivity, describing the radiation phenomena that occurred as the result of transformation processes, and subdividing radioactive rays into α radiation, β radiation and γ radiation. The particle nature of β-rays and α-rays was demonstrated in the studies by Becquerel and the Curies and in those by Rutherford and Geiger. Not until years later were X-rays characterized, like γ-rays, as electromagnetic radiation. Neutron rays were described as a further type of particle ray only much later, by James Chadwick in 1932.

The mechanism by which radiation works was initially unknown, so that the first applications, not only in medicine and technology, but also in everyday life, were frequently performed by a quite unsophisticated method of "trial and error." Few scientists seriously took up Albert Einstein's new ideas about light quanta, the photo-effect and the relationship between energy and mass of 1905 into their initial attempts to explain the mechanism by which radiation works. Arthur H. Compton's

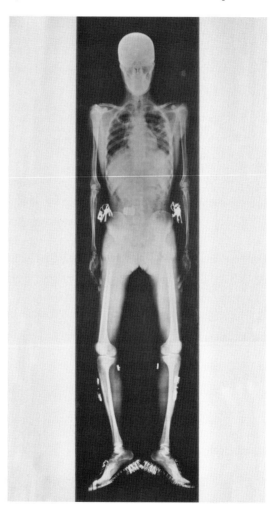

The first non-composite X-ray of the entire human body from Dr. Mulder, Bandung/Java, around 1925

studies of 1922 were the first to incorporate the general recognition of Einstein's ideas into an explanation of the direct interactions between X-rays and matter. The photo-effect and the Compton effect described two new kinds of X-ray interactions in addition to classical photon scattering, namely those of photon rays with matter. Later a further interaction of photons in the electromagnetic field of atomic nuclei was described, pairing, for which extremely high photon energies of around 1 MeV are required. For X-rays photon energies at a maximum of around 250 keV are applied, for γ-rays, on the other hand, energies of greater than 1 MeV can occur.

Hardly any other technology entered into practical everyday use as quickly as the application of X-rays and radionuclides in medical diagnostics and therapy. During this rapid entrance into a new technology, which was borne by great euphoria about technical innovations, the first negative effects of radiation were quick to appear. After using X-rays or radioactive rays carelessly and frequently, erythema, skin tumors and damage to the hemopoietic system were observed in the first patients, treating physicians and participating scientists.

Atom Bombs and Nuclear Technology

While the first years of using radiation in medicine and technology were marked predominantly by a view to the multifarious new possibilities and chances linked with its application, especially in medicine, the first atomic fission performed by Otto Hahn, Lise Meitner and Fritz Straßmann in 1938 and the atom bomb attacks on Hiroshima and Nagasaki in 1945 made visible to many another side of radiation and the technologies that brought it about.

After 1945, initially unreined development and armament with atomic weapons began, especially as a consequence of the Cold War between the Soviet Union and the U.S. In addition to building up

a vast potential for threat and destruction, the numerous aboveground and later underground bomb tests caused the local contamination of the test sites, as well as large-scale radioactive contamination of the atmosphere in their immediate surroundings. The public discussion about worldwide contamination through atom bomb tests – and the impression left by the Cuban Missile Crisis of 1962 – led to the 1963 Test Ban Treaty, which forbid nuclear weapons testing in the atmosphere, in outer space and under water. But France and China, which did not enter into this treaty, continued to perform aboveground testing of atomic weapons up until 1974 and 1980, respectively.

In the 1950s the development of nuclear reactors for the production of energy began in many industrial countries, embodying a "peaceful use of nuclear energy." The first research reactor in Germany went on line in Garching near Munich in 1957; the first nuclear reactor for energy production followed a few years later near Kahl am Main. Eighteen "power reactors" for energy production and four research reactors are currently on line in Germany. While energy production from nuclear power plants was initially supported to some degree by large sectors of the population in the context of the "economic miracle," the population's attitude to nuclear power changed over the course of the 1970s. The anti-atomic energy movement is linked with the names of communities like Wyhl, Brokdorf, Kalkar, Wackersdorf and Gorleben, in the vicinity of which nuclear technology projects were planned or operated. Major accidents at nuclear power plants, like those at Windscale in 1957 (today Sellafield, UK), at Three Mile Island in 1979 (near Harrisburg, PA in the U.S.) and not least of all at Chernobyl in 1986 (then Soviet Union, today Ukraine), which resulted in large-scale contaminations in Europe, influenced the attitude of society. Another event that must be mentioned here is the accident that occurred in a nuclear fuel reprocessing plant in Mayak (Russia)

Wilhelm Conrad Röntgen

in 1957. For many years this accident remained unknown in the West, but it resulted in large-scale contamination in Russia east of the Urals.

The Action and Risk of Rays

In the first years after the discovery of X-rays and radioactive rays, pathological changes were soon observed, the cause of which originated in the regular use of these rays, or their use at high intensities. Serious changes in the skin were observed (erythema, inflammations of the skin even extending to tumors). Similarly, acute, sometimes fatal, injuries to the hemopoietic organs were described. Soon a variety of acute injuries were described for various organs and organ systems, with the result that threshold values of exposure were determined below which these injuries did not occur. These kinds of injury are characterized as deterministic, because when exposure exceeds the threshold value, the occurrence of the illness is predetermined and foreseeable. As exposure to radiation rises, the intensity of the illness increases. The philosophy of the period was marked by the deterministic worldview derived from classical mechanics. In a world in which all forces effective at any given time are known, and the location of all parts with respect to each other known, events are completely predetermined and thus foreseeable. Heisenberg's uncertainty principle, described for quantum physics in 1927, was to change this deterministic image of science forever.

In addition to these deterministic injuries, quite soon scattered illnesses were observed in individuals who had been subject to regular exposure but exhibited no deterministic injuries. In particular, a high rate of leukemia was observed among the first pioneers of radiation applications in medicine and research. One of the best-known victims of this pioneer phase was probably Marie Curie, who died of the effects of leukemia in 1934. In particular it was the work of Hermann J. Müller

in 1927, who studied the mutations in the fruit fly Drosophila elicited by exposure to X-rays, which showed for the first time that it was possible to induce mutations with radiation.

Up to the 1960s, the scientific discourse about radiation risks concentrated on the genetic risks presented by radiation. Then around 1962 the first reliable data were published on the risk of leukemia among survivors of the atomic bombings in Japan. It soon became clear that other kinds of cancers also occurred with increasing frequency after exposure to radiation. According to today's knowledge, the primary risks of radiation are regarded as the risks of leukemia and cancer. This knowledge about radiation risks is based on the continued, longterm observation of cancer rates among the victims of atom bombs in Japan, of exposed population groups in the vicinity of atomic testing areas, of groups of patients from radiological diagnostics and therapy, and of persons exposed to radiation in their professions. The mechanism behind the inducement of leukemia and cancer after exposure to radiation is different in principle to that which causes deterministic injuries. This mechanism is characterized by its stochastic nature. The level of exposure to radiation determines the probability of an exposed person contracting leukemia or cancer. With increasing exposure the probability of illness increases. By contrast, the intensity of the illness is not dependent on the level of exposure. Overall the studies of exposed persons do not indicate the existence of any threshold for these stochastic illnesses. The relationship between exposure and cancers (dose-effect ratio) that concurs best with the available data is a linear dependency between the level of exposure and the probability of contracting cancer, without a threshold.

After the physical worldview of determinism had been shattered by quantum physics and its uncertainty principle, protection against radiation, which up to the 1960s had regarded deterministic

The explosion of the atomic bomb at Hiroshima, Japan on 6 August 1945

illnesses as radiation's only consequences for human health, was changed fundamentally by increasing knowledge about the stochastic nature of radiation effects, especially in cases of low exposure. This had corresponding effects for the view of radiation risks as well.

While a protection against radiation founded on a deterministic image of disease is oriented toward secure adherence to threshold values, a protection against radiation that presumes a stochastic image of disease – and thus the risk of radiation – must go beyond a pure discussion of threshold

Nuclear power
station
Biblis A and B

values. It is not sufficient to prescribe only that certain threshold values must be observed. Rather, even exposures below the threshold values must be avoided or reduced as much as possible by weighing out the circumstances of each individual case.

Is Radiation Healthy or Does It Make Us Sick?

Ionizing radiation must be categorized as carcinogenic and mutagenic. Its potential for causing illness is of stochastic nature, at least in the range of low exposures (below the effective threshold for deterministic injuries), i. e. every exposure to radiation is associated with an additional risk of illness. The level of disease risk depends on the level of exposure, that is, low exposure means a low additional risk of illness, while high exposure means a correspondingly higher risk.

Especially in the case of medicine, the risk of radiation is countered by a direct benefit for the affected patient. In this case rays at least can assist in the improved diagnosis of diseases, and, in many cases, can help to heal them. Thus rays

"Shamrock": warning signal for labelling radio-active materials and indicating potentially radioactive sites

may not be healthy, but when their application is medically justified, they are conducive to the process of recovery.

For other applications of ionizing rays, such as the production of energy in nuclear power plants, the monitoring of welded joints, and security checks of luggage in airports, generally no direct benefit to individuals exists. Yet, they are generally believed to benefit society. The risk-benefit calculations were and are assessed by society, or more precisely, through politics in the form of laws, regulations and guidelines.

Another factor that must be considered is the exposure to rays to which every person is subject: natural radiation. This exposure cannot be avoided at all, but its level can be influenced to some degree by individual behavior. Especially important in this connection is exposure to radiation from the radioactive noble gas radon in homes and other closed spaces.

The second part of the question posed above can be answered clearly. Ionizing radiation can be categorized as carcinogenic. Every additional exposure to rays is associated with an additional risk of disease. The first part of the question is difficult, however, because there is no clear answer. Ionizing radiation may not be healthy, but it can

be conducive to the recovery process when their application in medicine is justified. For applications outside of medicine, the risk of using ionizing rays often is countered only by a general benefit for society.

Klaus A. Vogel

The Revolution in the Image of the Earth

The Classical Model of the Spheres

From the periods of classical antiquity and the Arabian-Latin Middle Ages on, the residents of the Old World knew only the area of one or the other northern quadrants. The known world extended in the West-East direction from the Spanish penin-

described the ancient model of spheres. The spheres of the fixed stars and an outermost sphere in motion, the "primum mobile," formed the boundary of the universe. On concentric spheres within this moved the planets, the Sun and the Moon. In the center of the cosmos were the four spheres of the elements: fire, water, air

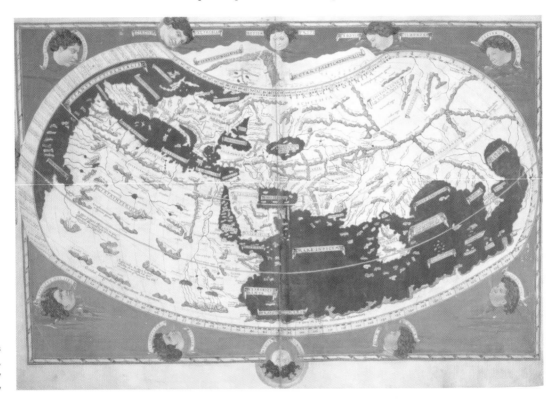

World map according to Ptolemy, from the 16th century

Alexandrinus Claudius Ptolemy, engraving by C.L.Bry

sula to Japan; in the North-South direction, from the North Sea to the equator.

A detailed portrayal of the known "ecumeny," the populated world, was presented in the work *Geography* by the scholar of late antiquity Claudius Ptolemy (1st century A. D.). In the scholarly tradition of the Western world, Ptolemy proceeded from the celestial nature of the cosmos and the Earth. Before him Aristotle, the father of the Western natural sciences, had already

and earth. Since Aristotle these were arranged according to their weight. In the middle of the cosmos was the sphere of the Earth, fixed, heavy and stationary.

How were the spheres of the earth and of water positioned in relation to each other? This question had been discussed by Greek, Arabian and Latin scholars since late antiquity. It proceeded from the arrangement of the spheres of the elements described by Aristotle, but could not be answered empirically. Since antiquity it had been assumed that ecumeny was surrounded by water. Only ocean was known to the west, to the north and to the east of the known world, ocean was alsopresumed to be in the south of Africa. But this did not clarify the relation between the sphere of earth and the sphere of water. Speculation was required to deduce the ratio of the size of the sphere of the Earth to that of the water and their relative positions.

In the 14th century the discussions intensified and the speculations became more differentiated. In a commentary to Aristotle's *De caelo et mundo,* the leading Parisian natural philosopher Jean Buridan (d. 1358/60) elaborated physical reasons why the sphere of Earth was permanently raised out of the surrounding sphere of water. He differentiated between the "lower" hemisphere of the Earth immersed in the sphere of water, and the "upper" hemisphere located above the surface of the water. The immersed "lower" part of the Earth sphere was heavier because of the water that had penetrated it. This is why the center of gravity of the sphere of the Earth is located in the immersed part. The above-water, "upper" hemisphere of the Earth is correspondingly lighter and is kept dry by the Sun. Thus while the center of gravity of the sphere of the Earth was located at the center of the cosmos, the center of its volume was not. Because of this, the eccentric position of the sphere of the Earth remained permanently stable. Up to the end of the 15th century, European scholars were convinced that the known, populated world was limited to one hemisphere, turned toward the sky and surrounded by ocean on all sides. In his text *De Mundo in universo* of around 1450, the church politician and humanist Aeneas Silvius, who was elected pope in 1458 and took the name Pius II, described the contemporary view of the world:

"Almost everyone is in agreement that the shape of the world is round. And they believe the same of the Earth, which rests in the center of all things, attracts everything heavy to it, and the majority of which is submerged in water."

The Discoveries of the Seafarers and the Columbus Project

While scholars speculated about the shape of the Earth and the relationship between the spheres, the Atlantic voyages of the Portuguese and Spaniards expanded the Old World's scope of

Representation of Aristotle's physics, with the four elements in the center, around 1450

experience. Starting in the second decade of the 15th century, the Portuguese Prince Henry the Navigator sent out ships to systematically explore the west coast of Africa. In 1444 Dinis Dias led an expedition that sailed around the green "Cape Verde"; in 1446 Nuno Tristâo reached the estuary of the Gambia; in 1460 Pedro de Sina sailed around the mountains of Sierra Leone. Just north of the equator the sailors had reached green, fertile land populated by black people, disproving the ancient and medieval theories about the insurmountable "burnt" or "withered zone" (*zona torrida*). In the year 1474 or 1475 Lope Gonçalves and Rui de Sequeira crossed the equator. In 1482–84 Diogo Çao reached the estuary of the Congo and Cape Santa Maria (13 degrees south). The coast proceeded southward. Bartholomeo Diaz returned to Lisbon four years later – he had pushed forward

Geocentric world-
view according
to Schedel,
woodcut, 1493
(photo: Lutz Braun)

far to the south, had sailed across the open sea and, in a heavy storm from the west, had sailed around a southern cape he christened the "Cape of Storms" (*cabo tormentoso*). The Portuguese King Joâo II (1481–1495) named it "Cape of Good Hope" (*cabo de boa esperanza*).

The navigators, after initially proceeding carefully along the African costs like coastal shippers, soon became more experienced and courageous on the open sea. Next in store was the circum-navigation of Africa. Yet the cosmographic knowledge of the seafarers and contemporary scholars had not changed. Then as before, it corresponded with the view of the populated hemisphere comprised of three continents, which were surrounded by an immense ocean.

Against this backdrop, Christopher Columbus' project of a journey westward to "India" all the way through the "lower" hemisphere must have appeared reckless to his contemporaries, even irresponsible. From 1483 until 1491 Columbus negotiated with Portuguese and Spanish scholars in Lisbon, Salamanca and Santa Fé (near Granada) about his project of a journey westward. The scholars rejected his project with good reason. In Salamanca, under the direction of the Archbishop of Granada, Fernando di Talavera, the following objections were raised:

1. A westward journey to Asia would take three years.

2. The western ocean was believed to be boundless and potentially not navigable. If Columbus reached the area of the antipodes located across from Europe, he would not be able to come back.

3. No antipodes existed, for, as Augustine said, the majority of the sphere of the Earth is covered with water.

This debate about Columbus' project was thus not only about the question of the distance between the western and eastern borders of the known world. At its center was the fundamental question as to the relationship between the elements earth and water and thus of the actual shape of the Earth.

In this context, the reference to Augustine had real weight. His name stood for Aristotelian philosophers like Buridan, whose differentiated conceptions were opposed to a westward journey on the ocean. Christopher Columbus, by contrast,

held a simple view of the shape of the Earth. In the volume of *Imago mundi* by Pierre d'Ailly that he annotated with his brother Bartholomeo, we find in the chapter *About the Four Elements and Their Position* the lapidary marginal note: "Earth and water together comprise a round body."

The Discovery of a "New World"

As we know, Columbus was ultimately successful in securing acceptance for his project. His western journey can be considered the first great experiment in the history of modern science. For the question of the actual shape of the Earth

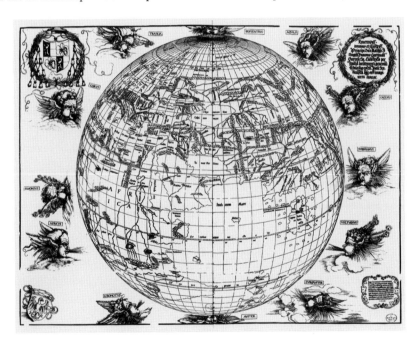

could not be decided by arguments alone. That Columbus did not give up and actually traveled westward shows not only his stubbornness and his daring, but also his practical-empirical sense. That he returned proves his skill as a navigator. In this Columbus and the sailors who accompanied him were very lucky. The experiment nearly failed on several occasions. And it supplied results that were unexpected and initially not even fully explicable.

The new image of the Earth. Planiglobus by Johannes Stabius and Albrecht Dürer, Nuremberg 1515

For Columbus did not reach "India" (as he believed up to his dying day), that is, China and the Far East of the known world. He found new, unknown islands, whose geographic position in relation to the known world was initially unclear. After the initial euphoria about the discovery of "new islands," reactions to Columbus' journey remained cautious. The Columbus letter of 1493, circulated throughout Europe, spoke generally of "many islands populated by countless people" in the "Indian Ocean." For the educated scientists of his day, this was too imprecise to clarify the question of the shape of the Earth. The experiment of the journey westward had provided surprising observations, but it did not achieve a breakthrough.

Yet Columbus' return gave new impetus to the overseas journeys of discovery. The Portuguese resumed their attempts to circumnavigate Africa. From 1497 until 1499 Vasco da Gama sailed around Africa, reaching Malindi on the east coast of Africa and Calicut on the west coast of India, thus opening up a new route to India by sea. For the first time since antiquity it was possible to establish direct contact with India and the Far East from Europe.

In terms of cosmography, the circumnavigation of Africa confirmed the traditional image of the known world consisting of the three continents Europe, Asia and Africa. But the discoveries continued. On the way to India in April 1500, 13 ships manned by over 1200 sailors under Pedro Alvarez Cabral's command chanced upon unknown land in the southwest Atlantic: what is today the coast of Brazil. The subsequent expedition along the newly discovered coast was accompanied by Amerigo Vespucci, a trained navigator and humanist from Florence. His report to Lorenzo di Pier Francesco de Medici, entitled *Mundus novus*, was a sensation for his contemporaries. Vespucci's cosmography was precise and did not shrink from criticizing the classics. Against the majority of

"the Ancients," proof had finally been found that the Earth was also inhabitable beyond the known world of the ecumeny:

"And this certainly transcends the knowledge of our Ancients, for the majority of them say there is no coherent land (*non esse continentem*) beyond the equator and in the southerly direction, but only an ocean they call the Atlantic. And if a number of them did agree that there was such land there, they denied on many grounds that it was an inhabitable Earth. But that their opinion is incorrect and completely contrary to the truth was revealed by this last journey of mine."

In few sentences Amerigo Vespucci made clear: in the southern antipodes a "New World" had been discovered. With this all conceptions that restricted the populated land to Europe, Asia and Africa and differentiated between an "upper" hemisphere (not covered by water) and a "lower" one (covered by water) were refuted. This was the cosmographic revolution. Contrary to all eccentric conceptions, the discovery of the antipodes proved the existence of the modern earth-water globe.

Vespucci and Copernicus

The discovery of the antipodes and the proof of the earth-water globe shook the authority of the classics. But "the Earth" still stood at the center of the ancient-medieval cosmos; the arrangement of the planets had not been affected. Yet it was now possible to recognize that the Aristotelian principle of gravity had a smaller scope than had been previously assumed. Apparently, earth and water were not arranged in the cosmos strictly according to weight, but together constituted a lighter, mixed body. This made more probable the considerations that the earth-water globe itself could be a moving celestial body.

This was an important point of departure for Nicolaus Copernicus. His heliocentric model of the world, which he formulated for the first time in the years 1507 through 1514, presupposed a relativization of the Aristotelian principle of gravity. For only if the gravity observed on the Earth was not a principle for universally arranging the cosmos was it conceivable that the Earth could be permanently shifted away from the center of the cosmos.

As a logical conclusion, Copernicus began his famous work *De revolutionibus*, printed in Nuremberg in 1543, the year of his death, with a refutation of the Aristotelian position. After two short chapters in which he reported on the medieval

textbook by Johannes de Sacrobosco, *De sphaera*, he wrote a chapter of his own entitled *How Earth and Water Constitute a Sphere* (*Quomodo terra cum aqua unum globum perficiat*). Concisely and pertinently, Copernicus emphasized that the earth-water globe had been proven according to the recent geographic discoveries. He expressly mentioned Amerigo Vespucci:

"That will become even clearer when those islands are added which were discovered in our age under the rulers of Spain and Portugal, and especially America, which is named after its discoverer, a navigator, and, despite the fact that its size has not yet been settled, is believed to be another world (*alterum orbem terrarum*), not to mention the many, previously unknown islands, so that it should come as no surprise that there are antipodes or antichthones."

In *De revolutionibus*, his *magnum opus*, Copernicus' argumentation proceeded not from the heavens, but from the earth. The overseas discoveries, and their proof of the earth-water globe, had clarified a decisive prerequisite of his cosmological model. The long-term effect of the Copernican model on the modern image of the cosmos is well known. Yet now we see that the famous scholar did more than observe the heavens and study the classics of astronomy. Nicolaus Copernicus had reacted to the cosmographic revolution that took place before his eyes, attentively, competently and creatively.

Hans Holbein the Younger, *The Ambassadors*, 1533

The Cosmographic Revolution

The changes in perception with reference to the earth, its surface and shape in the decade after 1500 were more unique and dramatic than ever before. The navigators' discoveries substantiated the image of the modern earth-water globe, previously considered a speculative concept, as the image of experienced reality. The globe, the new image of the Earth, provided a decisive prerequisite for Copernicus' cosmological conception and, together with the scientific and societal establishment of modern geography, had an exemplary effect on the development of early modern sciences.

Enrico Antonio Giannetto

Giordano Bruno and the Origins of Relativity

Giordano Bruno (1548–1600) was the first to propose a scientific theory of the relativity of motion, of space lengths and of time intervals in the period between 1584 and 1588–1591. This was not yet fully recognized essentially for two reasons: i) his heretical (in respect to the Catholic Church) theological ideas led him to the stake and his works were subjected to the so-called *damnatio memoriae* and were subsequently forgotten (some of his works are still not translated from Latin); ii) his conception of nature was not mechanistic and so for the dominant historiography, which identifies science and the mechanistic worldview, Bruno is not considered a scientist.

Giordano Bruno, presumably portrait from the *Livre du recteur* of the University of Geneva, 1578

At first, Bruno gave his assessment of the Copernican world system, but he went beyond Copernicus: he eliminated the solid spheres to which celestial bodies were considered to be related. He gave a physical basis to the new astronomical system: the alternative to Aristotle's physics was a mixed version of mediaeval impetus theory and ancient dynamical atomism, where atoms are not purely material and inert but full of power and form as in the original ideas of Demokritus, and, alternatively, the vacuum is not at all empty but full of aether. Atomism together with Christian theology led Bruno to the conception of an infinite universe made of infinite worlds. Since medieval natural philosophy, Christian theology produced a form of reasoning secundum imaginationem which involved a progressive deconstruction of Aristotelian physics and cosmology: it was the so-called argument *de potentia Dei absoluta* (on the absolute power of God). Scientific imagination was strictly bounded to theology. Since 1277, the year in which the Bishop of Paris, Étienne Tempier, condemned many points of Aristotelian physics, something similar to Christian natural philosophy had been tentatively constructed. Bruno introduced the idea that the power of God is infinite, and so also that creation must involve an infinite universe made of infinite worlds and infinite atoms, full of powers. This theological perspective was in opposition to the Calvinist idea, at the roots of the mechanistic conception of Nature as inert and passive matter, that the omnipotence of God would be limited by any power belonging to creation or to creatures be. However, at least in one respect, Reform theology was fundamental for Bruno: the potential for free interpretation of the Bible. Only the reference to the Bible had prevented John Buridan, Nicole Oresme, and Nicholas of Cusa from affirming the motion of the Earth. Bruno argued that the Bible gives only ethical indications and in no way scientific certitudes. The Joshua sentence contained in the Ancient Testament, "Stop Thou, Sun," gives no scientific assessment of the geocentric world system, but only an indication of the power of faith. Moreover, the revelation of the Gospel was that of a new earth and a new heaven, and Bruno felt himself to be one who had understood the hidden cosmological and physical content of this revelation: the new Earth and new heaven were an infinite universe.

Bruno's ideas were the basis of Shakespeare's famous verses in Hamlet, II, 2: "Doubt Thou the stars are fire / Doubt that the sun doth move / Doubt truth to be a liar / But never doubt I love." The doubt about the substance and the motion of the Sun, the doubt about Bible's sentences as scientific truths, leave only love as spring of certitudes for human life. The need for an infinite universe is to be ascribed to this same love, which is also written in the verses of Shakespeare's Anthony and Cleopatra (I, 1, 14–17): "Cleo. If it be love indeed, tell me how much. / Ant. There's beggary in the love that can be reckon'd. / Cleo. I'll set a bourn how far to be belov'd. / Ant. Then must thou needs find out new heaven, new earth."

Certainly, in Bruno's assessment of an infinite universe there were not only theological or ethical love reasons, but also scientific reasons: the scientific reasons of atomistic physics and cosmology, of the impetus theory, of the new astronomical observations of comets by Tycho Brahe, of the critical remarks by Bruno himself on the seeming fixity of stars, due to the distance from us, and on the

whole infinite space was for every body. Every body has a dynamical consistency and is in motion in infinite space where there is no body at rest as well as no mathematical or physical centre. The absence of any body at rest implies the impossibility to give an absolute measure of motion and consequently the necessary relativity of all motions. In Bruno's perspective, the rela-

Giordano Bruno burning at the stake of the Inquisition (Rome, 17 February 1600) Bronze relief, 1887, by Ettore Ferrari, Giordano Bruno Monument, Rome, Piazza di Campo de' Fiori

mathematical abstractions which can never correspond to physical measures and to physical reality.

It was the mathematical cosmological model which also remained dominant in Aristotle's physics and was destroyed by Bruno initiating a new non-hierarchical relationship between mathematics and physics.

In this way, Bruno broke the spheres and circles of the motion of celestial bodies: the motions of celestial bodies are completely free motions in empty (without matter) ethereal infinite space due to the impetus that every body has. No longer was there a natural space related to a body, but the

tivity of motion was not an argument against the reality of motion as it was for Parmenides, but, as for Heraklitus, motion and change are the fundamental characteristics of physical reality.

Indeed, for the first time in the dialogue, *La cena de le ceneri* (The Ash Wednesday Supper, IIIrd dialogue, 1584), which was written in England and published in London during a trip to escape the Catholic Inquisition, Bruno first drew upon the relativity of motion to give a proof of the Copernican world system by discussing a complex thought experiment on the motion of a ship, already used in a simpler form by Buridan, Oresme and Nicholas of Cusa, and then also simplified by Galileo. The relativity of motion considered by Bruno was based on the idea of the participation of all the things belonging to a system, to the motion of the system in such a way that any motion of translation without rotations (uniform or non-uniform, rectilinear or circular or along any curve) does not modify phenomena. Thus, one cannot perceive that the Earth has a circular motion around the Sun, because this motion does not affect phenomena on the Earth. Rotations would make a difference, but they are intrinsically relative motions among the different parts of the body. Here, Bruno also argued about the relativity of gravity, giving for the first time a sort of a principle of a dynamical general relativity.

Galileo Galilei, copper engraving, around 1793

Galileo, in his *Dialogo sopra i due massimi sistemi del mondo* (1632), repeated some of Bruno's arguments, but never quoted him because of the Inquisition. Thus, one has to speak of a Brunian and not Galilean "principle of relativity of motion." Interestingly enough, this has no relation with the so-called "principle of inertia," which the dynamical atomism of Bruno did not admit: it is only motion which can continue, whereas rest in the vacuum is unstable: by the so-called coincidentia oppositorum, rest is to be identified with motion at infinite velocity.

The relativity of time also follows from the infinite universe, but in some way it is independent of infinity. This consequence was discussed by Bruno in his *Camoeracensis Acrotismus* (1588), art. XXXVIII, and in his *De innumerabilibus, immenso et infigurabili; seu de universo et mundis libri octo* (1591, Liber VII, Chapter VII). One needs to remember the Aristotelian definition of time, which was also assumed in medieval natural philosophy: time was physically and cosmically given by the motion of the so-called eighth sphere, the sphere of fixed stars, because this motion was perfectly uniform, continuous and simple as required by a definition of time. However, if the universe is infinite there is no sphere and no privileged, perfectly uniform and continuous motion for the definition of time. There are infinite motions in the universe and every body could be used for the definition of time, and from the relativity of motions is implied relativity of time: thus, different motions define different, non-homogeneous, proper times, and so motion affects measures of time intervals which depend on motion.

The relativity of space is already implicit in its infinity without a centre, in the relativity of motion, but for Bruno it is also present in the relativity of space lengths and distances. He argued about it in the *Comoracoensis Acrotismus* (1588, art. XXVII, XXXII, XXXIV, XXXV e XXXVII), in *De innumerabilibus, immenso et infigurabili; seu de universo et mundis libri octo*, (1591, Liber IV, Chapter VI), and in a more profound way in *De triplici minimo et mensura ad trium speculativarum scientiarum et mulatarum activarum artium principia libri V* (1591, Liber II, Chapter V). Here, Bruno deduced the relativity of space lengths, starting from a radical epistemological critic of measurability and measures: motion affects measures and implies limits on the possibility to make exact measurements. From this point of view, measurements of space in different conditions of motion imply different space lengths.

As is well known, Galileo followed only Bruno's idea of a relativity of motion, limited to the case of uniform motions and considered as a kinematical relativity and not a dynamical one. Only Leibniz followed Bruno's idea of a general dynamical relativity of motion, time and space, giving it its first mathematical form. This idea had a

Scenographia Systematis Copernicani Copper engraving from the *Harmonia Macrocosmica* by Andreas Cellarius, Amsterdam 1708

complex and discontinuous story – which cannot be resumed here – within different conceptions of nature and of motion, space and time. Finally, it was reconsidered by Henri Poincaré, influenced by Leibniz physics, and through the influence of Poincaré and Mach by Albert Einstein who, at the time of the formulation of the special and general relativity theories, was not aware of this long story.

Jochen Büttner

Of Dwarves and Giants:
The Transformation of Astronomical Worldviews

"We are like dwarves on the shoulders of giants, so that is why we can see more than and further they ..." (Bernard of Chartres, ca. 1130)

Where did the giant, Newtonian mechanics, come from, whose shoulders the dwarf Einstein climbed upon to see the new science, and who, under the weight of Einstein's general theory of relativity, ultimately dropped to his knees and shrunk back to life size? How could Newton's mechanistic worldview develop, that makes the Earth a planet and the Sun a star, in a universe in which the movements of the celestial bodies were determined by the same laws that were responsible for the motion of bodies on the Earth?

Such a worldview actually contradicts the experiences of the observer on the Earth. For him everything seems to suggest that he himself, and thus also the Earth, are located at rest at the center of a moving cosmos, which regularly and incessantly revolves around them. On an Earth on which all observed motions exhibit the tendency to come to rest, such an observer moreover has hardly any reason to assume that the same laws determine both the self-depleting motions on the Earth and the eternal movements of the celestial bodies. More precise observations of the movements in the heavens show the observer that the fixed stars move the most regularly of all celestial bodies, as if they were affixed to a great, invisible sphere which revolves around the Earth once every 24 hours. Of similar regularity is the movement of the Sun, which, however, lags slightly behind the movement of the fixed stars, and is overtaken by this movement over the course of a year. The observed movements of the planets, finally, exhibit scarce if any regularity; at times, some planets even change their direction.

In order to describe and interpret observations of this kind, the mathematicians and astronomers of Greek antiquity developed models, according to which the irregular movements could be understood as the result of regular circular movements

Christian interpretation of the Aristotelian worldview, 15th-century codex

superimposing each other. In order to make such a superposition possible, Eudoxus in the 4th century B. C. introduced a model of nested "homocentric" shells. The celestial bodies were attached to the spheres, at whose center the Earth was

located. Each of these spheres had its own motion, but at the same time each also took on the movement of the spheres that surrounded it.

The way this model functioned becomes especially clear in the example of the motion of the Sun. The Sun moves around the Earth daily and, at the same time, in an annual motion relative to the fixed stars. This double motion is explained in Eudoxus' model by the fact that the shell of the Sun is pivoted in the sphere of the fixed stars, and is thus compelled to participate in the daily revolution of sphere of the fixed stars' from east to west. Hence observer on the Earth sees the Sun rise in the east and set in west each day. At the same time, however, the Sun's sphere also revolves around its own axis, which is tilled with respect to the sphere of fixed stars. Therefore the Sun's path in the sky, that is, its position with respect to the fixed stars, shifts slightly from one day to the next, thus causing the change of seasons.

the universe. According to Aristotle's system, the stationary Earth was surrounded by no less than 55 crystalline spheres, holding the Moon, the Sun, the planets and the fixed stars. Although many of these spheres were superfluous in the mathematical sense, they were introduced by Aristotle in order to construct a mechanical connection, such that the spheres located on the outside could drive the inlying spheres by friction. As part of his attempt to organize all knowledge in the system of the natural philosophy, Aristotle developed his model of celestial movements into the most comprehensive cosmology of antiquity.

Almost simultaneously to this cosmology, an alternative system describing the celestial movements was developed, oriented more strongly toward the demands of astronomers. In this "system of epicycles and deferents," too, it is the superimposing of regular circular movements that produces the seemingly irregular movements of the planets. Here small circles, the epicycles, move on larger circles, the deferents, at the centers of which the stationary Earth rests. The planets themselves revolve on the epicycles, so that their motion results from the superposition of this rotation and the movement of the epicycles on the deferents.

Based on such a system, and on the most accurate astronomical data of his age, Ptolemy in the 2nd century B. C. created a complex model of the motion of all celestial bodies around the Earth and published it in his *Almagest*. The Ptolemaic system rendered the movements of the celestial bodies more precisely than the Aristotelian one and proved to be better adaptable to the astronomic data, becoming ever more comprehensive and precise over time. In some

Aristotelian worldview, representation from the 16th-century

Still in the same century, the basic idea of this model was taken over by Aristotle, who developed from it his own influential conception of

respects, however, it was also inferior to the Aristotelian alternative with its crystalline spheres, for instance as concerns the question of the causes and the mechanism of conveying motion.

As different as the Aristotelian and Ptolemaic systems were in detail, in the ways they were integrated into systems of knowledge and in terms of their audiences, they still can be understood as little more than distinct variations of the same basic idea. They embody a geocentric worldview, that is, the image of a resting Earth around which

so revolutionary in retrospect, which put the Sun at the center of a stationary universe and made the Earth a planet, revolving just like the other planets, around the stationary Sun. As radical as this new heliocentric worldview was in many respects, it nevertheless maintained many aspects of the old system, such as for instance the crystalline shells.

Copernicus left a difficult legacy for his era. Even half a century after the publication of his epochal work, only few scholars had grappled with the de-

Schematic representation of the way the Ptolemaic system works, 16th-century

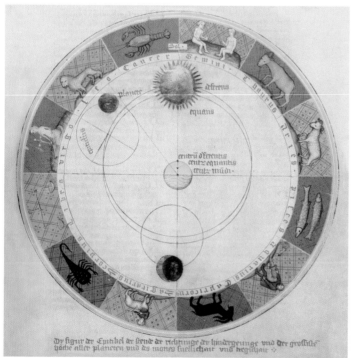

Alhazen's (Ibn al-Haytham's) interpretation of the Ptolemaic system, 15th-century codex

all celestial bodies move. Over the course of time, both systems were criticized, modified and ultimately even harmonized with Christian doctrine. As variants of a geocentric worldview, however, they remained practically without competition for two millennia.

Serious competition for the geocentric worldview did not emerge until the year 1543, when Copernicus presented the astronomical work *De Revolutionibus*, on which he had worked almost his entire life. In this work he introduced his worldview,

tails of his new system, among them the excellent astronomer Tycho Brahe. Tycho recognized the potency of the Copernican system over the Ptolemaic, yet he preferred a geocentric worldview for physical as well as theological reasons. Therefore, starting from his extremely precise and comprehensive astronomic observations, he constructed yet another version of the worldview.

This Tychonic, or "geo-heliocentric" worldview restored the Earth to its stationary position at the center of the universe. As in the geocentric model, the Sun revolves around the Earth, but, as opposed to the purely geocentric system, the planets revolve around the Sun (see below).

Tycho's system found many disciples among the 17th century Ptolemaic astronomers, since it

With regard to their success in foreseeing the motions of celestial bodies, however, they differed little or not at all.

In the course of the 17th century, however, new observations and arguments in favour of the

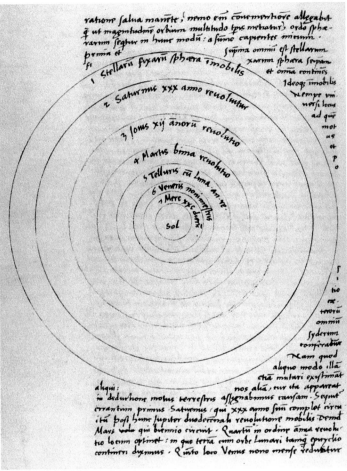

Nicolaus Copernicus' representation of his worldview in a handwritten version of his *De Revolutionibus* (1520–1540)

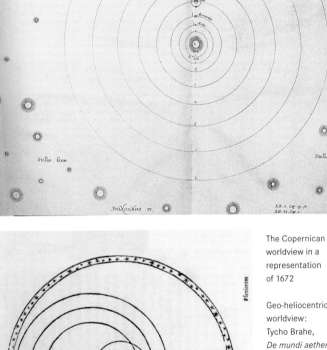

The Copernican worldview in a representation of 1672

Geo-heliocentric worldview: Tycho Brahe, *De mundi aetheri recentioribus phaenomenis*, 1588

avoided the ideological problems associated with Copernicus' setting the Earth in motion (see above). However, it soon became apparent that Tycho's system also involved considerable problems. For instance, because planetary orbits intersected in this system, it was out of the question that crystalline spheres could serve as the mediators of motion.

Initially it was exceedingly difficult to decide in favor of either of the competing systems. Each of the worldviews had strengths and weaknesses.

Copernican system accumulated. Thus Galileo, using the telescope which he had further developed, confirmed that Venus, similar to the Moon, exhibits phases, a phenomenon that had been predicted by the Copernican, but not by the Ptolemaic system. However, in its complete lack of a mechanics of the heavens, i. e., an explanation of the causes as well as the mechanism for conveying the motions described, the Copernican system shared a fundamental deficit with its competitors.

The assumption of elliptical orbits, on which the planets were supposed to move – at changeable speeds, moreover – could of course not be reconciled with the prevalent interpretation of movements in the heavens, which assigned the regular circular motions of the celestial bodies a special status with regard to all earthly motions. In the

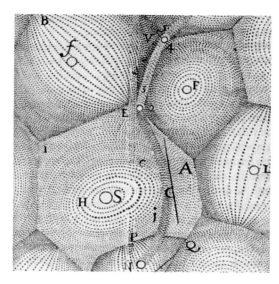

René Descartes' mechanistic vortex theory to explain the celestial motions, in his *Principia philosophiae*, 1644

search for a new physical explanation for the celestial movements, 17th-century scholars developed a variety of different approaches. Kepler himself proposed the sunlight or a magnetic force as the cause of planetary motion; René Descartes attributed it to the collisions of small corpuscles filling up the universe.

At the end of the 17th century, based on these proposals, Isaac Newton developed his own explanation for the motions in a Copernican universe Kepler had described. According to Newton, planets fall to the Sun, just like an apple falls to the Earth, and are thus held in their orbits. The acting force is identical in both cases: a universal gravitational force, acting between heavy bodies which depends only on their masses and their mutual distance. With this theory Newton had finally breached the gap between heavenly and earthly motion and had subjected the universe to uniform mechanical laws of motion. Newtonian mechanics

Copper engraving for the title page of Johannes Kepler's *Mysterium Cosmographicum* of 1596

The resolution of this deficit, i. e., the mechanization of the Copernican system, was to proceed along intricate paths. In his *Mysterium Cosmographicum*, published in 1596, Johannes Kepler imagined a god who had used the five simplest regular polyhedra to fix the distances between the Copernican celestial spheres in order to create a perfectly harmonic universe. However, intensive study of Tycho's astronomic data led Kepler to abandon this idea and instead assume that the planets move around the Sun on elliptical paths, whereby their movements were supposed to obey regularities, which he formulated in his three laws of planetary motion.

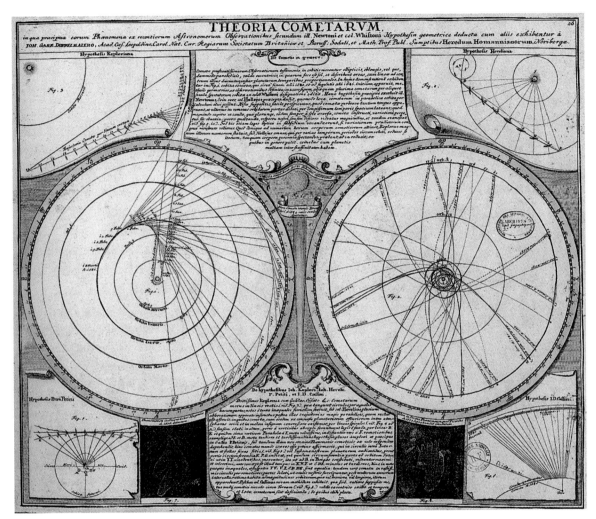

Comet orbits according to Newtonian theory, in an atlas of the heavens, 1742

subsequently became the exclusive instrument for the solution of astronomical problems and time and again served to generate new knowledge about the way the universe is constituted. Newton picked up Chartres' metaphor and called himself a dwarf standing on the shoulders of giants. Although born a dwarf, Newtonian mechanics, over the course of the next two centuries, with the help of generations of researchers, grew up to be a giant itself, whose shoulders were to be scaled at the beginning of the 20th century by another dwarf again: Einstein.

Matthias Schemmel

Curved Universes Before Einstein:
Karl Schwarzschild's Cosmological Speculations

Albert Einstein's theory of general relativity of 1915 represented a revolution of fundamental concepts in physics. Gravitation, previously a force that diverted material bodies from their natural path, was now supposed to be a deformation of space and time that guided bodies in their natural paths. Space and time, previously a stage on which physical events took place, were now supposed to be actors themselves: dynamic spacetime not only influences the motions of matter, but is also bent and curved by matter.

While Einstein's new theory thus demanded fundamental rethinking on the part of physicists, the empirical utility of the theory was marginal. All the observed phenomena that were described correctly by Einstein's revolutionary theory could be described equally well by Newton's established theory, the mathematics of which, moreover, was much easier to handle. It was just one small and up until now unexplained deviation of the motion of the planet Mercury from the orbit predicted by Newtonian theory that Einstein was able to explain with his theory. Even as late as 1917, the renowned physicist Max von Laue commented that this fact of "agreement between two single numbers" was hardly reason enough "to change the entire physical worldview in its essentials, as Einstein's theory does."

Among the astronomers too, a cautious, and even negative attitude to the new theory prevailed. In fact, all the spectacular astronomical objects that are described so successfully today on the basis of the theory of general relativity – supermassive stars, black holes, galactic nucelei and quasars, and in fact the expanding universe as a whole – were not objects of research, indeed, their exis-

Karl Schwarzschild in 1900 published a paper about non-Euclidean cosmology. His works on the general theory of relativity were written in early 1916 when he served on the Russian front. There he contracted a skin disease, which caused his death a few months later

tence was entirely unknown! Yet a few astronomers were among the most important pioneers of the theory of relativity. One such exception was the German astronomer Karl Schwarzschild (1873–1916).

Even before completion of the theory of general relativity, Schwarzschild's work concerned the possible consequences this theory entailed for astronomical observation. In 1913 he started a series of observations of the solar spectrum to investigate the redshift of the spectral lines of the Sun, which Einstein had predicted on the basis of his equivalence principle. Just a few weeks after Einstein had introduced the successful calculation of the anomalous motion of Mercury on the basis of his new theory in November 1915, Schwarzschild published the first nontrivial exact solution of Einstein's equations that describe the gravitational field and its interaction with matter in the universe. The solution was essential to the further development of the theory of relativity, in which it continues to play a central role today. Schwarzschild presented a second exact solution in a further publication, which introduced a quantity for the first time, the "Schwarzschild radius," which later was to play a major role in the theory of black holes.

Why was Schwarzschild's reaction to the theory of relativity so different from that of most of his colleagues in physics and astronomy? How could he recognize the importance of the theory of general relativity at such an early stage? In pursuing these questions it becomes apparent that long before the emergence of the theory of relativity, Schwarzschild was concerned with problems that could only later be dealt with adequately within the framework of this theory. The following example serves to illustrate this.

As implied above, space and time are "curved" in Einstein's theory of general relativity. This is to be understood in the sense in which the surface of a sphere is curved relative to a flat surface. This means, for instance, that – in opposition to the flat, so-called Euclidean geometry usually taught in school – the sum of the angles in a triangle no longer amounts to exactly 180°. The mathematics of curved spaces had already been developed in the course of the 19th century. The mathemati-

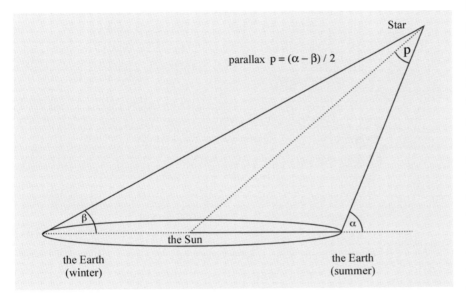

cian Bernhard Riemann (1826–1866) brought this development to a certain completion, which is why it is also refered to as Riemannian geometry. The knowledge of the mathematical possibility of non-flat spaces naturally brought with it the possibility to contemplate whether these are actually manifested in nature. Carl Friedrich Gauss (1777–1855) was one of the first to consider this possibility.

What Schwarzschild did was to combine the mathematical knowledge about curved spaces with his physical and astronomical knowledge about the cosmos, as notes from 1899 and a publication from 1900 show. He assumed that light is always propagated along what are known as geodesic lines, the shortest distances between two points. In flat space these lines are straight, but in curved spaces they generally are not. Geodesics were to play a central role in Einstein's theory as well: Not only the propagation of light, but also the inertial motion of matter particles follows these lines according to this theory.

In his publication of 1900, Schwarzschild posed the question as to the extent to which, in view of our knowledge of astronomy, it is possible for our world to deviate from flat space. Obviously, the deviation could not be very great, for all terrestrial measurements of the angles in triangles obtained by connecting any three points in space with straight lines always produce a total of 180°. But astronomy offers much larger structures, and Schwarzschild looked at the largest triangles constructible at that time through observation. The corners of this kind of triangle were given by the position of the Earth at two points in time six months apart and by the position of a distant star. The baseline of this triangle is thus the diameter of the Earth's orbit around the Sun. The other two sides of the triangle are given by the rays of light that emanate from the star and strike the Earth at the two terrestrial positions. Half of the difference between the angles at which the star is seen from the Earth at intervals six months apart is also known as the parallax (see Figure 1). Presuming flat space, the parallax is a measure for the star's distance from the Earth: If the baseline of the triangle (the diameter of the Earth's orbit) and the two adjacent angles are known, then it is possible to construct the entire triangle. However, with the aid of this triangle it is not possible to determine whether the space is truly flat. For this one would have to know the angle of the triangle at the star and calculate the sum of all three angles. It is possible, however, to consider what happens to the triangle if one assumes space to be curved.

As an alternative to Euclidean, flat space, Schwarz-schild considers two types of curved spaces, the hyperbolic and the elliptical. Two-dimensional surfaces may serve as graphic illustrations of the three types of space (see Figure 2). Here flat space

Figure 2:
Triangles on a flat
(a), a hyperbolic (b)
and an elliptical (c)
surface

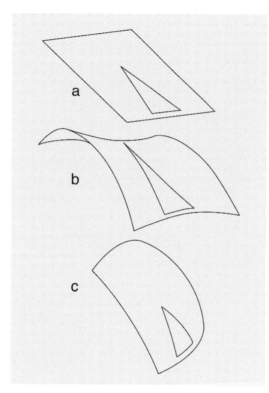

is represented by a plane (a), hyperbolic space by a saddle-shaped surface (b), and elliptical space by a spherical surface (c).

In hyperbolic space the sum of the angles of a tri-angle is less than 180°. The sides of the parallax triangle representing the geodesic lines along which the light runs from the star to Earth are thus curved inward (see Figure 2b). For an equal parallax in hyperbolic space, the star is thus fur-ther away than in flat space. In fact, even stars that are infinitely distant still have a parallax in hyperbolic space. The more that space is curved, the greater the parallax. It is of course impossible to determine with infinite accuracy the angle at which one observes a star. Therefore the parallax can only be perceived when it is greater than the

minimum value dictated by measurement preci-sion. From the fact that for many stars no paral-laxes are observed, Schwarzschild infers the per-missible curvature of a hyperbolic universe: the curvature must be so slight that the parallax of a star located infinitely far away is less than the minimum value measurable at the time.

In elliptical space, the sum of the angles of a trian-gle is greater than 180°. The sides of the parallax triangle given by the geodesic lines on which the light from the star runs to Earth are thus curved outward (see Figure 2c). Hence, for an equal paral-lax in elliptical space, the star is closer than in flat space. In order to estimate the permissible degree of curvature in elliptical space, Schwarzschild makes use of a property of this space that distinguishes it from flat and from hy-perbolic space. For in contrast to these kinds of space, elliptical space is finite, as one may see by considering a closed spherical surface. For any given parallax value, all stars with a smaller par-allax thus must be located in a finite volume. The larger the curvature of the elliptical space, the smaller this volume. In a test calculation with a certain curvature value, Schwarzschild comes to the conclusion that the stars distant from the Earth were separated from each other by an aver-age of only 40 times the radius of the Earth's or-bit. "It is quite out of the question," Schwarzschild concludes, "that the stars could be located so close to each other without their mutual physical influences" – namely, their mutual attraction – "being revealed." From considerations like this Schwarzschild derives a value for the permissible degree of curvature of elliptical space.

Schwarzschild's application of non-flat geometry to the universe is different from that of the theory of general relativity in many respects. For instan-ce, Schwarzschild considers curved three-dimen-sional space, whereas the theory of relativity deals with curved four-dimensional spacetime. In Schwarzschild's considerations, space remains the rigid stage for all events concerning matter and radiation, whereas in Einstein's theory, space-time itself is subject to changes. Moreover, our image of the cosmos has changed profoundly since Schwarzschild's day: While cosmological considerations at that time were usually related to the system of fixed stars – the part of our galaxy known then – today we know that our galaxy is only one among billions, which are organized in clusters and superclusters and fill a space several powers of ten larger than the space known at the time. Correspondingly, Schwarzschild's maximum curvature values are large compared with the values considered in contemporary cosmology.

But despite these differences, Schwarzschild's earlier considerations influenced his reception of Einstein's revolutionary theory in such a way that, for him, in contrast to many of his colleagues, the idea of a curved spacetime was no longer strange. These earlier considerations even made him capable of recognizing the cosmological consequences of the new theory long before Einstein was able to. Thus Schwarzschild, in a letter to Einstein, was the first to articulate the possibility of an elliptical universe as the solution to Einstein's equations.

Equally seminal considerations are found in Schwarzschild's early work, which from today's perspective belong to the problem complex of relativity theory. These questions deal with subjects such as the relativity of motions, the origin of inertia, the validity of the Newtonian law of gravitation, and the anomaly of Mercury's motion.

All of these examples from Schwarzschild's work show that there was more than one path to the insights of general relativity. Indeed, indications about the character of such a theory were to be found on many paths for those who left the narrow lanes of specialized disciplinary discourse and dared to approach the overarching questions that concern the foundations of physics and its neighboring disciplines.

Eberhard Knobloch

Truth and Freedom in Mathematics: The Emergence of Non-Euclidean Geometry in the 19th Century

Albert Einstein's geometrical description of space is based on a form of non-Euclidean geometry, as it was propounded particularly in the 19th century by such mathematicians such as Carl Friedrich Gauss (1777–1855), Johann Bolyai (1802–1860), Nikolai Lobatschewski (1793–1856), theory. Why did this first occur in the 19th century, two thousand years after Euclid (approximately 300 B.C.) had formulated what was to become the Euclidean parallel postulate? In order to answer this question, a few terms must be explained.

Six of the seven free arts. From top: dialectics, rhetoric, geometry, arithmetic, music and astronomy, around 1250

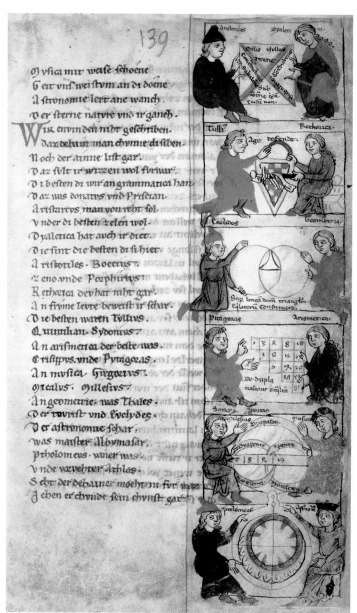

The Euclidean Parallel Postulate P: old and new

Euclid's parallel postulate states: If two straight lines a and b are intersected by another straight line h, and if the sum of the internal angles α and β on one side is less than two right angles, then the straight lines will meet at a point P if they are extended towards this side.

This formulation allows for only one negation: there exists at least one pair of straight lines that, intersected by a third, form angles α and β with a sum smaller than 180° without there being an intersection of the two lines. The geometry elaborated by Gauss, Bolyai, and Lobatschewski is based on this negation. Since Felix Klein (1849–1925) we speak of "hyperbolic geometry."

The Euclidean parallel postulate as formulated today states: For every point P not lying on a straight line l, there is at least and at most, that is exactly one line, h that goes through point P, and which does not intersect line l.

Bernhard Riemann (1826–1866), which, after overcoming considerable resistance, finally entered into the body of accepted mathematical

This formulation allows for two negations: There is not just one such straight line, i. e., there are more such straight lines (hyperbolic geometry).

Carl Friedrich
Gauss (1777–1855),
mathematician
and astronomer,
painting from
around 1850

the theorem *P* was replaced by its logical negation, the consistency of the new geometrical system did not imply the truth of *P*.

While usually accounts in the history of science
claim that, after countless futile attempts since
antiquity to solve the parallel problem, the insight that one had to accept the true *P* theorem
as an additional postulate found increasing recognition. It shall be shown in the following that the
solution to the parallel problem was neither necessary nor sufficient for the emergence of non-
Euclidean geometry. In order to see this, one must
first know that *P* is equivalent to the statement:
The sum of the angles of a (flat) triangle amounts
to 180°.

For the following discussion we will differentiate
between the three possible cases:

P is true, so non-*P* is false: Euclidean geometry;

P is false, so non-*P* is true: anti-Euclidean geometry;

P is true and non-*P* is true: Euclidean and non-
Euclidean geometry can exist side by side.

The solution of the parallel problem is not sufficient …

Aristotle (384–322 B.C.), Euclid.

With Euclid's decision to include the parallel
postulate in his geometry, he had solved the
parallel problem. But this decision between two
equally possible solutions was insufficient for
the advent of non-Euclidean geometry. It was an
either / or decision: for *P*, against non-*P*. In the
writings of Aristotle we find traces of a previous
discussion about the two conceivable possibilities.
In the *Nicomachean Ethic* (VI 5, 1140b 13–15)
it is stated that, for example, one's desire or lack
thereof cannot influence the decision that the
sum of the angles of a triangle equals or does not
equal the sum of two right angles. In the *Posterior
Analytics* (90a 31–34; 93a 32–35) the question is
repeatedly dealt with as to which of two contra-

There is not at least one such straight line, i. e.,
there are no such straight lines. This geometry is
free of parallel lines, or in other words, any two
straight lines cross (Riemann). Since Felix Klein
we speak of "elliptical geometry." With such a
formulation of geometry we enter the realm of
Einstein.

The Parallel Problem and the Emergence of Non-Euclidean Geometry

This parallel postulate *P* has long been accompanied by the parallel problem; namely, the question of whether the true theorem *P* is a theorem
of absolute geometry; that is to say whether the
parallel postulate can be deduced from the logical
combination of all the unprovable geometric principles, excepting the parallel postulate and its logical negation.

While the independence of *P* within the framework of absolute geometry was provable, its
"truth" was not. Since from the 18th century on,

dictory theorems is correct, as for example the theorem that the sum of the angles of a triangle is equal to two right angles or not. In any case Aristotle leaves the answer open in light of two equally justifiable possibilities.

The solution of the parallel problem is not necessary
Thomas Reid (1710–1796)

In 1764 Thomas Reid suggested a geometry of "elliptical" type. This had no connection with the parallel problem that was at the time still considered to be unsolved. Reid was a philosopher. In his work *An Inquiry Into the Human Mind* he wanted to contest the assertion of George Berkeley (1685–1753) and show that a creature without the ability to move was able to pursue a consistent geometry. The creatures he presented (the so-called Idomenians) could only move their eye, but in all directions. One has to imagine this eye as being at the centre of a sphere. Their range of

vision was the surface of a sphere in Euclidean geometry. What Reid described as projection of the Euclidean space on the surface of the sphere, results in a spherical geometry, in which the great circles play the role of Euclidean straight lines. Since, by definition two great circles intersect, it was a parallel free geometry; a non-Euclidean geometry of "elliptical" type.

The Essence of Mathematics

If we now try to answer the question of why we first saw the emergence of non-Euclidean geometry in the 19th century, it has now become clear that the answer is related to the concept of mathematics, and not to the parallel problem. In the 19th century mathematicians freed themselves from the most diverse limitations valid up until that time:

Operations do not have to be commutative, as is the case with the addition and multiplication of numbers (Gauss, William R. Hamilton (1805–

Left:
Georg Cantor
(1845–1918),
mathematician,
1910

Right:
Albert Einstein,
portrait with pipe

RIEMANN-GEOMETRIE
MIT AUFRECHTERHALTUNG
DES BEGRIFFES DES
FERNPARALLELISMUS

VON

A. EINSTEIN

SONDERABDRUCK AUS DEN SITZUNGSBERICHTEN
DER PREUSSISCHEN AKADEMIE DER WISSENSCHAFTEN
PHYS.-MATH. KLASSE. 1928. XVII

BERLIN 1928
VERLAG DER AKADEMIE DER WISSENSCHAFTEN
IN KOMMISSION BEI WALTER DE GRUYTER U CO.
(PREIS ℛℳ 1.—)

1865)), spaces need not necessarily be at most three dimensional (Hermann Graßmann (1809–1877), Hamilton), sets need not necessarily be finite (Georg Cantor (1845–1918)).

There is not necessarily only one geometry, the Euclidean geometry. The mathematician is free to choose consistent and coexisting axiomatic systems; he is free in deciding whether he takes P or non-P as a basis. "The essence of mathematics resides in its freedom," as Georg Cantor, the founder of the theory of infinite point sets, stated in 1883 in an unsurpassable way.

Albert Einstein:
Riemann-Geometrie mit Aufrecht-erhaltung des Begriffes des Fernparallelismus.
Title page of the special printing: *Sitzungsberichte der preußischen Akademie der Wissenschaften,* Berlin 1928

Volkmar Schüller

Newton's Worldview

After Johannes Kepler (1571–1630) discovered the three laws, named after him, on the movements of the planets, it became possible to describe the correct mathematical dependence on time of the planetary places. In particular it was now known that the planets moved in ellipses around the Sun, and that the Sun was located at one of the focal points of these ellipses. Kepler was not yet able to explain in physical terms why the planets moved along such paths. He had to be satisfied with the idea that the planets moved around the Sun under the influence of some kind of forces that had their source in the Sun itself. Ever since then the search for the correct physical explanation for the movements of the planets was one of the most important tasks in physics and astronomy. One famous attempt at such an explanation was made by René Descartes (1596–1650), who maintained in his *Principia philosophiae* of 1644 that the heavens were completely filled with aether in which the planets were floating. In his opinion this aether moved in whirls around the Sun and in this way transported the planets around the Sun. In his paper *An attempt to prove the motion of the Earth from observations,* dated 1674, Robert Hooke (1635–1703) announced he would soon explain the structure of the planetary system with the help of the following three assumptions: 1. All celestial bodies possess a force of attraction or gravitation that is directed towards their center. 2. All bodies that are set in a straight uniform motion would persevere in this motion until they were detracted by another force and compelled to make a different movement, for instance in the form of a circle or ellipse. 3. The force of attraction or gravitation will be greater, the closer the attracted bodies are to the centre of the force. But because Hooke's mathematical skills were insufficient to develop the correct mathematical laws for the force of attraction or gravitation from Kepler's three laws, he was unable to provide the announced explanation of the structure of the planetary

Johannes Kepler

system, despite his three physically very reasonable assumptions.

The secret of gravitation and the subsequent correct physical explanation of planetary motion were finally revealed by Isaac Newton (1642–1727) in his *Philosophiae naturalis principia mathematica* published in 1687. Newton had realized that he could only solve the problem of gravitation by first developing a general theory of mechanics suitable for describing not only the motion of the planets

but also other physical phenomena. The mechanics he formulated were then mathematically and physically developed by later generations into the theory that since the beginning of the 20th century is named Newtonian mechanics. Irrespective of the forces with which bodies may influence each other, the basis of Newton's mechanics is formed by the three famous Newtonian axioms. According to the first axiom (the law of inertia) every uninfluenced body moves uniformly in a straight line. The second axiom (the law of force) says that the change in the quantity of motion of a body is proportional to the force influencing it and that the change will occur in the direction of the influencing force; and the third axiom says that when one body acts on another, it will suffer an equal reaction from the other body. Whenever someone wishes to solve a concrete problem, they have to be conversant with the law of force. As Newton wrote in the first preface to his *Principia*, we have "to discover the forces of nature from the phenomena of motions and then to demonstrate the other phenomena from these forces." In the case of the planetary movements the phenomena of motion are given in the laws of Kepler from which Newton deduced the mathematically correct law of attraction or gravitation within the framework of his mechanics. Newton recognized that the force with which the Earth attracts a body on the Earth (i. e. the weight of the body on earth), and in turn the body on the Earth attracts the Earth, is of the same nature as the force that keeps the Moon in orbit around the Earth, and conversely with which the Moon also attracts the Earth. By generalizing his results Newton developed his famous law of gravitation: two bodies, irrespective of whether they are on the Earth or in the heavens, whether they are in the state of motion or rest, always mutually attract each other with a force directly proportional to the product of their mass and inversely proportional to the square of their mutual distance.

According to this law of gravitation, if the mass of a body changes, this change is immediately noticeable, i. e. without any time delay, as a change in the force of attraction with which this mass attracts another distant mass. A change in the gravitational force thus has an immediate effect in another place in space, or, to put it another way, it spreads into space at an infinitely great speed. This means that the gravitational force between the bodies is not produced or transferred by an intermediary medium (aether), but that it makes a 'timeless' leap through space as a force at a distance. This is the weak point of Newton's theory, because it means that something is simultaneous-

ly having an effect somewhere, where it is not present, and this is a physical impossibility. Newton was criticized by his contemporaries, especially by G. W. Leibniz (1646–1716), for this conception of gravitation as an action at a distance. But even Newton himself found this idea absurd, and for this reason he endeavored in vain throughout his lifetime to find a physical theory in which gravitation is not an action at a distance and yet correctly describes how and why gravitation comes

Isaac Newton

about. It was Albert Einstein who eventually taught us why Newton, and all of his successors in quest of such a theory, failed to solve this problem. Despite the weakness in Newton's law of gravitation, it still provides an excellent basis for explaining and calculating the planetary movements. When Newton concluded from this law that the comets repeatedly return and thus belong to our planetary system, it was a spectacular discovery at that time. The comets proved to be a special kind of planet orbiting the Sun in extremely eccentric ellipses.

The following generations of physicists found it difficult to detach themselves from Newton's ideas on space and time, despite the fact that G.W. Leibniz had vehemently criticized them and had pointed out the contradictions involved. Newton was convinced that an absolute time existed, that it flowed uniformly in itself and of its own nature without reference to anything external, and that

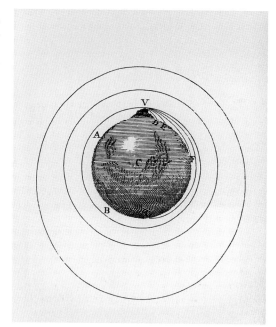

Representation of the effects of centripetal forces on the Earth (after Newton)

Using a telescope, a nobleman explains to a lady Newton's worldview. Copper engraving, around 1745

there was also an absolute space that of its own nature without reference to anything external always remained homogeneous and immovable. Here, absolute space and absolute time represent, so to speak, the vessel in which all happenings take place and in which Euclidean geometry is always valid, irrespective of the way the masses are distributed within. Newton did not see the absolute space as an ideal thing, but he was convinced that he had proved the existence of the absolute space with the help of the bucket experiment he had described in his *Principles*. It took the successes of the theory of relativity to finally banish Newton's concept of absolute space and absolute time from physics. Interestingly, absolute space and absolute time were not only of physical signif-icance to Newton, they also had a fundamental metaphysical and theological meaning in his worldview. He maintained that God, who exists always and everywhere, had constituted absolute time and absolute space, and he once even called this space "God's sensorium." To later generations of physicists, and to us, giving God a place in physical theories seems quite strange, but to Newton and many of his contemporaries it was normal in physics to refer to God in the context of natural phenomena.

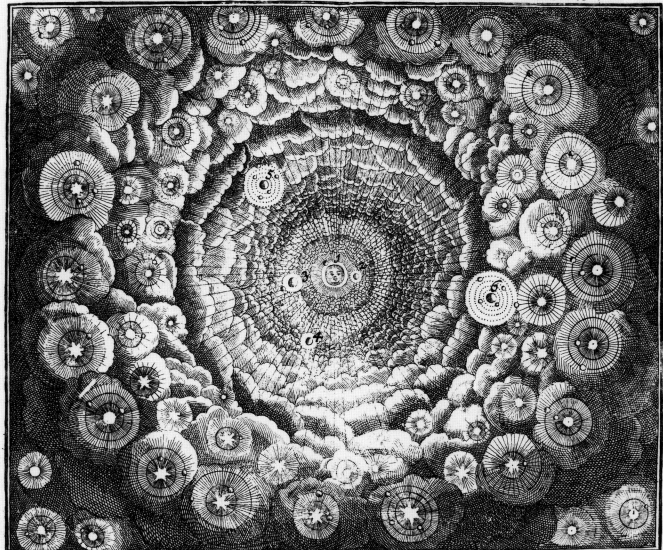

Newton believed our planetary system was unstable and that one day it would disintegrate. John Flamsteed (1646–1719) had observed that the two planets, Jupiter and Saturn, not only influence each other's orbits, but that this perturbation was increasing with time. As a result Newton believed that this perturbation would escalate uncontrollably and that the planetary system would one day disintegrate. God would then have to intervene with a supernatural act to restore order to the planetary system. It was not until 1786 that by means of Newton's law of gravitation Pierre-Simon Laplace (1749–1827) was able to provide exact mathematical proof that the Jupiter-Saturn perturbation runs through a cycle of approximately five hundred years and that Flamsteed had made his observations at a time when this perturbation was on the increase, rather than at a time when it

The Universe after Descartes

was declining. That meant that the stability of our planetary system had been saved, and Laplace's explanation of the Jupiter-Saturn perturbation marked a further major triumph of Newton's law of gravitation.

As in the case of scientists before him, Newton was convinced that matter consisted of non-divisible atoms and that it was one of physics' major tasks to explain the physical properties of matter by means of the properties of atoms. Newton's predecessors had been intent on deducing the properties of matter from the expansion and impenetrability of atoms, in processes where the atoms were thought to mutually influence each other only when coming into direct contact. But he was of the opinion that the single atoms also possess forces of attraction enabling them to interact with each other over relatively large distances. This concept helped Newton explain such things as Boyle's gas law (pV = constant, at a constant temperature), and it also helped him derive the barometric formula.

Newton also made exceptional contributions in the field of optics. He split a ray of white sunlight into its composite colors by letting a ray of sunshine pass through a narrow slit into a darkened room and then through a prism inside the room. Brightly colored strips in the colors of the rainbow appeared on a screen that was set up behind the prism. When he selected any one of these colors and let it pass through a second prism onto a second screen, he found that the individual color remained unchanged. He deduced that the original ray of light had been split into its constituent parts and was no longer divisible. Newton concluded from this experiment that a ray of white light was composed of different rays of light, each of which corresponded to a color of the rainbow, and that these rays could be split no further. When, in another experiment, he took all of the rays corresponding to the colors of the rainbow, then bundled and mixed them, the result was

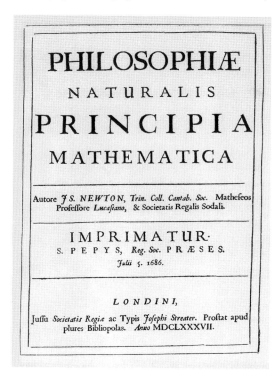

Title page of Newton's *Principia*, 1687

again white light. Newton thought of the different components of light as different sized and rapidly moving light particles. When passing through a prism he supposed that the larger particles (belonging to red light) suffer the least deflection, while the smaller particles (belonging to violet light) suffer the greatest deflection. Newton expressly stated that the colors should not be ascribed to the individual light rays as properties. He maintained that humans only ascribe a particular color to the different types of light rays, because the particles of these rays induce certain stimuli when coming into contact with the retina of our eyes and this process results in the corresponding color sensations in the brain. Newton's particle theory of light, which he described in detail in his *Opticks* of 1704, was at first unable to assert itself against the competing light wave theory of his Dutch contemporary Christian Huygens (1629–1695) who interpreted light as the propagation of waves in an experimentally indeterminable aether. It was not until Einstein correctly ex-

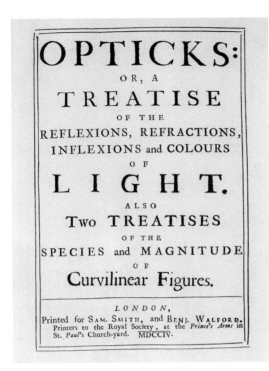

calculated by Albert Einstein within his theory of general relativity, was confirmed by astronomical observations, thus providing significant experimental proof of the theory of general relativity, the value contained in Newton's theory was far too small.

Not only did Newton's outstanding scientific achievements decisively influence developments in physics, his practical research methods were exemplary, and his physics proved increasingly successful in the explanation of natural phenomena. In time physicists were increasingly prepared to adopt his methodology as a theoretical scientific framework. In his famous quote "hypotheses non fingo" (I do not feign hypotheses) Newton did not mean that he made no hypotheses, but that he did not simply invent them, i.e. that he derived them from the natural phenomena in order to then test them in experiments. In his opinion his research method consisted of two parts. In the first part, which he called analysis, inferences as to causes were made from observations of natural phenomena which one then attempted to observe in subsequent experiments. In the following second part, which he called synthesis, inferences were made from the observed causes as to the effects, which are a result of the causes. Newton's research method was the first consistent attempt to pursue physics as a pure science based entirely on empirical experience.

plained the photo effect by his light-quantum theory that physicists realized both Newton and Huygens had succeeded in correctly describing different essential aspects of light with their particle and wave theories. Newton's successors actually contemplated the possibility within the framework of Newton mechanics, that light, i. e. light particles, are deflected from their straight path when in the gravitational field of a very large mass, e.g. the sun. While the value for the deflection of light in a gravitational field, as

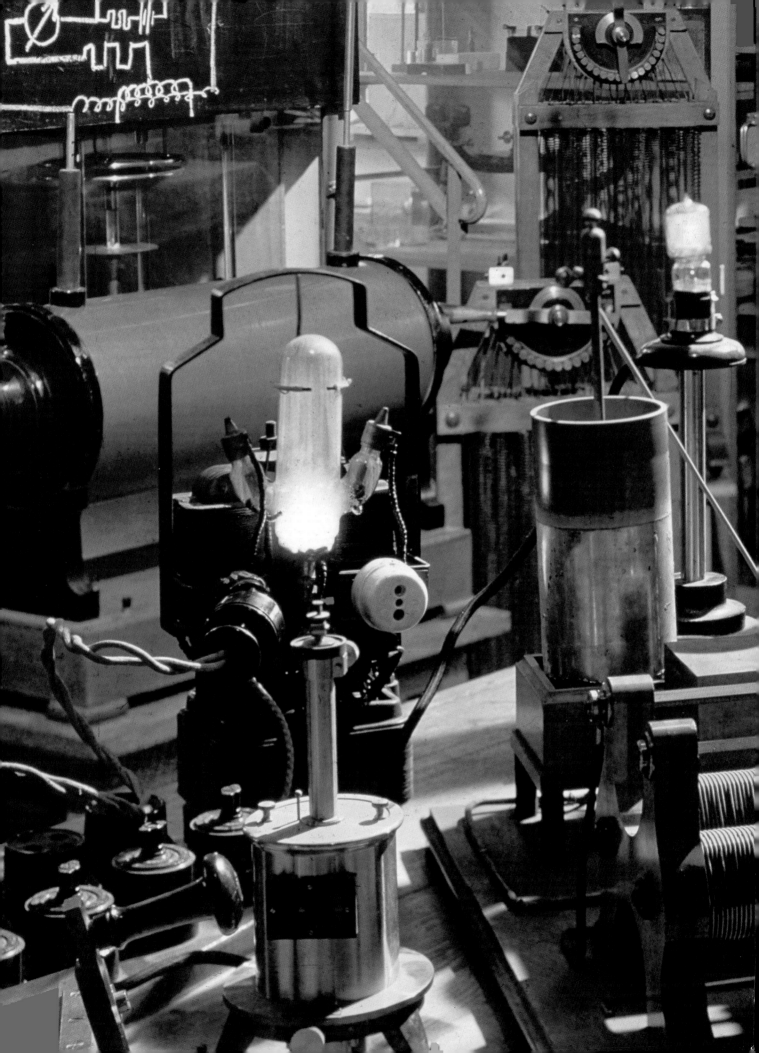

Thomas de Padova

Riding on a Beam of Light

Science journalists love Einstein. We can't help ourselves. Just as the admirers of the Argentine tangos still make their pilgrimage to the grave of Carlos Gardel in Chacarita cemetery in Buenos Aires, to light one cigarette after the other for his life-size statue to keep his vital spark from ever

Albert Einstein in the post office on Berlin's Winterfeldstraße at the live radio broadcast of the Golden Jubilee of the invention of the electric light by Thomas Edison, 22 October 1929

Pages 104–105: Laboratory in Zurich at the time Einstein worked there, reconstruction from 1970

dying, we still decorate our newspapers and magazines with Einstein photographs and quotes, to keep the small flame burning with which we can always ignite the public's interest in science.

At the same time Carlos Gardel made the tango song popular, Albert Einstein helped science gain unforeseen popularity. Gardel touched the soul of the people with sad lyrics and melodies that revolved around the loneliness and uprootedness of big-city inhabitants. For his part, Einstein touched the soul of the people with sad knowledge that revolved around the loneliness and uprootedness of mankind as inhabitants of the universe. Of a cos-

mos in which there is no more absolute space and no time to depend on. Einstein laid the foundations for this exactly one hundred years ago, with his special theory of relativity, which he expanded to a more general worldview a few years later. Comparing Einstein with a tango icon may seem strange, but it is no accident. Einstein is the only scientist of the 20th century whom everyone knows. He, too, became a pop icon.

While a scientist's fame usually amounts to little more than being quoted by those working in his field, the specialized colleagues, after the end of World War I, Einstein was catapulted into the public spotlight as a result of a media chain reaction, after a British research team established that beams of light do in fact, as he predicted, move along curved paths in the vicinity of the Sun. He seized this chance with passion and courageous meddling, with a non-conformism occasionally bordering on obstinacy, dressed carelessly and with a violin case in his hand.

Little of this is visible yet in his miracle year 1905. Only the world of physics becomes aware of him. But even this happens rather hesitantly, despite the fact that Einstein's activity during this year brought forth findings in a wide variety of fields of research, which condensed into a new worldview: nearly simultaneously he, first, makes the existence of atoms perceptible to sensory experience, and, second, declares that light consists of individual packets of energy, the quanta. Third, he revolutionizes our perception of space and time and, in an afterthought, jots down the most famous physics formula ever: $E = mc^2$. This very formula expresses the powerful energy contained in a single gram of matter, the energy contained in thousands of tons of coal.

His mental explosion in 1905 is unparalleled. His new insights, coupled with his fundamental critique of the concepts of classical physics, put him as far ahead of his time as Galileo and Newton had been in theirs. And if we honor him with

dors of Zionism, collecting money for a Hebrew University in Jerusalem.

In the face of bloody confrontations between Israelis and Palestinians, the cosmopolitan Einstein then turned sharply against Jewish nationalism as well. After World War II the Israeli presidency was offered to him on account of his fame. He rejected it.

Einstein sat on the fence all his life. Politics made him suffer; he saw himself confronted with historical constellations that were constantly shifting. He rejected authority, even scholastic authority.

But even at the age of twelve he was enraptured by mathematics, especially geometry – an early enthusiasm he shared with Galileo and Newton. He also occupied himself with current issues in physics at a very young age. When something really interested him, it continued to interest him his whole life long.

For instance, the question of whether it is possible to ride a beam of light and how light would be perceived in such a case. Even his intuitive answer to this question contained an essential fundamental thought of the theory of relativity: namely, that all physical laws remain unchanged as long as a system is in homogeneous motion.

He was a master of questioning. And his genius, often described as naive, was based in large part on this: on simply asking, posing questions Galileo and Newton before him had also concerned themselves with in another manner. He stood on the shoulders of these giants, yet as a child of his age. And this age had elaborate physical and mathematical tools ready for him, which he only needed to make proper use of.

ceremonies, congresses and exhibitions in 2005, this is because in many respects, his work continues to guide us even today.

For instance, the gravitational waves he predicted have yet to be discovered, although physicists are long since convinced of their existence. And if someday we irradiate the smallest structures of matter with an atom laser instead of light, this will be primarily because researchers in the 1990s pursued a new kind of state of matter which Einstein had predicted: the Bose-Einstein condensation, which has since been discovered.

So who was this Einstein?

"Albertle," as his mother called the boy born in Ulm in 1879, came from a family of Jewish businessmen that relocated frequently. He was raised with no special emphasis on religious customs. Not until he experienced anti-Semitism ever more strongly in his later years did he join forces with Judaism, which he did more closely than would have accorded with his nature. In the 1920s he traveled through the United States with ambassa-

"The famous" were under constant observation by photographers even back in 1922. As Lord Haldane's guest, who had a political mission to fulfill, Einstein doesn't seem satisfied as much as resigned to his fate. (*Berliner Illustrirte Zeitung*, June 1922)

Of course, other scientists were also making these attempts. Like the Dutchman Hendrik Antoon Lorentz, some had even brought to paper important formulas of the theory of relativity. However, they had not recognized their actual value.

That Einstein was initially positioned on the margins of organized science was not a disadvantage for him. In 1905 he held down a job at the Patent Office in Bern, where he was an expert, third class. This work left him enough time to read journals and study physics in its entire breadth. He was a regular member of a small circle of intellectuals, profited especially from discussions with his engineer friend Michele Besso, read Kant, Hume and Mach. Not least through them he learned to scrutinize concepts for their essential meaning. One

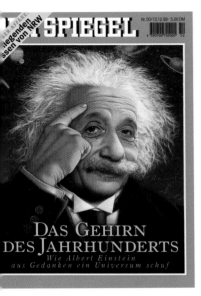

Cover picture of the magazine *Der Spiegel,* 50/1999

can only agree with a contemporary physicist like the Austrian Anton Zeilinger when he regrets that most students of natural science subjects today lack such philosophical training.

Einstein, we can perhaps summarize, posed the right physical questions and then never let go. This is how it occurred to him that the speed of light is something absolute and of an equal magnitude in all systems. He picked the special theory of relativity like a ripe fruit from a tree – complete with all of the peculiar consequences it entails. For example, that every system possesses its own time: time stretches for anyone who moves. Einstein in a speedy spaceship lives longer than Newton sitting under an apple tree.

Right after he had formulated the special theory of relativity and while he was still fighting for its acceptance, he took the next revolutionary step, now turning the Newtonian idea of gravity upside down – his greatest, his own characteristic

achievement. From then on, space and time were changeable phenomena, and the matter distributed throughout the universe curved space and time around it.

Einstein finished work on the general theory of relativity in Berlin, as World War I was already raging. Shortly before the outbreak of the war, Max Planck and other established scientists had secured him a position at the Royal Prussian Academy of Sciences without any obligation to instruct: a dream job for the young researcher. And this in the midst of that day's stronghold of natural sciences, which Berlin remained up until the forced emigration of Einstein and numerous other Jewish scientists in the 1930s.

Einstein set great store in the close specialized exchange of thoughts with Planck and company. Yet his move to Berlin was initially combined with a bitter disappointment on the human level. Barely arrived, he was an unwilling witness to how his scientific colleagues chimed in with the patriotic war cries by signing the proclamation "To the World of Culture." Several of them volunteered to fight on the front; Walther Nernst and above all Fritz Haber soon performed laboratory experiments at the Academy to prepare for the military implementation of poison gas, to which thousands of soldiers fell victim near Yper in Belgium in April 1915.

In the midst of this battle noise, Einstein raised his voice as a pacifist for the first time. He signed an "Appeal to the Europeans" against the war and tried to convince his colleagues to take his side and maintain contact to foreign countries. These courageous actions contributed decisively to his popularity outside of Germany.

After the war, when the general theory of relativity had already been confirmed by experiments and the press had become a constant, unwelcome companion, he traveled through Europe, to the U.S. and to Japan. There he was celebrated as an international scientist, rather than as the cultural ambassador of Germany his native country so liked to display.

His appeals to pacifism and conscientious objection to military service brought him a worldwide following – which he later disappointed, in one of

the many historical twists he faced. He was prompted to these appeals by the anti-Semitism and nationalism that grew constantly during the Weimar Republic, and to which Einstein himself was subjected at numerous events. At a very early stage he recognized in these movements a threat for all of Europe, against which he believed that states like Belgium and France had to protect themselves.

Once Hitler had come to power, Einstein went a step further. Shortly before the outbreak of World War II, he warned U.S. President Roosevelt that the Germans could build an atomic weapon. By this time he had turned his back on Germany forever and settled in Princeton, New Jersey, in a true ivory tower of research.

At this point in time Einstein had long since taken leave of many current issues in the field of physics. He had ignored nuclear physics, obstinately closed his mind to the successes of quantum theory, and engrossed himself unsuccessfully in the search for an even more general, overarching global theory. He had become the cranky outsider of his guild.

The formula he discovered, $E = mc^2$, already contained the entire explosive power of the bomb. He did not participate in the protracted research of nuclear fission, the discovery of the chain reaction or the "Manhattan project," which ultimately led to the atomic bomb. But in view of the National Socialist threat he advocated the bomb's construction. He later regretted this step and directed his efforts to opposing the buildup of nuclear arms. This twist, too, was characteristic for him. All his life he maintained his intellectual independence. We associate with his rebellious spirit and his brave political commitment, his frequent mistakes, his increasing obstinacy and his later "cosmic" religiousness anything other than a sharp division between objectivity and subjectivity. This very haziness makes him, as compared to other scientists of the Modern Age, the ideal link between two cultures often regarded as opposites: science and art.

After a dinner for an American-Palestinian campaign in the Hotel Astor in New York, 4 March 1931. Seated: Felix Warburg, Albert Einstein, Elsa Einstein. Standing: Robert Szold, Morris Rothenberg, Rabbi Wise, Jefferson Seligman

Gereon Wolters

Albert Einstein and Ernst Mach

At first glance, Einstein and Mach do not appear to have any relationship to each other at all, or only a negative one. Einstein is the creator of the theory of relativity and never expressed doubt about the value of atomic theories or the reality of atoms. This is apparent, for instance, in his work on Brownian movement in the anniversary year of 1905. Mach, on the other hand, or so it ap-

pears, brusquely rejected the theory of relativity, and had always been against atomic theories and disputed the existence of atoms. Notwithstanding these weighty divergences, Einstein venerated Mach highly and with good reason. Moreover, today it is almost certain that the texts in which Mach appears to dismiss the theory of relativity (without any argument, by the way) are posthumous forgeries. And there is reason to believe that Mach changed his opinion about the existence of atoms late in life.

Mach was an experimental scientist. For lack of money, he started out working in the field of sensory physiology. Among his achievements here are the discovery of what are called "Mach bands" (sharpening contrast at the boundaries between fields of different color) and the localization of a special sensory organ for angular acceleration movements in the semicircular canals of the hu-

Albert Einstein and Ernst Mach towards the end of 1910

man inner ear. As an experimental physicist he became known primarily for his explanation of the Doppler effect and the investigations on shock waves at supersonic speeds based on this effect. As impressive as Ernst Mach's experimental science achievements may have been, his importance for Einstein lies in an entirely different area. When Albert Einstein was born (1879), the Austrian Ernst Mach (1838–1916) was already 41 years old and had been a duly appointed university professor for 15 years, starting as a professor for mathematics in Graz in 1864, and from 1866 for physics as well. In 1867 Mach took over the chair for physics in Prague, which he held for 28 years until 1895. At the age of 57 he accepted a chair created for him in Vienna, for "Philosophy, especially the history and theory of the inductive sciences." It was in this field, that is, in the "historical-critical studies" on the history of physics performed by Mach from the epistemological perspective, that he became important for Einstein. These works essentially begin with *History and Root of the Principle of the Conservation of Energy* (1872) and continue with the great work, published in so many editions, *The Science of Mechanics: A Critical and Historical Account of Its Development* (1883). These were followed by other books with a similar conception, on the history of the theory of heat (1896) and – posthumously – of optics (1921).

Einstein and Mach met personally on one occasion. More precisely: in late September 1910, Einstein sought out Mach at his house, where he had been largely confined since a stroke in 1898 left him lame on one side and speech handicapped, but mentally alert.

The conversation with Mach concerned atomic theory. Even in his last interview (1955) Einstein expressed his pleasure at Mach's willingness to accept the atomic hypothesis without further ado for reasons of "economy of representation," if it allowed the relationships between observations to be depicted more accurately. The interviewer I. B. Cohen continues, "Einstein was quite satisfied,

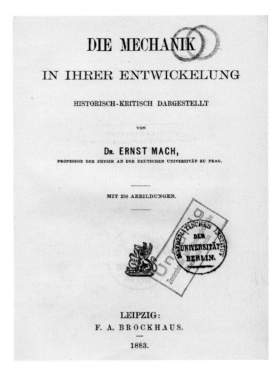

Cover page of Ernst Mach's *Die Mechanik in ihrer Entwickelung*, Leipzig 1883

in fact, more than a little pleased. With a serious expression he told me the whole story once more to make sure that I understood it completely." There are a number of further indications that Mach changed his opinion around the end of 1910. For instance, the physicist Stefan Meyer remembers him calling out, on the occasion of the demonstration of the spinthariscope in his *Radiuminstitut* in Vienna, "Now I believe in the existence of atoms!" Mach also completely ceased his polemics

against atomic theory in his writings after 1910. Further, in a letter in the year 1914 he characterizes his earlier position (and only his attitude to atoms can be meant here) "as not corresponding to reality, as overcome and eccentric in view of the progress of science."

When Einstein speaks of scholars who influenced his path to the theory of relativity, the name Mach is almost never mentioned. Yet it is known that Mach's mechanics was one of the books fervently studied at the *Akademie Olympia* in Bern.

But Mach was not a theoretical physicist from whom Einstein could have learned, of a sort like Maxwell, Abraham and Lorentz. Let us listen to what impressed him. In his sympathetic, frankly tender obituary for Mach – the special and general theories of relativity were well under way – Einstein writes in 1916 that Mach "had the greatest influence on the epistemic orientation of natural scientists of our age." This influence consists of two components: 1) Mach's critique of "the mechanistic world view" and 2) his demand for scrutinizing the basic concepts of physics.

Fragment of a letter from Ludwig Mach to Joseph Petzoldt, 26 December 1919

The mechanistic world view consisted in the agenda of conceiving all physical phenomena or even all phenomena of nature as ultimately mechanical (moved material) and depicting them with the means of mechanics. In his *Autobiographical Notes* of 1946, written at the age of sixty-seven, Einstein states, "It was Ernst Mach, who, in his *History of Mechanics* (*Geschichte der Mechanik*), shook this dogmatic faith (in mechanics as the definitive basis of all physical thinking). This book exerted a profound influence on me in this regard [...]" As a matter of fact, the theory of relativity can be regarded as the proof of the untenability of the mechanistic world view.

Mach's critique of the basic physical concepts was more concrete and no less important. In his obituary for Mach (1916) Einstein writes that, "Concepts which have proven to be useful in ordering of things easily attain such an authority over us that we forget their earthly origin and accept them as unalterable facts. They are then branded as "intellectual necessities," "*a priori* givens" and the like. [...] Through such errors the path of scientific progress is often made impassable for long periods of time." Einstein's "anti-authoritarian" antidotes, taken over from Mach: "show the circumstances upon which their justification and utility depend, show in detail how they grew out of the circumstances of experience." This demand for the 1) operative (how and what am I measuring exactly?) characterization and 2) empirical reference (what does the concept correspond to in reality?) of basic physical concepts is the epistemological basis for both the special and the general theories of relativity.

Mach's critique of the Newtonian bucket experiment became important in the general theory of relativity. In this critique Mach points out that inertia – as opposed to the way Newton conceived it – could constitute a property of mass which is relative to the great mass of stars in space. With this, the basic Newtonian concepts of "absolute

space," and of "absolute velocity" and "absolute acceleration" (especially in the form of absolute rotation) were shaken. Einstein delivered the fatal blow. From Mach's idea of the relativity of inertia, Einstein developed "Mach's principle" as part of his general theory of relativity. Furthermore, an exact reading of Mach's *Mechanics* shows that

Cover page of Ernst Mach's *Prinzipien der physikalischen Optik*, Leipzig 1921

Mach, in asides, questions and wishes for the future of physics, had practially drafted a research agenda. Besides the relativity of inertia, this agenda contained three more important components of the general theory of relativity:

1) general covariance, that is, the demand that the laws of physics be formulated such that they are equally valid in any system (even in systems accelerating with respect to each other) ("general principle of relativity")

2) the equivalence of inert and gravitational mass ("principle of equivalence")

3) a theory of the local effects of gravitation. Einstein apparently must have failed to notice this

somewhat tangentially discussed agenda when he wrote in his obituary: "His observations about Newton's bucket experiment show how close the demand of relativity in the general sense (the relativity of accelerations) was to his intellect. However, what is missing here is a lively consciousness that the equivalence of inert and gravitational mass of bodies categorically demands a relativity postulate in a wider sense."

For the development of the special theory of relativity, Mach's demand for an operative characterization of concepts was the guide for Einstein's new definition of the concept of simultaneity, and ether (as the carrier of propagation of electromagnetic waves) could not stand the test of Mach's demand for an empirical reference. Thus it is not surprising when Einstein writes in the obituary, "It is not improbable that the theory of relativity would have occurred to Mach, if the question as to the meaning of the constant speed of light had already been of interest to physicists when his intellect was in the bloom of youth. But because this [...] impulse was missing, not even Mach's critical need was sufficient to awaken the feeling that a definition of simultaneity of spatially distant events was needed." And thus in a certain sense it is not an exaggeration when Einstein closes a card to Mach (1909) with the phrase: "Your reverent pupil A. Einstein."

Due to his age and illness, Mach himself was able to take notice of the theory of relativity only peripherally, but he expressed only positive remarks about it, and saw in this theory the realization of his own concepts of physics. All the more astonishing were the anti-relativistic invective in the foreword to the posthumously published

The Principles of Physical Optics (1921) and the alleged quotes in the foreword to the ninth edition of *Mechanics* (1933). Today it is practically certain that these texts were forgeries by Ernst Mach's son Ludwig Mach (1868–1951).

Ludwig Mach was a medical doctor with a series of failures under his belt. As a laboratory assistant to his father he showed good manual and technical skills, especially in interferometry, but he understood almost nothing about physics and had no command of differential and integral calculus. Nevertheless, after the disaster of the First World War, in which he had lost his entire fortune, he believed he could somehow follow in his father's footsteps. On top of all this he had also become a drug addict (morphine, cocaine) and attempted over a period of many years to acquire financial support with the promise that he had to perform more experiments in order to take a position on the theory of relativity "in accordance with his father's wishes."

In the context of the public, often anti-Semitic discussions about the theory of relativity from 1920, Ludwig Mach soon noticed that such funds were to be expected above all from Nationalist and anti-Semitic sources. This was perhaps the most important reason for him to invent Mach's anti-relativism.

This twist of the matter had its own irony, as Mach and Einstein were in complete agreement as anti-nationalists. In his obituary, Einstein called special attention to Mach's "kind, humanitarian and optimistic disposition." "This disposition also protected him from the disease of the times, from which few today are spared [1916], from national fanaticism."

Anne J. Kox

Hendrik Antoon Lorentz and Albert Einstein

Hendrik Lorentz (1863–1928) can be characterized as one of the last great classical physicists. He completed the building of classical, 19th century physics in essential ways and thereby prepared the way for the revolutionary developments in physics in the 20th century.

Lorentz was born in Arnhem in the Netherlands in 1853. In 1875 he obtained his doctorate at the University of Leyden and only two years later he was appointed professor of theoretical physics at the same university. His chair was one of the first in theoretical physics in Europe. Lorentz would spend his whole career in Leyden, in spite of several tempting offers from other universities in Europe. At first he led a relatively secluded life, but towards the end of the 19th century he started attending physics meetings outside of the Netherlands and soon he became one of the world leaders in theoretical physics, whose work and views were admired and respected by many.

Albert Einstein with Henrik Antoon Lorentz, around 1920

Although Lorentz made important contributions in many fields of physics, his most famous and influential work was done in the theory of electromagnetism. He succeeded in clarifying and expanding Maxwellian electrodynamics by taking a consistent microscopic approach. For him, the world was built of small particles, some of which, later called electrons, carry electrical charge. Thus, for instance, an electrical current, a highly mysterious phenomenon in Maxwell's theory, could simply be viewed as a material current of charged particles. In addition to these material particles, Lorentz assumed the existence of an electromagnetic aether as the medium in which electromagnetic phenomena take place and in which, for instance, electromagnetic waves propagate. Lorentz's theory, which became known as the "electron theory," made a fundamental distinction between this aether and matter: the charged matter particles were the sources of electromagnetic action, whereas the ether functioned only as its "carrier." This separation turned out to be very fruitful in the further development of the theory of electromagnetism.

In the course of the years Lorentz refined and expanded his theory, until at the beginning of the 20th century it was a consistent whole, capable of explaining all known electrodynamical phenomena. As he developed his theory, Lorentz modified his views of the nature of the aether. At first, he tried to treat the aether as a mechanical substance, obeying the laws of mechanics, but over the course of the years it gradually lost its mechanical properties, until only the property of immobility was left. The final version of the electron theory predicted that it was impossible to find experimental proof of the existence of the aether. Nevertheless, Lorentz never abandoned it, because he felt some kind of medium was needed to function as the seat of electromagnetic waves and electromagnetic field energy. As a consequence, the aether represented a privileged reference system, even though it could not be detected by physical means.

What was the relation between Lorentz's theory and Einstein's special theory of relativity, which appeared a year after Lorentz had completed his electron theory? As it turns out, the empirical consequences of the two theories are the same. Furthermore, there are many similarities in the formalism, for instance in the form of the transformations employed in connecting moving systems. But the foundations of the two theories are totally different: the inability to detect the aether in Lorentz's theory implied the experimental equivalence of all uniformly moving systems. In Einstein's theory, on the other hand, the equivalence of these systems (and thus the dispensability of the aether) was postulated from the beginning. No-one has better expressed the differences in the theories than Lorentz himself in his book *The Theory of Electrons*:

"Einstein simply postulates what we have deduced, with some difficulty and not altogether satisfactorily, from the fundamental equations of the electromagnetic field."

ments on the personal relationship between the two men. In spite of their disagreement on very fundamental points in physics, Lorentz and Einstein had a very close relationship. Lorentz admired Einstein for his bold ideas and revolutionary mind: witnesses describe their interaction as one between a teacher and a brilliant and promising student. Einstein, on the other hand, admired Lorentz for his vast knowledge and grasp of physics as well as for his personal character-traits: mild in his opinions, utterly dependable and always ready to help others.

In an oft-quoted letter Einstein called Lorentz a "work of art" and in 1953, 25 years after the death of Lorentz, he wrote: "On the personal level he was more for me than all others I met on the path of life." In the same article he gives Lorentz's work the ultimate praise by saying that the new generation of physicists "have absorbed Lorentz's fundamental ideas so completely that they are hardly able to realize to the full the boldness of these ideas and the simplification which they brought to the foundations of physics."

James Clerk Maxwell, stipple engraving

Hendrik Antoon Lorentz, around 1920

One might say that Lorentz had pushed classical electrodynamics as far as it would go and that physics needed an Einstein to make new progress. It is interesting to make a few concluding com

Clayton Gearhart

Black-Body Radiation

Hot objects radiate light and heat. And they do so over a range of wavelengths or colors (See Figure 1). Our Sun, for example, is brightest in the yellowish-white part of the spectrum. But one needs only to pass sunlight through a prism to see that its radiation extends from long wavelength red and infrared light at one end of the spectrum, to short

Max Planck

Graph of black-body radiation for a range of temperatures, from the data of Otto Lummer (1860–1925) and Ernst Pringsheim (1859–1917), published in 1899. Wavelength is plotted on the horizontal axis, and radiation intensity on the vertical. Note the characteristic peak in brightness – in the infrared, for these data – trailing off at both shorter and longer wavelengths. These data were the first to suggest a deviation from Wien's law: The solid curve represents the data, and the dashed curve was calculated from Wien's law

wavelength violet and ultraviolet light at the other. Even a red-hot iron bar, brightest at red and infrared wavelengths, emits some radiation towards the violet end of the spectrum. Physicists call this phenomenon "black-body radiation," defining a "black"object as one that emits and absorbs radiation most efficiently.

By the late 1800s, black-body radiation was a lively research topic for both theoretical and experimental physicists. Samuel Pierpont Langley (1834–1906) in the United States, and an exceptional group of experimental physicists in Germany centered around the *Physikalisch-Technische Reichsanstalt* (PTR) in Charlottenburg, had developed sophisticated techniques for studying

this radiation. Part of their motivation was practical – establishing better absolute temperature scales, and measuring light intensities, at high temperatures.

In nearby Berlin, starting in the mid-1890s, the theoretical physicist Max Planck (1858–1947) had begun to formulate a new theory of black-body radiation. By late 1899, Planck thought he had succeeded. He was already an authority on thermodynamics, the new theory of heat developed over the course of the 19th century. In a series of lengthy papers, he combined his thermodynamics with the new electromagnetic theory

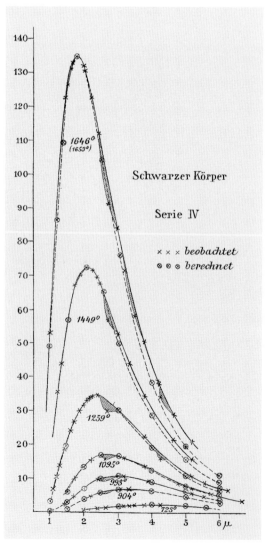

of James Clerk Maxwell (1831–1879). This path led him to a new and more rigorous derivation of Wien's law, an equation describing black-body radiation discovered in 1896 by his friend and colleague Wilhelm Wien (1864–1928), and seemingly in good agreement with experiment. Planck put

Fig. 526.

Fig. 527.

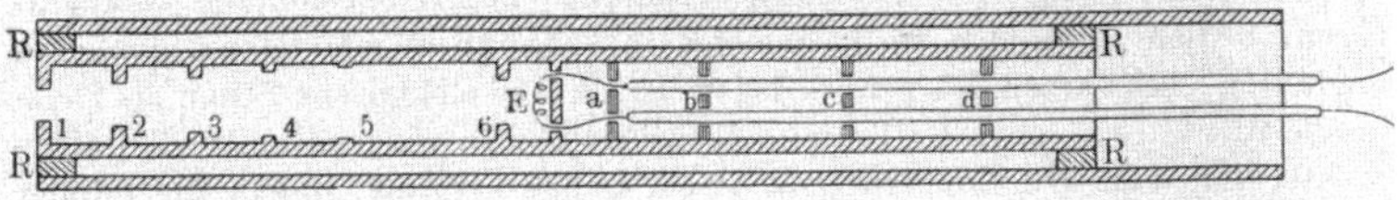

great stock in his new derivation, proclaiming that Wien's law, was a "necessary consequence [...] of the second law of thermodynamics."

Strong words indeed! At almost the same time, however, physicists at the PTR had found systematic deviations between Wien's law and the latest experimental data (fig. left). Planck went back to work. In December 1900 and January 1901, he published three short papers in which he derived a new radiation law – one that ever since has given excellent agreement with observation. But his derivation was mysterious. It relied on a little-known statistical method introduced in 1877 by the Austrian physicist Ludwig Boltzmann (1844–1906). And it went Boltzmann one better – Planck could make his derivation work only by taking

the additional step of introducing what he called "energy elements" – finite chunks of energy that, unlike Boltzmann, he could not make arbitrarily small. Today we call them "quanta," and over the last century, physicists have developed the strange new theory called quantum mechanics to describe and explain nature at the atomic level. But in 1900, all this was yet to come. The "energy elements," whatever they might be, had no obvious interpretation in the physics of the day. Planck himself said virtually nothing about how to interpret them physically, and both his contemporaries and later historians found it difficult to grasp his meaning. Among Planck's early readers was Albert Einstein, a recent graduate of the *Eidgenössische Technische Hochschule* (ETH) in Zurich who was seeking to make his mark on physics. His own first paper on molecular forces had appeared within 30 pages of Planck's January 1901 papers in the *Annalen der Physik*. A few months later, he alluded to Planck's earlier work on black-body radiation in skeptical terms in two letters to Mileva Marić, his fellow student at the ETH and future wife. Nevertheless, in the same letters he announced his intention of reading Planck's "most recent theory." Apparently it made an impression. We do not know in any detail how Einstein's thinking developed. We do know that in three papers published between 1902 and 1904, he developed his own version of "kinetic theory" – a molecular theory of heat – in which he independently rediscovered many of the results of Boltzmann and the American theorist Josiah Willard Gibbs (1844–1906). It was just the tool he would need for his own investigations, and in the third paper, in 1904, he showed how his methods could be used to investigate energy fluctuations in black-body radiation. A year later, he proposed a spectacular new view

Illustration of a black body from the *Physikalisch-Technische Reichsanstalt* (Fig. 526) and schematic longitudinal section (Fig. 527)

of black-body radiation. In 1905, in a paper he called "very revolutionary" in a letter to his friend Conrad Habicht (1876–1958), he suggested as a "heuristic point of view," that light consists of "a finite number of energy quanta that are localized in points of space, move without dividing, and can be absorbed or created only as a whole." Einstein pointed to several experimental effects that, he argued, could best be explained by these particle-like light quanta. But in Maxwell's theory, light was a wave, and many other experiments could be explained only by wavelike behavior. Einstein's "heuristischer Gesichtspunkt" was revolutionary indeed – so much so that it took nearly two decades to win widespread acceptance. Planck himself, though an early supporter of Einstein's theory of relativity, could not bring himself to accept this "very revolutionary" step. Whatever Planck may have thought of his energy elements, they involved only the exchange of energy between radiation with matter; light itself remained a wave.

Moreover, however important Planck's work may have been in arousing Einstein's interest, he apparently remained skeptical of Planck's approach in 1905. Einstein's innovative plausibility argument for light quanta was based not on Planck's law but Wien's older one, which worked well at short wavelengths. Although he showed a close acquaintance with Planck's work, he referred to it only briefly, almost dismissively, when he noted that Planck's two new physical constants – among his most important results – were "to some extent independent" of his theory. And in an earlier section, he had used one of Planck's 1899 results to show that Maxwell's theory led to an impossible prediction: the intensity of black-body radiation should become infinite at short wavelengths! Earlier drafts may have been more pointed. In a 1928 letter, Michele Besso (1873–1955), Einstein's friend and colleague at the Swiss Patent Office, reminded Einstein that "On my side, I was your public in

the years 1904 and 1905; if by the phrasing of your communications on the quantum problem I deprived you of a part of your fame, in return I made a friend for you in Planck."

A year later, Einstein was more complimentary. In a review, he praised Planck's 1906 book that drew together and extended all of his work on black-body radiation, and in a second *Annalen* paper – perhaps Besso's letter referred to this one – he announced that Planck's work and his own were after all in accord: Planck's theory, he said, differs from Maxwell's precisely because "Planck's theory makes implicit use of … light quanta." Planck, as we have seen, would not have agreed! Neither he nor most other physicists accepted light quanta until Arthur Compton's (1892–1962) experiments made them inescapable in the early 1920s. Nevertheless, black-body radiation and its connection to light quanta remained at the center of Einstein's thoughts. In 1909, for example, they were at the heart of his address to a meeting of the *Gesellschaft Deutscher Naturforscher und Ärzte* – the first scientific conference he attended, and the first time he met Planck and many other prominent physicists. By considering "fluctuations" – variations in energy intensity – he showed that not only did light behave sometimes like a wave and sometimes like a particle, but that the dual wave and particle nature of light was inescapable – he spoke of "a kind of fusing of the wave and emission theories of light." In 1916, he found a new derivation of Planck's radiation law, involving assumptions on the "stimulated emission" of light that set down the underlying principles of the laser, not invented until decades later. And in 1924, he understood immediately the significance of a paper sent to him by the then-unknown Indian physicist Satyendra Nath Bose (1894–1974), who had found yet another derivation of Planck's law – one which implicitly suggested that Einstein's light quanta were not independent particles, but somehow mysteriously influenced one

Satyendra Nath Bose and Max von Laue after a lecture by Bose at the Fritz Haber Institute

another. This strange phenomenon was not fully understood until the advent of modern quantum mechanics. Einstein translated Bose's paper into German and arranged for its publication. He also saw its implications for the seemingly unrelated topic of quantum ideal gases, and published the papers describing what is now known as Bose-Einstein condensation, experimentally confirmed only within the last decade.

In short, black-body radiation was not only the starting point for Einstein's exploration of the new quantum universe in 1905; it remained for him a continuing source of inspiration and new discoveries for the next twenty years.

Charlotte Bigg

Brownian Motion

Consider this figure: a sheet of squared paper on which three broken lines have been drawn, reminiscent of the familiar join-the-dots game, though no pattern is obviously recognisable. No scale is inscribed that might provide clues about the nature of the phenomenon represented here. No numbers, letters or symbols to tell the viewer how to hold the figure, or in what direction the lines run. Yet show this figure to a physicist and the response will be immediate: this is Brownian motion.

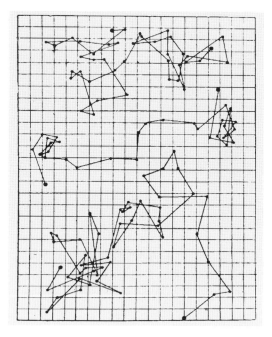

Diagram of Brownian motion

This diagram, published in 1909 by French chemist Jean Perrin (1870–1942) on the basis of Albert Einstein's work, has become an icon for the physical sciences. In the words of physicist Max Born (1882–1970), it stands for "...the reality of atoms and molecules, of the kinetic theory of heat, and of the fundamental part of probability in the natural laws."

Brownian motion, the irregular and perpetual motion of particles suspended in a gas or a liquid, was of considerable interest to scientists throughout the 19th century, but its origin remained a puzzle. The first publication on the phenomenon, by botanist Robert Brown in 1828, excluded bio-

logical causes by showing that the motion also animated inorganic particles – an important result for microscopists liable to confuse it with the proper motions of minute organisms. Among the remaining explanations, careful experiments in the following decades further excluded convection, evaporation in the liquid, or surrounding vibrations. Samples of particles floating in liquid sealed and left to rest for years displayed undiminished activity, prompting scientists to attribute Brownian motion to internal causes: the play of chemical and/or physical forces inside the liquid itself.

In the late 19th century, proponents of the kinetic theory – according to which gases are made of minute spheres, or molecules, in constant, rapid and random motion – ascribed the particles' dance to their collisions with the molecules. Experimental evidence in part supported this explanation, notably the motion's temperature-dependency. The kinetic theory indeed states that temperature is proportional to the average kinetic energy of a molecule: the higher the molecular velocity, the hotter the gas. However, the theory predicted values for individual molecules' weight and velocity that seemed to preclude any such visible motion: even the smallest particles were so massive compared to the molecules that they would barely be affected by molecular collisions. The kinetic theory taps the resources of mechanics to understand the properties of gases. It enables the behavior of individual molecules to be calculated, but also that of the great number of molecules making up the volume of gas under study. The kinetic theory thus provides a means of tracing macroscopic, perceptible phenomena, such as a gas's temperature or pressure, to the mechanical behavior of its constituent molecules. Conversely, macroscopic properties supply information on the molecules' characteristics – though of course only average molecular characteristics can be obtained on this basis. This had wide-ranging consequences. The kinetic theory stimulated the development, notably by James Clerk Maxwell (1831–1879) and Ludwig Boltzmann (1844–1906), of statistical mechanics: the combination of mechanics, the study of forces acting on individual bodies, with statistics, mathematical tools for dealing with large numbers of agents (such as populations).

Note that the kinetic theory postulates that gases are made of molecules, but it provided no positive evidence as to their actual existence. The theory's successes in predicting the actual behavior of gases was reason enough for some scientists to endorse molecular reality. But others objected to this approach on several grounds. At the epistemological level, the phenomenologists headed by Ernst Mach (1838–1916) argued that science should proceed from perceptible phenomena, and not from hypotheses about unproven, invisible entities such as atoms or molecules. At the pragmatic level, many scientists agreed that the kinetic theory was of value, but more conclusive evidence was required in support of the basic atomic postulate. At a more fundamental level, this was a dispute over whether matter was ultimately made of discrete entities (atoms), or of energy. Many scientists investigating phenomena from the perspective of thermodynamics, e. g. focussing on energy processes, in particular the energeticists Wilhelm Ostwald (1853–1932) and Georg Helm (1851–1923), opposed the kinetic theory for this reason. In this debate that lasted for decades, scientific method, the kinetic theory, and atomic reality were closely intertwined.

In the early 1900s, the relatively marginal phenomenon of Brownian motion was turned into a crucial test to resolve this dispute. Albert Einstein played a determinant role in this development. In the introduction to his 1905 paper *Über die von der molekularkinetischen Theorie der Wärme geforderte Bewegung von in ruhenden Flüssigkeiten suspendierten Teilchen*, Einstein stated that: "In this paper it will be shown that according to the molecular-kinetic theory of heat, bodies of microscopically-visible size suspended in a liquid will perform movements of such magnitudes that they can be easily observed in a microscope, on account of the molecular theory of heat. It is possible that the movements to be discussed here are identical with the so-called 'Brownian molecular motion'; however the information available to me regarding the latter is so lacking in precision that I can form no judgement in the matter. If the movement discussed here can actually be observed (together with the laws relating to it that one would expect to find), then classical thermodynamics can no longer be looked upon as applicable with precision to bodies even of dimensions distinguishable in a microscope; an exact determination of actual atomic dimensions is then possible. On the other hand, had the prediction of this movement proved to be incorrect, a weighty argument would be provided against the molecular-kinetic conception of heat."

The botanist Robert Brown

In other words, if Brownian motion corresponded to the motion which Einstein said was predicted by the kinetic theory, then classical thermodynamics was no longer to be found valid at the molecular level; if it did not, classical thermodynamics was safe but the kinetic theory and atomism were in trouble. Thermodynamics entered the argument in that its second law, the law of increasing entropy, postulates the impossibility of a perpetuum mobile. Einstein pointed out here that, if it could be proven to exist, the perpetual motion of suspended particles constitutes such a perpetuum mobile, and is an exception to this law; such that this law must be seen no longer as absolutely, but as statistically true: at the macroscopic scale of the liquid the law holds, but at the atomic or molecular scale contrary events might occur.

Albert Einstein, portrait taken in Prague, 1912

What Einstein proposed to do in the first place was to decide between classical thermodynamics and the kinetic theory, between energy and atoms. But this was in some sense a rhetorical question. In fact, Einstein sought to reconcile the apparently contradictory premises of statistical mechanics and thermodynamics. To this end, he reinterpreted the second law of thermodynamics as being statistically valid. This also involved doing considerable violence to the kinetic theory of gases, whose domain of validity Einstein extended by various means to liquids and to suspended particles.

How does this relate to the diagram? Einstein's 1905 paper was the first attempt to produce a mathematical theory of the Brownian motion of particles suspended in a liquid. Part of this work involved developing a new method of measuring the motion of individual particles. On the basis of the kinetic-statistical conception of diffusion (e. g. that it consisted in the displacement of a great number of molecules), Einstein found that "the mean displacement is therefore proportional to the square root of time". The relevant quantity to measure when assessing the motion of individual particles was average displacement, not velocity (distance covered over time interval), as earlier experimentalists and critics of the kinetic theory of Brownian motion had believed.

This amounted to say that the variation in velocity and direction in the motion of individual particles is so rapid and irregular that it is meaningless to try observing and measuring it. Instead, Einstein proposed to measure the position of individual particles at regular intervals of time and to average these displacements. The diagram is a visual rendering of this new method of measuring the motion of microscopic particles. It stands simultaneously for the realisation that the laws of physics function somewhat differently at the molecular scale (velocity measurements are meaningless, statistical law of entropy), but it also stands for the solution to the problem of how to connect the macroscopic and the microscopic dimensions: the Brownian motion of suspended particles reflects molecular agitation. Einstein's displacement method now enabled Brownian motion to be observed properly, quantified and made to fit with the kinetic theory, a prediction eventually confirmed by the experiments of Jean Perrin in the late 1900s. In this way, Einstein's search for a higher physical unity that might integrate the insights of statistical mechanics and thermodynamics brought in the same crucible atoms, the role of probability in physics, the kinetic theory – and Brownian motion.

In its enigmatic simplicity the diagram further stands not just for the motion of a single particle

and each particle's displacements during two consecutive intervals are also independent of each other, the size of the time interval chosen is unimportant. As Perrin put it, "one of these drawings features fifty consecutive positions of a single particle. It gives only a weak idea of the prodigious entanglement of the actual trajectory. If indeed one marked [the particle's position] second by second, each one of these rectilinear strokes would be replaced by a polygonal outline with thirty sides as relatively complicated as the drawing reproduced here, and so on." In other words, if one could zoom into these lines and see the intermediary motions between each point, one would always obtain similar, broken lines. Perrin compared these "curves devoid of tangents" to Brittany's coastline. As Einstein's theory was developed over the 20th century, Brownian motion was indeed found to be but one instance of a family of widely recurring phenomena. Today we would call them random fractals, and we find them all around us in the shape of tree branches, mountain crests, distant galaxies, but also the patterns produced by stock market fluctuations or the variability of our heartbeats.

but, because of its statistical character, all the possible motions of suspended particles. It is also valid at any scale: because the movements of individual particles are independent of each other

Jordi Cat

Einstein and James Clerk Maxwell: Unification, Imagination and Light

To pay simultaneous attention to two scientists can be insightful when one of them can be shown to have, in some way, engaged the other, and when each can be used as an optic lens to explore the other: namely, when perceived similarities open the door to differences and, in turn, noted differences open the door to similarities, and when this dialectic of reflection and refraction expands and brings into light other issues and explanatory aspects.

In 1931, on occasion of the 100th anniversary of Maxwell's birth, Einstein wrote that Maxwell's "change in the conception of reality is the most profound and fruitful one that has come to physics since Newton." Einstein expressed dissatisfaction with Newton's mechanics as a picture of a world of material points and with the fact that quantum theory was only a direct representation of statistical probabilities of particles' behavior. Instead, Einstein concluded, "we shall then, I feel sure, have to return to the attempt to carry out the program which may be described properly as the Maxwellian – namely, the description of physical reality in terms of fields which satisfy [like de formable bodies] partial differential equations without singularities." Einstein was on his way in the search for a unified theory of continuous fields of matter and forces. But his emphasis on the "program" signals a gap between that Maxwellian ideal and Maxwell's and Einstein's own actual accomplishments.

Both Maxwell and Einstein shared a drive toward a unified picture of physical reality. The concept of energy was a big part of it. Yet, at another level, both left us with a temporary dualistic picture, in terms of particles and fields, dynamical equation and statistics. To illuminate this point, we can take a look at their reflections on light.

In the early 1860s Maxwell reached the following conclusions: (1) electric and magnetic phenomena are inseparable and their mathematical representation obeys the connected equations of his electromagnetic theory; (2) electric and magnetic forces form continuous fields

Albert Einstein, Berlin 1930

that are most intelligible as mechanical and energy properties of a continuous aether which pervaded the universe like a cosmic muscle and a machine, and he pictured its hidden structure in his imaginary models; (3) perturbations in these fields travel like elastic waves in the ether; (4) these waves travel with a constant velocity dependent on the constant properties of the aether; (5) light waves travel with roughly the same measured velocity; (6) therefore, light must be electromagnetic in nature and optical phenomena could be explained by electromagnetic theory. During the same time, Maxwell started to develop his kinetic theory of heat and gases. In it he ex-

Mechanical model of the aether, according to Maxwell

James Clerk Maxwell

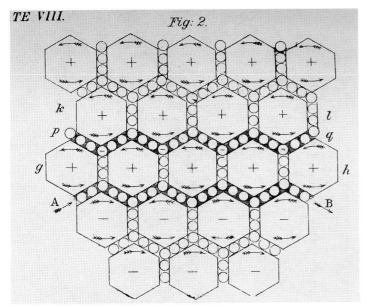

plained a number of macroscopic properties of gases as statistical distributions and averages of the mechanical and energy properties of a large number of molecules. His imagination now moved from the mechanical models of the hidden machinery of aether to thought experiments in the microscopic world of gas molecules. There lived Maxwell's fictional demon, who taught him that the new, statistical, second law of thermodynamics merely expresses our ignorance of the behavior of individual molecules.

Finally, we should note that Maxwell's thoughts were not isolated from Victorian culture. The artistic family background, the varied romantic philosophical influences, the industrial culture of steam and machines and electrical culture of global telegraphic signaling found their way into Maxwell's imagination and researches: in the formulation of theory, its concepts and vocabulary, in the determination of the value of the velocity of light, and in the role of the imagination and its fictions, in the images of the geometrical lines in his portrayal of fields and diagrams, and in his imaginary mechanical models of the luminiferous and electromagnetic aether.

With no weaker powers of imagination than Maxwell, at sixteen Einstein tried to imagine what it was like to travel sitting on a beam of light and looking at its electric and magnetic oscillations. In 1905 light figured centrally in Einstein's novel ideas.

In the dominating version of Maxwell's theory, due to Lorentz, Einstein and others noted that there was an asymmetry in the theoretical description of the interaction between electricity and magnetism and that it corresponded to no observable differences. There were also problems modeling the interaction between the matter and aether. In addition, the aether represented absolute space and a standard of absolute velocity, with no theoretical and empirical value. Einstein blamed all the difficulties on the assumption of the aether and its properties. And he concluded that a general, simple, unified, coherent, relativistic, empirical picture of the physical world could not include Maxwell's beloved mechanical aether – fields cannot be machines after all. Instead, Einstein replaced it with one of its absolute properties: the constant velocity of light, now constant for every system in uniform motion. With constant velocity, light signals can be used to establish a network of synchronized time measurements, basic to his special theory of relativity.

Here, too, we can say that Einstein's imagination was of his time and place. The culture, Jewish and German, of unified, harmonious, symbolic representations, the electro-technical culture of his family background, his job at the patent office, and the role of electrical and light signaling to synchronize clocks, made their way into Einstein's researches.

In the same year, Einstein announced additional results of far-reaching significance. A number of phenomena involving electromagnetic radiation and its interaction with matter were best explained in terms of light particles, or quanta, and their statistical properties. Like Maxwell, also at the same time Einstein published a number of papers with new results for thermodynamics and the statistical molecular theory of heat in gases and liquids.

Despite the underlying role of energy, matter and radiation, fields and particles yielded a fatally dualistic picture of physical reality. To Einstein's chagrin, in the 1920s the results of the new quantum theory would only extend and reinforce this duality.

The drive toward a simple unified picture of reality led Einstein to extend his thinking about continuous fields and energy in his theory of general relativity. With another thought experiment Einstein eliminated the concept of gravitation as a force acting at a distance between discrete bodies. Instead, he reduced it to the local curvature of a "field" of space-time occupied by matter. And matter was now conceived as a local distribution of energy. He was on his way to carry out the machine-free "Maxwellian program."

Stefan Siemer

"In the Brightest Arc Lamps and Incandescent Lights": The Electrical Factory *Jakob Einstein und Cie.*

In 1930, Einstein wrote retrospectively: "I never stopped occupying myself with technical things. This also has advantages for performing scientific research." Thus he developed a new kind of airfoil during World War I, made important contributions to the construction of gyrocompasses well into the 1920s, and in 1927, along with his colleague Leo Szilard, registered a patent for a new kind of cool-

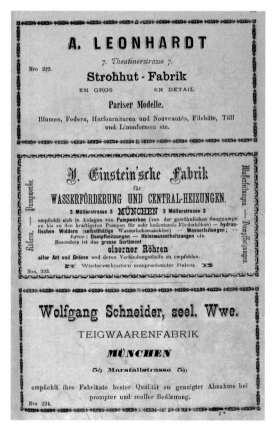

Advertisement for the Factory for Water Conveyance and Central Heating founded by Jakob Einstein in 1876 (*Munich address registry*, 1878)

ing process for refrigerators. For him, "technical things" were both the point of departure for theoretical considerations and the occasion to test out theoretical views in the field of practice. Einstein owed his exact knowledge in this field not least of all to his family home. Between 1885 and 1894 his father, Hermann, along with his uncle Jakob, ran the *Elektrotechnische Fabrik Jakob Einstein und Cie.* in Munich. For a time Albert was even supposed to take over the company, which is why he was sent to Zurich to study

electrical engineering. This never came about due to the company's bankruptcy in 1896, and because of Einstein's fundamental aversion to technology as a livelihood. "The very thought," he wrote in a letter in 1918, "of having to use the power of ingenuity on things that are supposed to make every day life even more sophisticated, with the objective of the dreary drudgery of capital, was unbearable to me." However, at this point in time he had already found his "livelihood": in the summer of 1913 he had been elected to membership in the Prussian Academy of Sciences, which he took up on 1 April 1914. "For at Easter I am going to Berlin as an Academy person without any kind of obligation [...] I am looking forward to this difficult profession," was his self-deprecating remark on the appointment.

Einstein und Cie. in Munich

Ever since Werner Siemens had introduced his electrodynamic machine for the first time in 1866, ever more powerful dynamos were making it possible to install public lighting systems in large cities all over the country. Since the seventies, strong arc lamps had lit construction sites, theaters, train stations, parks, and outdoor restaurants. Electric light incessantly forged ahead into public space, replacing gaslight lamps, and ultimately, with the invention of the light bulb, making its way into private households.

The companies *Siemens und Halske* in Berlin and *Schuckert und Co.* in Nuremberg emerged as the first large enterprises of this new branch of industry. This was also the age of industrial inventors, who placed their technical skills entirely at the service of progress and the improvement of everyday life, without losing sight of their own concrete, financial interests.

How much contemporaries were fascinated by this new technology is reflected in the *Memories of the Paris World Exposition of 1881* by Oskar

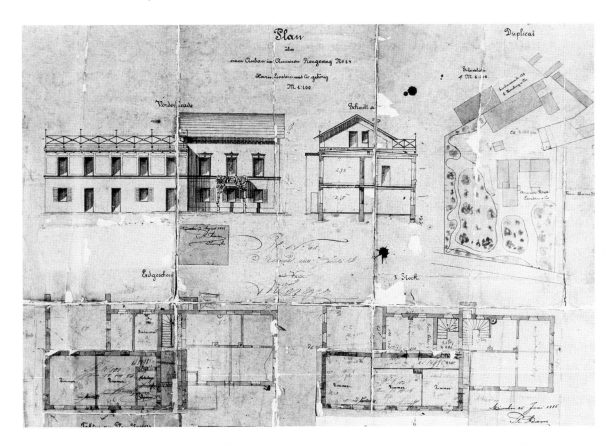

von Miller, an electrical engineer from Munich. He writes: "But the greatest sensation of all was a light bulb by Edison, which could be lit and extinguished with a switch; hundreds of people stood in line for the chance to operate this switch themselves just once." By the time the major "Electricity Exhibition in the *Glaspalast*" was organized one year later, the Bavarian capital had advanced to become a center of electrical engineering.

At the early date of 1883 an electrical experimental station was founded by the Munich Polytechnic Association. In addition to extensive activity in writing expert opinions, the association also held courses, including instruction on how to position lightning rods. Its director Friedrich Uppenborn was also the editor of the mouthpiece of this new technology: the *Centralblatt für Elektrotechnik*, the first issue of which appeared in 1879. Last but not least, the entries in the Munich address books of the 1880s provide information about the boom of the electrical industry in Munich. The classified directory of 1886/87 lists just five companies under the category "Electric Lighting." More than five years later the number of listings had multiplied by a factor of three. The number of private lighting systems increased rapidly as well: While there were only a total of thirty in 1885, three years later this number had risen to onehundred and sixteen.

The *Elektrotechnische Fabrik Jakob Einstein und Cie.* had a significant share in this boom. Its founder was Albert Einstein's uncle, Jakob, who moved from Stuttgart to Munich in 1876 as a certified "civil engineer" and in that same year opened a business on Müllerstrasse for water installation, the *J. Einstein'sche Fabrik für Wasserförderung und Central-Heizungen*. In 1879 Albert's father, Hermann followed him to Munich and joined the company as a partner.

Plan for an extension to the Einstein family house in Munich, June 1886

Title page of the festival program for the opening of the electric lighting system in Schwabing, 1889

Due at least in part to the impression of the above mentioned Electricity Exhibition in Munich, in 1882 the brothers concentrated themselves exclusively on the manufacture of electrical products. They moved to a location near Sendlinger Tor and in 1885 built a two-story factory building on a block between what are today Lindwurmstrasse and Adlzreiterstrasse. The Einstein family resided at the address Adlzreiterstrasse 14 (No. 12 today), right next to the company grounds.

At the end of the 1880s, the company employed almost two hundred workers. The core of production was a dynamo developed by *Einstein und Cie.*, whose trademark soon adorned both advertising and the letterhead. Jakob Einstein was the undisputed technical head of the company. He had invented and patented a new kind of electricity meter and an improved self-regulating arc light. Telephone receivers and light bulbs manufactured under license rounded out the product range.

Festival Lighting

The electrically illuminated *Oktoberfest* (postcard from around 1895)

Effective public displays were necessary to achieve commercial success. An opportunity for such a display was offered by the International Electrical Exhibition in Frankfurt am Main in 1891, where *Einstein und Cie.* was a prominent participant. The exhibition catalog reports that an "Dynamo made by Einstein, with 75,000 voltampere power, coupled directly with a 125-horsepower standing steam machine" served "to illuminate the Pfungstadt beer hall, the café, the maze, and the shooting range." The focus of the exhibition, however, was the problem of transmitting electricity over long distances, which was just starting to be addressed at the time. That

such transmission was possible was demonstrated brilliantly by a 170-kilometer-long electric cable from Lauffen am Neckar to Frankfurt, which powered an artificial waterfall like the one at the Electricity Exhibition in Munich.

ry grounds on Adlzreiterstrasse. Electric lighting for the fairgrounds appears to have become standard from this point on, as is documented in pictures and postcards. From 1892 the first carnival rides were operated electrically.

The Schwabing power station in 1889 (*Centralblatt für Elektrotechnik*, XII, 1, 1889)

Opportunities arose again to present the company and its products, especially in Munich. In 1885 *Einstein und Cie.* installed 16 arc lamps at the October Festival, supplying them with electricity over a 6.5-km-long overhead cable from the facto-

The seventh German Gymnastics Festival in 1887, too, was not least a demonstration of the phenomenon of electrification. This time *Einstein und Cie.* were supposed to install a system of fifty-five arc lamps and sixty-six incandescent lamps on the Munich Fairgrounds. This was not only a lucrative contract, but was awarded to the company despite powerful competitors like the *Allgemeine Deutsche Elektrizitätsgesellschaft* (AEG).

Dynamo by the Einstein company (*official catalog of the Power and Labor Machine Exhibition*, Munich 1888)

Public Lighting for Schwabing

For a limited period of time, *Einstein und Cie.* was able to demonstrate impressively the possibilities of electrical lighting systems at festivals of this kind. The permanent lighting of entire streets was an entirely different challenge, however. Since 1882 the electrification of city lighting had been

Design for the building of the Schwabing power station from the year 1888

supplied by Einstein. The exterior system comprised eight arc lamps and 170 incandescent lamps, connected to each other by overhead cables, since laying underground cables would have been too expensive. With this Schwabing took on, as Friedrich Uppenborn wrote in the *Centralblatt für Elektrotechnik,* "a completely American appearance."

The inauguration of the new lighting system on 26 February 1889 at 7.00 p.m. was a social event celebrated by the city in the fashion of the age, with fanfares and speeches followed by a parade and Bengalese lighting. "All at once," the *Schwabing Gemeindezeitung* reported, "the festival square and the streets of Schwabing shone forth with the brightest arc lamps and incandescent light, greeted by those present with lively applause." Just before, Jakob Einstein had officially handed the system over to the city.

However, the lucrative follow-up orders the company had hoped for failed to appear. Ironically, when the system was to be expanded at the beginning of the 1890s, the city gave preference to the more economical gas lighting. Yet the lighting in Schwabing had made the company well known in the region. *Einstein und Cie.* designed a number of lighting systems for other Bavarian cities; however, they realized only one of these, for Pfarrkirchen in 1891.

The End of the Company

The rapid decline of *Einstein und Cie.* began after 1891. In the long term it was not possible for a family enterprise of this kind to hold its own against large companies like *AEG, Siemens* and *Schuckert*. In 1892 the Einsteins lost out to *Schuckert* from Nuremberg in the competition for a contracted lighting system for the city of Munich. It never fully recovered from this setback. In the same year the brothers attempted a new start in Italy, but in 1896 the company went

a matter of discussion in Munich, and locations for hydraulic power stations had already been selected. Yet protracted negotiations with the gas company competing for the city's lighting contract were necessary before the first public invitation of tenders in 1891.

Between 1887 and 1891 the situation was different in Schwabing, which was an independent community on the outskirts of Munich. In October 1888 the city council resolved to set up an electric lighting system. *Einstein und Cie.* was awarded the contract for the electrical part of the system. The electricity was generated by a power station in which a 40-horsepower, economical gas motor by the Deutz company drove the two dynamos

Interior view of the Elektrotechnische Fabrik J. Einstein und Cie. in Munich (Julius Kahn: *Münchens Groß-industrie und Großhandel,* Munich 1891)

into bankruptcy for good. Jakob Einstein died in Vienna in 1912 as the director of an electrical company; his brother Hermann followed in 1920. The factory building in Munich was torn down in 1959, after having served the local Israelite Cultural Community as its last refuge from Na-tional Socialist persecution from 1938. The Einstein's first home on Müllerstrasse was destroyed in World War II. The house on Adlzreiterstrasse survived the war and post-war period. Today a memorial plaque points calls attention to its most famous resident in the years 1885 to 1894.

Peter Galison

Einstein's Compass

At the beginning of 1915, Albert Einstein found himself engaging more and more in politics; he started to protest the militarism that had plunged Europe into a devastating war. That year also marked a significant change in the path of his long life in science. Collaborating with mathematician Marcel Grossman, Einstein was scrambling to learn all he could about a new kind of geometry, heretofore almost entirely unknown to physicists, that might aid him in characterizing the bending of space-time. The stakes, he realized, were vast: could special relativity be generalized into a theory of gravity? Could the Newtonian cosmos of distant inverse-square forces be scrapped in favor of one based on the equivalence of mass and energy with fields of curved space and time? In November 1915, after the most intense intellectual struggle of his life, Einstein was finally able to reveal general relativity to the world. His gargantuan effort was no less than a triumph of theory, reason, and abstraction.

Yet from the start and through much of that eventful year, Einstein had stepped back from the Platonic reaches of tensors and coordinate transformations to focus on bench experiments involving gluing quartz fibers to mirrors and pulsing electric currents through electromagnets. As he wrote to his best friend, Michele Besso on 12 February 1915: "The experiment will soon be finished. [...] A wonderful experiment, too bad you can't see it. And how devious nature is, if one wants to approach it experimentally! I've gotten a longing for experiment in my old age." Working with Hendrik Lorentz's son-in-law, W. J. de Haas, they undertook

Hans Christian Ørsted, woodblock 1906, from a contemporary portrait of 1830

an experimental challenge that had stumped some of the most adept lab hands of all time – explaining the mechanism responsible for magnetism in iron.

The basic concept was simple. An electric current traveling in a loop makes an electromagnet. Einstein wondered whether magnetized iron might not also owe its capacity for magnetization to a similar phenomenon, as André Marie Ampère and his successors had long speculated. Einstein asked whether, at the atomic or molecular level, there were many such current loops all oriented in the same direction. If so, there might be just one kind of magnetism. He said :

"Since [Hans Christian] Ørsted discovered that magnetic effects are produced not only by permanent magnets but also by electrical currents, there may have been two seemingly independent mechanisms for the generation of the magnetic field. This state of affairs itself brought the need to fuse together two essentially different field-producing causes into a single one – to search for a single cause of the production of the magnetic field. In this way, shortly after Ørsted's discovery, Ampère was led to his famous hypothesis of molecular currents which established magnetic phenomena as arising from charged molecular currents."

A Simple Thought Experiment

Reducing two causes to one: here was quintessential Einstein. He had begun his work on special relativity with the usual understanding of James Clerk Maxwell's equations must be very wrong, because it seemed as if there were two explanations for why current was produced when a coil approached a magnet. If the coil was moving and the magnet still, the standard story held that this was because the charge in the coil was moving (along with the wire) and so was pulled around the loop by the magnetic field. If the magnet was moving toward the coil, then, according to the conventional view, the growing magnetic field near the coil was produced by an electric field that drove charge around the coil. Einstein's special theory of relativity accounted for both phenomena by reassessing the meaning of space, time and simultaneity.

In his 1907 principle of equivalence, Einstein had objected to the previously unchallenged claim that there were two kinds of mass – gravitational mass (responsible for the weight of a lead ball) and inertial mass (the resistance of a mass, say a lead ball, to acceleration, even far out in space). Instead, Einstein stated that there was just one kind of mass. There was no way to distinguish the behavior of mass pressed to the floor of an accelerating rocket ship and that of mass pulled to the

floor of a stationary room in a gravitational field. So Einstein likewise believed deeply that there was but one kind of magnetism and that it was caused by the aligned orientation of tiny magnets–current loops formed by electrons as they raced around atomic nuclei. The question was: how could one test this idea? Suppose that you are standing on a lazy Susan with a gyroscope in each hand, each with its axis pointing away from you and spinning clockwise from your point of view. The gyroscopes' angular momenta are oriented in opposite directions, so the system's total angular momentum adds to zero. Next, say you raise your hands above your head so the gyroscopes are now both pointing up. This means their angular momenta are both aimed in the same direction, so they sum to a nonzero value. But because the angular momentum in a closed system is conserved (stays the same), you begin to rotate on the lazy Susan, in this case to counter the angular momenta of the gyros. Einstein imagined this scenario in miniature, inside an iron bar. Suppose that an unmagnetized iron cylinder was suspended by a fine, flexible fiber and that suddenly a strong magnetic field was applied, enough to magnetize the cylinder by orienting all the little electron-orbits. If he was correct, many of the little randomly oriented electron orbits would than be aligned. Their angular momenta would suddenly add instead of canceling. And again, just as the lazy Susan did, the cylinder would rotate to compensate. This was the notion behind the experiment.

Fascinated by Magnetism

In time, amazingly, Einstein and de Haas succeeded in eliciting results from the remarkably delicate apparatus they built subsequently. But from where did this concept come, and why just then in 1915, amid the worst war and his own high-stakes struggle to define general relativity?

Albert Einstein, 1916

Gyrocompass
by Einstein, 1926

For an answer, one must look back to the period
after Einstein's graduation from the Zurich Polytechnic in 1900, years during which he found
it difficult to find gainful employment. Rejection
letters piled up until mid-1902, when he finally
received a very welcome job offer from the Bern
Patent Office. Although Einstein had battled with
one teacher after another during his school years,
he admired and learned much from the head of
the patent office, Friedrich Haller. Einstein loved
Haller's injunction to "remain critically vigilant,"
to view suspiciously the inventors' claims. Einstein learned to adhere strictly to Haller's injunction to "remain critically vigilant" – to view inventors' claims with skepticism. Einstein loved machines and corresponded with other enthusiasts
about them; he even built new ones in his apartment. Over the years he patented refrigerators, invented new electrical measurement devices and
advised his friends about machinery. Indeed, his
father and uncle had long run an electrotechnical
business and had patented their own inventions.
Sadly for us, nearly all of Einstein's patent evalua-
tions were, by law, destroyed, but a few remain –
in particular, those that made their way into court
proceedings. That is because Einstein soon became one of the most esteemed technical authorities in the patent office and thus a much appreciated expert witness.

Herein lies the key to understanding Einstein's
fascination with magnetism. In the early 20th century the tried-and-true magnetic compass began to
suffer difficulties. It worked poorly on new ships,
which were becoming metallic and electrified,
and functioned badly inside submarines or near
the earth's poles. And the standard compass was
problematic in aircraft because its directional in-
dicator led and lagged during turns. Two companies took up the compass problem, one headed
by the American inventor and industrialist Elmer
A. Sperry and the other by his German archrival,
Hermann Hubertus Maria Anschütz- Kaempfe.

The solution was to convent powered gyroscopes
into compasses. Anschütz-Kaempfe cleverly built
the casing of his gyroscope so that it would pre-
cess (slowly cycle its axial orientation) in such a
way that its axis lined up with the rotational axis
of the Earth. Soon afterward, Sperry produced a
similar instrument. Anschütz-Kaempfe promptly
sued for patent infringement. Sperry mounted the
usual defense: he was merely following an older,
preexisting idea. In mid-1915 Einstein was called
in to serve as an expert witness. His testimony
showed, to the court's satisfaction, that the earlier
gimbaled gyroscopes could not possibly have worked as compasses, because they were designed to
move only within a very tight range inside their
casings – a ship's slightest pitch and yaw would
render them useless. Anschütz-Kaempfe won the
case. Einstein went on to become sufficiently expert in gyrocompass technology to collect royalties for his work in this field for decades to come.

Only a Theoretician Believes his Theory

Einstein's royalties in the science physics proved
to be even greater. On 27 January 1930, he wrote
to Professor Emile Meyerson in Paris: "I was led

to the demonstration of the nature of the para-magnetic atom through technical reports I had prepared on the gyromagnetic compass." He saw that just as the Earth's rotation oriented a gyro-compass, a cylinder of iron could be made to ro-tate by orienting all the little atomic gyroscopes inside it. The experiment turned out to be a spec-tacular success. Einstein and de Haas had demon-strated an effect so subtle that even the great James Clerk Maxwell had failed to discern it. But this story has a twist. The two physicists showed excellent agreement between the theory (ferromagnetism caused by orbiting electrons) and their experiment. Unfortunately, their striking re-sult soon came under attack – cautiously at first, then with growing insistence. It seemed that their measurement of magnetism per unit of angular momentum was off by a factor of two, a difference no one could adequately explain until much later, after the development of quantum mechanics and the concept of electron spin. It seems that Ein-stein's commitment to a particular theoretical model had cut two ways. On the one hand, it had given him real conviction about how to organize and conduct the experiment – specifically, where to look for the effect. Maxwell and others who had failed before had no feeling for the magnitude of the phenomenon. On the other hand, the theo-retical model Einstein chose made it easy to accept an experimental answer when blackboard calcu-lation and laboratory results agreed – despite the

existence of many potentially interfering factors, which included such things as the effect the Earth's magnetic field and the vagaries of the fragile lab apparatus itself.

The tale reminds me of one of Einstein's wonder-ful sayings: "No one but a theorist believes his theory; everyone puts faith in a laboratory result but the experimenter himself."

Wander Johannes de Haas in front of a large electromagnet manufactured by *Siemens & Halske*

Lea Cardinali

A Fifteen-Year-Old With Very Clear Ideas:
Albert Einstein

It must have been a very pleasant surprise for his mother, Pauline, father, Hermann, and uncle Jakob to see young Albert suddenly appear in front of them in Milan that day in the winter of 1894/1895. You see, they thought he was 500 kilometers away, diligently studying at the *Luitpold Gymnasium* in Munich where they had left him two months earlier.

But let's take things chronologically. How come the brothers Hermann and Jakob Einstein were in Milan with their families? Not everyone knows the story of the Einsteins in Italy...

ing Jewish parents, Albert had been brought up with few deep-seated rules and a lot of intellectual freedom. He was fascinated by all mysterious aspects of nature and by scientific literature and felt insufficiently challenged by his studies. He also disliked the authoritarian Prussian style of teaching, so Albert quit his studies and suddenly appeared in Italy intending to turn his life around. He had planned everything: he left classes with a medical certificate documenting a nervous breakdown and with a reference from a teacher which would get him into another school; he intended

Left:
The Einstein-Garrone factory in Pavia (left-side of building complex)

Right:
Title page of the product catalogue of the Einstein-Garrone company in 1896

The Einstein brothers, who worked in Munich from 1885 as electrical system installers, had direct connections with the engineer L. Garrone (1857–1911) from Turin. When their Munich company failed, the Einsteins decided to open a new one in Italy. Thus the *Officine Elettromeccaniche Nazionali Einstein-Garrone* opened for business in March 1894. The two Einstein families moved to Milan in the autumn, while they were building a big factory in Pavia next to the canal that flows into the Ticino. It was completed that year. Meanwhile, Albert, left alone in Munich, was very unhappy. As he was of middle-class, non-practic-

to renounce his German nationality and thus avoid the nightmare of military service, so he was evidently a rather difficult character who perhaps lacked adaptability but certainly not determination.

"When I crossed the Alps to Italy, I was so surprised to see how the ordinary Italian, the ordinary man and woman, uses words and expressions of a high level of thought and cultural content, so different from the ordinary German. This is due to their long cultural history [...] The people of northern Italy are the most civilized people I have ever met." (Admittedly, it is rather difficult

Left:
Courtyard
(colonnade) of the
Einstein residence
in Pavia

Right:
Frescoes in the
home of Albert
Einstein's family
in Pavia (photo with
kind permission of
F. and G. Torti)

technicians had come from the firm which was previously in Munich.

Albert initially wrote about Pavia: "The city's soul could be defined in mathematical terms as roughly (1) the sum total of the ramrods the various gentlemen and ladies have swallowed, and (2) the mood created in the observer by the uniformly filthy walls and streets everywhere. The only beautiful aspects are the delightful, graceful little children." In fact, Albert always had pleasant memories of his months in Pavia, spent in the solitude he loved, with his scientific studies, improving his violin-playing and enjoying the open-air life by the Ticino and in the countryside. He was encouraged to frequent the family factory – and solved a problem for his Uncle Jakob with a machine design that had caused difficulty for several of the company's engineers. Although initially a little shy, Albert turned out to be sociable, ironic and witty. It was Ernesta Marangoni (1876 – 1972) who saved the records of this period. They didn't come to light until 1955 when she published the letters she had received from Albert and her close friend, his sister Maja, in a local newspaper. In 1895, Ernestina was studying science, could speak German, and loved playing Wagner and Beethoven – an ideal companion for the Einsteins!

In the autumn of 1895, Albert sent his favorite maternal uncle, Caesar Koch, his first scientific article: *Über die Untersuchung des Aetherzustandes im magnetischen Feld,* sometimes called the "Pavia Paper." Albert was only sixteen and the subject chosen ("The Function of Aether as a Medium for Electromagnetic Action," which was seriously debated in the scientific world at the time) touched on mechanics and electromagnetism and foreshadows a question that was to interest Albert all his life: how to connect these two fields more

to see what he based this judgment on, given that it was the first time he had traveled and Albert didn't understand the language!)

Albert's time in Italy was a positive one: "The happy months of my sojourn in Italy are among my most beautiful memories [...] Days and weeks without anxiety and without worries." He reassured his parents by saying he intended to work on his own to get into Zurich Polytechnic. He was thus able to devote himself to his studies. In Milan he bought the university-level volumes of a physics book by J. Violle and underlined passages as he studied them.

From the beginning of 1895, the Einsteins were in Pavia, a town with a predominantly agricultural economy and a population of about 30,000, whose cultural and social life centered around the ancient university. Albert's family lived in a smart house, where the famous poet U. Foscolo (1778 – 1827) had lived, and got along well with the local bourgeoisie.

Engineer Jakob Einstein, well known for his low-current parasite dynamo, obtained patents for certain arc lamps and measuring instruments. The Einstein-Garrone works specialized in "producing electrical dynamos with direct and alternating currents, arc lamps, measuring instruments, electric light systems, the long-distance transportation of power, electric heating and repairs."

Detailed regulations laid down working hours and formalities, weekly wages and contributions for sick workers. In early 1896, there were 80 workers, including 3 women and 2 girls. Some German

closely and more organically. The concluding statement – *"Die quantitativen Forschungen über dem Äther können erst beginnen, wenn qualitative Resultate existieren, die mit sicheren Vorstellungen verbunden sind"* ("quantitative research on aether cannot begin until there are qualitative results related to well-founded conceptions") – also indicated his methodological approach to science. Albert failed to get into the Polytechnic in autumn 1895, so he went to Aarau for his final school year. There is a second work dating from this period: *Über Elektrizität und elektrische Ströme* ("On Electricity and Electric Currents"). Ernestine received a copy in 1898, but only the title remains in her records. One year later, Albert, now in Zurich, asked his friend to return this article because he was no longer convinced about the contents. The whole thing then vanished for good. What we do know is that Albert already disclosed his intention of "taking a look at the mysterious nature of electric current" in the "Pavia Paper." It is now impossible to say whether this meant investigating the structure of matter or the theory of the electric field. (It is well known that at Aarau he was thinking about Maxwell's electromagnetic theory and the ideal experiment for following a ray of light at the same speed.)

While Albert was completing his studies at Aarau in autumn 1896, the factory at Pavia, although apparently well-organized, went into liquidation. It was quite a mysterious situation, but perhaps the Einsteins went to Pavia with the specific aim of supplying the town with electricity. In fact, Italian friends, partners and the factory itself supported the cooperative company A. Volta in its attempts to raise the required capital. Unfortunately, when these two enterprises had to come to an agreement, they were no longer on good terms, and in mid-1896 the Einsteins finally failed to get the Pavia lighting contract. A few months later, their factory went into liquidation. As a result of this failure, Jakob stopped working as a businessman.

It is little known that between 1897 and 1899 he was employed by the Italian-German firm of Ruggeri-Koppel, responsible for building the electric tram line from Lace to S. Cataldo. First he worked as an electrical engineer and then as the technical director. Hermann, however, tried something different. He founded another firm in Milan, helped by his cousin, brother-in-law and indefatigable financier Rudolf Einstein. The Einstein family moved to Milan into a sumptuous apartment which had originally belonged to the Trivulzio princes. Very little is known about this new firm, but it, too, closed down within two years.

While at the Polytechnic, Albert remained in contact with Italy. He went through bouts of depression because of his father's business failures. He returned to Milan during the holidays but stayed in contact with his friends from Pavia. It was probably during this time that he asked Ernestina's father for directions and organized a 150 kilometer trip from Voghera to Genoa. About four days' walk brought him to his uncle J. Koch. This was a strange destination, given that Albert didn't like his aunt much, "a real monster of arrogance, insensitivity and formalism." He wasn't enthusiastic about his cousin either; she was "a very spoilt, snotty-nosed brat." He was probably duty-bound to visit his uncle in recognition of the fact that that he paid 100 francs a month for Albert's studies in Zurich.

In 1898, Hermann was at Canneto sull'Oglio (MN), where he won the contract to provide the town with electric lighting. This was installed in 1899 using a dynamo in the so-called Mill of the Madonna. The mill concession cost 2,500 lire a year. To compensate, the council would pay 720 lire per annum for public lighting, to be supplied from 5 a.m. to dawn and from sunset to midnight, from 1 September to 30 April. Among the private users, the local hospital paid an annual subscription of 60 lire for 3 five-candle-power lamps! In April 1900, Hermann won the contract in Isola della

Scala (VR) to install a dynamo for public lighting in the ancient Palasio mill. The commune promised to pay 2870 lire a year for fifty-two lamps, mostly for street lighting. Older residents tell us that Hermann stayed in the village for several months.

His son was with him on one of his stays, and they then went on to Venice. In fact, Hermann wanted Albert to work with him and eventually to succeed him. Legal deeds show that Hermann was still running both concessions personally in the spring of 1900, although he eventually passed them on to his cousin Rudolf. In 1902 Hermann died unexpectedly of heart problems in Milan and was buried in the Monumental Cemetery.

With Hermann's death, the Einstein family's Italian parentheses closed. Wife and daughter moved to Switzerland a few weeks later, although they stayed in touch with their relatives and friends still in Italy.

Albert's connection with Italy diminished with his father's sudden death in 1902. His life now became increasingly centered on Switzerland. The combination of being together with Mileva and looking for a job to make a living had created so much tension in the family in the final years that it was only shortly before his father's death that the latter agreed to Mileva and Albert getting married. But that is another story.

And so the child Albert, trembling with emotion at the mystery of a magnetic needle moving, became self-taught and intolerant of any teaching imposed on him. During the Italian period, a strong and deeply non-confessional faith and the need for firm ideas turned him into a scientist who searched, even against the mainstream, for a single theory to explain the inner mysteries of nature – an engineer of the universe, not of the Einstein Electric Works, "because the thought of applying my inventiveness to things that would make everyday life even more sophisticated, with the aim of piling up capital, was unbearable to me. Thought for its own sake, like music!"

Rudolf Einstein's company invoice to the community of Canneto sull'Oglio for electric lighting, June 1900

Fabio Bevilacqua
Stefano Bordoni

Einstein's 1895 Pavia Paper

"At its inception, an electric current sets the surrounding aether in a kind of momentary motion whose nature it has not yet been possible to determine with certainty. Depite the continuance of the cause of this motion, i.e., the electric current, the aether remains in a potential state and forms a magnetic field."

Albert Einstein's first scientific essay was written in Pavia in 1895. Soon after the departure of his family from München he left the Liutpold Gymnasium in December 1894 and joined them in Milan, on physics and mathematics and with its cabinet of physical instruments going back all the way to Volta.

Einstein's proficiency in physics was remarkable: in October, despite his failed admission to the Zurich Polytechnic, he was invited by the physics professor Weber to follow his second year courses. Albert sent his first paper to his Uncle Caesar Koch, in Brussels, as a sign of his own desire to pursue scientific subjects. He wrote this paper at the age of sixteen and, in contrast to his later

Left:
Caesar Koch,
Albert Einstein's
uncle, around 1910

Right:
Herrmann Einstein,
Albert Einstein's
father

where he bought some physics textbooks that are still part of his library. To be closer to the electrotechnical factory owned and run by his father and his uncle, in spring the whole family moved to Pavia, where Albert prepared privately for the entrance examinations to the Zurich Polytechnic, to be held in October. The Pavia context was stimulating: Albert's uncle was the owner of a number of patents in the field of electrotechnology and there are memories of Albert's involvement with the factory's problems. Moreover, only a few hundred meters away from the Einstein's family house on via Foscolo, was the Pavia University with its marvelous library full of German texts

writings, a physicist today would find it difficult to understand. The subject is closely connected to the debate on ether which took place in the last decades of the 19th century. Over three pages some words are quoted repeatedly: *Bewegung* ("motion"), *Äther* ("aether"), *Kraft* ("force"), *Feld* ("field"). Aether has the highest number of occurrences. This debate was placed on the borderline between mechanics and electromagnetism. Despite its cautious preliminary remarks, the paper is not without interest. Einstein asserts that there is an elastic aether, where magnetic phenomena and propagation of light take place. The magnetization of the aether amounts to the storage of potential energy in the form of strains of its structure that modify density and elasticity (that electric and magnetic fields be related to the energy of the medium had been a precise suggestion very interestingly, makes some remarks on the energy required by the current to overcome the inertia of the aether structure to achieve the magnetization. In other words, part of the kinetic ener-

First page of Albert Einstein's first scientific essay, 1895

by Maxwell, but in Einstein's paper the identification seems different). He predicts that the velocity of light propagation would be affected by the magnetization of the aether and suggests that experiments be performed. A connection is thus assumed between magnetism and optics, and also between electromagnetism and mechanics of continuous bodies. Einstein explicitly quotes Hertz's "marvelous" (*wunderbaren*) experiments and, gy of the current is stored as potential energy of the aether, a possible clue, as noticed by A. Pais, of an independent discovery of self induction. How does Einstein's first paper relate to existing knowledge in the field? How much of this knowledge was Einstein aware of? The general reference frame for the whole paper seems to be, first, the conceptual model of the contiguous action of Maxwell's theory as interpreted by

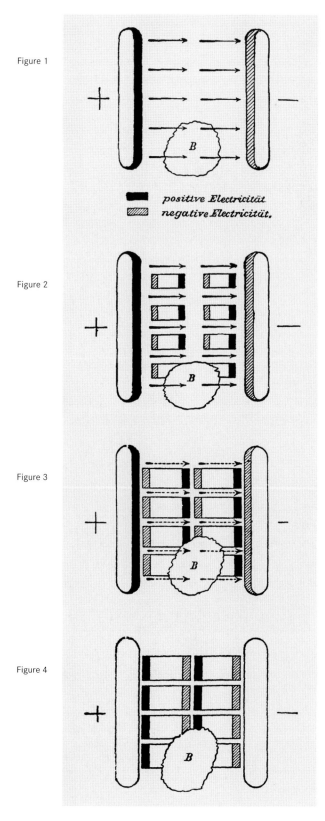

Figure 1

Figure 2

Figure 3

Figure 4

Hertz and, second, a mechanical model of aether. This conception, considered by some scholars as being "somewhat antiquated by the time the essay was written," belongs to the struggle between mechanical and electromagnetic foundations of aether.

A good summary of the views on electromagnetism is given in Hertz's famous introduction to his *Electric waves* of 1893. First of all, he shared with his master Helmholtz some basic assumptions. Among them:

a) electric polarisation in dielectrics and electric currents in conductors give similar electromagnetic effects

b) air and empty space behave like all other dielectrics.

Within the landscape of electromagnetism, he distinguished four main standpoints which he depicted with the help of a couple of conducting plates (a "condenser"), the electric charges on them and the medium between them:

1. the standpoint of action at a distance between at least two bodies, the same conception of the well-known Coulomb's law

2. the standpoint of German "potential theory," where every body is shown as a centre – site and source – of force; removing part B of the aether, the electric forces are unaffected

3.a) the standpoint of forces inducing changes in the surrounding medium (no empty space!) and those giving rise to new distance-forces: nevertheless the energy is stored in the electric bodies; removing part B of the ether, the forces would remain but the polarisation would disappear

3.b) the standpoint of forces inducing changes in the surrounding medium and those giving rise to new distance-forces but, unlike the previous case, the energy is stored inside the medium; removing part B of the aether, only vanishingly small distant-forces would remain

4. the standpoint of the contiguous action through a medium, where the medium polarisation is the

only thing really operating: no forces at a distance and the energy is stored inside the medium; removing part B of the aether, every electric effect would disappear.

Einstein seems to incline towards the approach described by Hertz as 3.b, the energy mostly in an aether, but the primary quantities are still the conduction charges on the condenser's plates. This model, that appears in Maxwell conflated with model 4, where the primary quantities are the polarizations, was the model of Helmholtz's theory.

The paper shows a specific attention to mechanical and electromagnetic features that would have survived the disappearance of the aether as a material body in Lorentz's theory and the disappearance of the aether itself in Einstein's special theory of relativity: the contiguous action, the magnetic energy, the increased inertia of the moving charges, and, the velocity of light. Ten years later what was described in mechanical terms in this first paper would have been explained as the result of adopting a different frame of reference. We could add an historical curiosity to the scientific topics of the paper. Ms. Marangoni, a friend of Einstein's from the time Einstein's family lived in Pavia, in the north of Italy, in a letter quoted a contemporary (1895/6) paper with the following headline: *Über Elektrizität und elektrische Ströme*. A Pavian physicist, B. Bertotti, wrote to Helen Dukas, Einstein's secretary, requesting further information, but Ms. Dukas answered that she had never been acquainted with the existence of the corresponding paper.

Volker Barth

Universal Exhibitions and the Popularization of Science in the 19th Century

The second half of the 19th century can be seen as science's golden age of popularization. The founding of museums, various scientific journals and public libraries made the knowledge of the age accessible to broad sectors of the general public. Increasing numbers of citizens were able to

The Universal exhibitions were the focal point and symbol of these trends. Starting with the Great Show of the Works and Industries of all Nations in London in 1851, more than 20 international exhibitions were staged by 1900 in Europe, Australia and the USA. Despite their many differ-

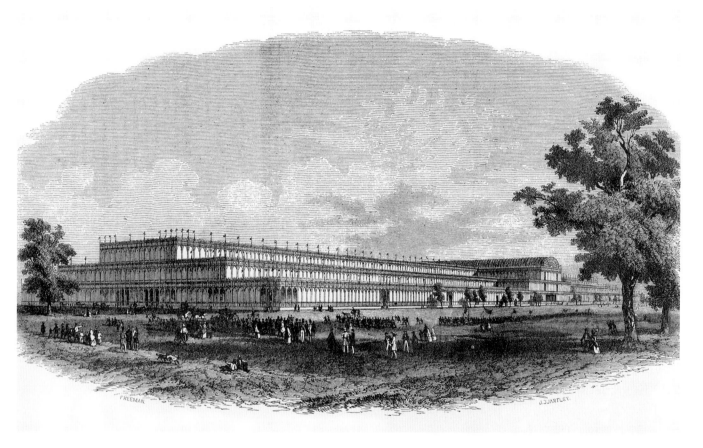

Crystal Palace,
London, 1851

inform themselves about the latest scientific achievements. In Germany the term *populärwissenschaftlich* (popular scientific) was first used in its modern sense from the 1850s onwards. From then on it signified the communication of scientific knowledge to an ever increasing lay public. There were similar developments in other European countries, such as France or Austria, with the continuous establishment and expansion of these types of popularizing institutions between 1850 and 1900.

ences, their common denominator was their distinctly encyclopedic character. Their aim was to present all the products of human creative power, free of any thematic restrictions, to a public of millions from all over the world.

The concept of progress served as the ideological base. The accumulation and increase of knowledge was seen by the exhibition organizers as the basic principle of human civilization. It was the task of each and every individual to contribute to this social cause through disciplined and committed labour. The idea was to develop universal laws for constant perfection in all areas of human activity. The extensive state-of-the- art presentation of current scientific achievements in a world exhibition acted as a documentation of progress that not only demonstrated the paths already covered but also helped to signpost the major tasks of the future. Making established knowledge accessible to a millionfold public was thus an important evolutionary step.

The exhibitions quite emphatically stressed the complete scientific objectivity of the achievements and insights on show. The exhibitions were not designed as a forum for the debate and comparison of different concepts. Nor were they there for people to discuss unsolved questions. The whole idea was to present recognized scientific findings to a broad lay public. Consequently, this type of popularization presented the exhibited results as significant, definite and unalterable.

	exhibition year	exhibitors	visitors
London	1851	14.000	6 Mio.
Paris	1855	23.954	5,1 Mio.
London	1862	ca. 30.000	6 Mio.
Paris	1867	52.200	11 Mio.
Wien	1873	53.000	7,2 Mio.
Philadelphia	1876	30.000	10 Mio.
Paris	1878	52.835	16 Mio.
Paris	1889	61.722	32 Mio.
Chicago	1893	70.000	27,5 Mio.
Paris	1900	83.047	50,8 Mio.

But above all, the Universal exhibitions of the 19th century were an impressive scientific spectacle. Financial considerations compelled the organizers to think about how to attract as many visitors as possible to the exhibition grounds. So obviously, they were not interested in presenting an array of dry and boring academic research. They were more intent on transforming science into a real experience where the interface between science and entertainment was imperceptible. In this respect the exhibitions in the United States proved particularly pioneering, as the name "World Fair" already indicates.

A glance at the impressive statistics of these huge spectacular events shows just how effective and important the world exhibitions were in the popularization of science:

The exact number of objects put on show was so enormous, that they were not even comprehensively listed in the exhibition catalogs. So it would be a futile pursuit to try to discover exactly how many scientific achievements and innovations were presented to the public.

In the various sections of the exhibition visitors were able to view the latest developments which were all too often de-

The *Menschen-hebemaschine* (people-lifting-machine), Paris 1867

L'Exposition universelle de 1867, in Paris, aquarelle with a view of the exhibition compound on the Champs-de-Mars

clared as pioneering innovations. The comparatively under-represented areas of the humanities look somewhat lost amongst this impressive wealth of products. All the more lively was the peaceful competition for the much sought-after medals, which were awarded by an international jury to exhibitors of industrial, applied arts, medical and natural science objects of all kinds. The spectacular displays of heavy engineering and mechanical engineering were particularly awe-inspiring and the public was appropriately surprised and impressed. And even though the contemporary viewers were unaware of the fact

at the time, some of the exhibited appliances were destined to change the world for ever. This applies less to the inaugural exhibition of the Bessemer process or the gas-driven engines shown by Prussia in 1867. The guests at the first exhibition in the USA in 1876, held in Philadelphia to celebrate the centenary of the Declaration of Independence, had the opportunity of admiring the world debut of the telephone which is now considered to be a basic household appliance. The same can be said of the 1878 presentation in Paris of innovative cooling systems, even though at the time they bore no resemblance whatsoever to our present-

Corliss-Dampf-maschine, 1400 PS, at the universal exhibition in Philadelphia, 1876

of industrial products without actually having to move themselves. This experiment was not only repeated at the *Exposition universelle de 1900,* it was perfected in the shape of a 3.4 kilometre long moving sidewalk beside the Seine. The organizers' idea was to let the visitors feel and experience science and so memorize as much as possible. This principle had already created incredible excite-

Ferris Wheel, Chicago 1893

day refrigerators. And at the end of the 19th century the presentation of the automobile at the *Exposition universelle de 1900* in Paris heralded the advent of one of the great symbols of the 20th century.

The decisive popularization factor at the Universal exhibitions was the way in which science and technology were displayed as some kind of magic door to the future. In 1889 in Paris, for instance, the organizers were not satisfied with simply exhibiting the variety of electric lamps. They decided instead to focus on the Eiffel Tower, which was nothing but the enormous entrance to the exhibition. They installed a 3,000 ampere floodlight at the top of the tower and bathed the French capital in light. The effect was stunning since at that time only few lamps of a meagre 100 ampere were in use. At the same exhibition an electrical conveyor belt was installed inside the engine hall so that the visitors could be transported through the show

ment in 1867 when the public had the chance to be transported to the roof of the huge exhibition building by means of the first hydraulic elevator ever to be exhibited in France.

But the efforts to popularize the items at the Universal exhibitions did not exhaust themselves in the staging of stunning events. Visitors also had other opportunities to absorb the latest knowledge of the day. There were numerous accompanying conferences and congresses which became an established part of every exhibition from 1867 onwards. In that particular year a total of twenty-four conferences were included in the programme. Among these the ones focussing on medicine, anthropology and biology brought the natural sciences in particular to the forefront. In 1878 a total of eight organization committees concentrated exclusively on staging public congresses to accompany the exhibition. The comparison between the number of congresses held in the year prior to the exhibition and those held in the exhibition year provides excellent insights into the exhibitions' popularization potential. There were only 22 congresses in the year prior to the Universal exhibition of 1878 but 65 in the exhibition year. In the exhibition year of 1889 the number rose to 111 (38 in the previous year), in 1893 in Chicago there were 95 congresses (52 in the previous year), and in the year of the 1900 exhibition in Paris there

Le Triomphant,
Paris 1900

were 232 (66 in the previous year). At the 1900 Paris exhibition 48 per cent of the 70,000 congress participants came from abroad. In this way the Universal exhibitions of the 19th century became an international showcase of the sciences, and in Paris, which hosted five Universal exhibitions, there were seven times more congresses than in Berlin which never hosted one.

Finally, it is important to mention the enormous output of scientific and popular scientific texts surrounding the exhibitions. The great number

Trottoir roulant at the Parisian Universal Exhibition, 1900

Illumination of
the Eiffel Tower,
Paris 1889

of official reports, publications of conference contributions, commented exhibition catalogues and
assessments of the juries – *the Rapport général
sur l'Exposition universelle de 1889* in ten volumes is a perfectly normal example – is vastly
exceeded by the reports in the national and international press. For months surrounding the 1867
exhibition *Le Figaro* devoted a daily column to
the topic in which objects of all kinds and origins

were described in detail together with the way
they functioned.

The Universal exhibitions presented a spectacular
wealth of exhibits, a huge variety of congresses
and a mass of publications spreading scientific
knowledge from all spheres to a public of millions
of which most 20th century exhibitions can only
dream.

David Kaiser

Einstein's Teachers

Albert Einstein had two types of teachers in his early years: those with whom he interacted directly in the classroom, and those from whom he learned at a distance, poring over their publications in devoted self-study. With the first group, Einstein's relationships were at best ambivalent and at times antagonistic. It is from the second group that some of his most profound influences derive.

Einstein's difficult relationships with his teachers began at the tender age of five, when his family hired a private tutor for him. She didn't last long, as Einstein's sister Maja later recalled: one time, when dissatisfied with the lesson, "he grabbed a chair and with it struck the woman tutor, who was so terrified that she ran away in fear and was never seen again." Nor did Einstein have particularly high regard for his teachers in elementary school or at the *Gymnasium* in Munich. Both groups struck him as far too militaristic: the former group were too much like "sergeants," while the latter were like "lieutenants," he later explained to his friend and biographer Philipp Frank. The feeling was apparently mutual: the director of the Munich *Gymnasium*, Dr. Joseph Degenhart, informed the teenaged Einstein that he "would never get anywhere in life," adding that "Your mere presence here undermines the class's respect for me." Degenhart later asked Einstein to leave the school two years before completing his degree – a request to which Einstein gladly acceded. Despite his high school teacher's dour view of Einstein's prospects, the young student had already shown great promise in mathematics and physics. His preparation in other subjects, such as foreign languages and literature, however, was far less solid. No *Gymnasium* diploma in hand, he wanted to forget about the drudgery of secondary schooling and jump into university studies. His one hope was to enter the *Eidgenössische Technische Hochschule* (ETH, or Swiss Federal Polytechnic in Zurich), for which no formal high school degree was required. Despite months of careful self-study, he failed the ETH entrance examination in 1895 – his high marks on the physics and mathematics portions were insufficient to balance his poorer showing on the other subjects. Upon advice from the ETH's physics professor, Heinrich Friedrich Weber, Einstein went to study for one year at a small cantonal school in Aarau, Switzerland, before re-taking the ETH entrance examination. Einstein did well in the tiny school's liberal atmosphere – a far cry from the "sergeants" and "lieutenants" he had left behind in Germany – but he hardly aced all of his classes there, either. His father reacted with equanimity to Einstein's grades half-way through his Aarau studies, writing to a family friend that "I have always been used to Albert bringing home, alongside some very good grades, also some poorer ones, & I am not disconsolate."

Einstein re-applied to the ETH and was accepted, enrolling in October 1896. Although the Polytechnic had some great mathematics professors, such as Hermann Minkowski, Einstein failed to take advantage of the opportunities to learn from them, often skipping their lectures altogether. He got through these courses thanks in large measure to his new friend Marcel Grossmann, who took immaculate notes in lectures and graciously shared them with his less responsible classmate. (Grossmann went on to a career in mathematics and became a kind of teacher to Einstein himself when Einstein was working on his general theory of relativity and needed a crash-course in some of the advanced mathematics topics that he had ignored as a student.) Einstein was more impressed by Weber's lectures on thermodynamics – the science of heat – eagerly attending these classes and taking careful notes during his third year at the ETH. Yet he continued to find even his physics professors lacking. He wrote to his fellow student and future wife Mileva Marić that "Unfortunately, no one here at the *Technikum* [ETH] is up to date in modern physics & I have already tapped all of them without success." No one, that is, taught subjects that Einstein most wanted to study: the latest developments in theoretical physics. Still worse, in Einstein's view, was Professor Jean Pernet's

continued to search for an academic position without success, he came to suspect (probably incorrectly) that his former teacher had been blackballing him. Einstein complained to Grossmann: "For the past three weeks I have been here [in

physics laboratory course, which Einstein also took during his third year at the ETH. The brash young student skipped so many sessions of the class that Pernet had Einstein officially reprimanded, and gave him the lowest possible grade (1 out of 6).

Although Einstein had found Weber's lectures on thermodynamics worthwhile, he had not really impressed Weber with his overall record. When the time came for Weber to hire assistants among the recent ETH graduates, he passed over Einstein in favor of two engineering students. As Einstein

Milan] with my parents and am trying from here to find a post as an assistant at some university. And I would have long found one if Weber wasn't double-crossing me." Several years later, after Einstein's paper on special relativity had been published, his former mathematics professor, Minkowski, found himself surprised that the work could have come from such a mediocre student. As Einstein recalled late in life, he had learned quickly how best to spend his time at the ETH: "Some lectures I would follow with intense interest. Otherwise I 'played hookey' a lot and studied

Albert Einstein at the *Luitpold-Gymnasium* in Munich, 1890 (front row, third from right)

Hermann
Minkowski

Hermann Ferdinand
von Helmholtz,
around 1880

the masters of theoretical physics with a holy zeal at home." This "holy zeal" of self-study proved far more influential for the young Einstein than most of his classroom teachers ever were. In fact, his habit of self-study began long before he entered the ETH. When the Einstein family still lived in Munich they followed the Jewish custom of inviting a local student to come to dinner at their house once a week. Max Talmud, a medical student, began to join the family dinners when Einstein was ten years old. Talmud often brought popular science books for the young Einstein to read, including Alexander von Humboldt's famous five-volume series, *Kosmos: Entwurf einer physischen Weltbeschreibung* ("The Cosmos: Attempt at a Description of the Physical World"). Talmud also gave Einstein a geometry textbook around Einstein's twelfth birthday; the following year, he introduced his precocious friend to Immanuel Kant's *Kritik der reinen Vernunft* ("Critique of Pure Reason"), and the two delighted in discussing the famously ponderous volume together. Among the works that Einstein studied so intensely on his own while a student at the ETH were Ludwig Boltzmann's writings on statistical mechanics. Boltzmann's work aimed to provide

a first-principles foundation for the subject that Weber taught in the classroom. The goal was to derive the science of heat – thermodynamics – from the motion and scattering of tiny atoms. Einstein similarly threw himself into the works of Hermann von Helmholtz and of Helmholtz's student, Heinrich Hertz, thereby gaining his first introduction to the science of electromagnetism. From Hertz in particular Einstein learned about James Clerk Maxwell's unification of electricity, magnetism, and light – but, crucially, separate from the British Maxwellians' tethering of their equations to the straps-and-gears mechanical models of an underlying aether. Another important figure, from whose writings Einstein learned a great deal, was August Föppl. Föppl's textbook, *Einführung in die Maxwellsche Theorie der Elektrizität* ("Introduction to Maxwell's Theory of Electricity"), published in 1894, helped to plant key ideas in the student's head about how to think about relative versus absolute motion and how to handle the electrodynamics of moving conductors. To a student like Einstein, who had been clamoring for just these sorts of subjects in his classroom lectures – but who had been disappointed time and again by their absence – a fresh, new book like Föppl's was sure to be influential. Several echoes of Föppl's text reappear throughout Einstein's early work on special relativity. Einstein's longtime friend Michele Besso, whom Einstein met early in his student years at the ETH, introduced Einstein to the writings of another of the teachers who would exert a strong influence on him: Ernst Mach. Einstein's careful inspection

argued forcefully, and Einstein appreciated the lesson. Einstein wrote to Mach a few years later that "I know, of course, your main publications very well, of which I most admire your book on Mechanics. You have had such a strong influence upon the epistemological conceptions of the younger generation of physicists [...]," among whom Einstein certainly included himself.

Heinrich Rudolph Hertz

Ernst Mach, 1900

of Mach's works continued after he moved from Zurich to Berne to begin work as a patent clerk. In Berne he joined with two friends who shared his intellectual curiosity, Maurice Solovine and Conrad Habicht, forming what they jokingly called their "Olympia Academy" to read and discuss works on physics and philosophy. In these years Einstein was especially taken with Mach's positivism: a true science of physics, Mach had urged, should be based entirely on observable quantities, on objects of positive experience. Unobservable or empty metaphysical concepts – such as the aether – should be banished, Mach

Among these two types of teachers – those Einstein encountered face to face, such as Degenhart, Weber, Minkowski, and Pernet, and those from whom Einstein learned at some remove, such as Boltzmann, Helmholtz, Hertz, Föppl, and Mach – the most lasting intellectual influences most certainly came from the latter group. Neither a model student nor a complete rogue, the young Einstein charted his own intellectual course, often preferring to explore the exciting new worlds of theoretical physics alone, at home, with a good book.

Robert Schulmann

Einstein's Swiss Years

Of the three countries in which Einstein spent most of his years, it is only Switzerland to which an inner voice drew him. As a fifteen-year old and without even bothering to consult his parents, he set out to seek his fortune there. Neither fate nor external inducements played a role. Contrast this with his native Germany which he left as a teenag-

Moving from Munich to Italy to Switzerland, all before the age of sixteen and all on his own initiative without the security of emotional and significant financial support from his immediate family gives some indication of the steely resolve in Einstein's character. Though citing a reference to the young man as a "Wunderkind," the rector of the

The Polytechnic, later the *Eidgenössische Technische Hochschule* in Zurich, around 1890

er and to which he returned after receiving an irresistible offer of academic employment; or with the United States where he sought refuge in middle age from the inhospitable climate in Germany. At various times a citizen of all three countries, Einstein was connected by a special bond to Switzerland. At the very end of his life in the United States he poignantly recalled the strength of that bond. In bidding farewell to his attending physician (Rudolf Nissen), who was returning to Basel, he remarked "You go now to the most beautiful parcel of earth that I know."

Swiss Polytechnic (ETH) urged him to finish a final year at the Aargau cantonal school in Aarau, a small town west of Zurich after he failed his entrance examination in 1895. The failure came most likely because of an insufficient knowledge of the French language and literature. In Aarau Einstein boarded in the home of Jost Winteler, a history and language teacher. Einstein's affinity for the gentle yet fiercely democratic landlord extended to other members of the Winteler clan. He referred to Jost's wife Pauline as his second mother. With daughter Marie he developed a more complex relationship. In 1897, when breaking off his relationship with Marie, he wrote to mother Pauline with an indirection and hyperbole that revealed an individual willfully distancing the self from engagement with the world around him. This psychological device, reinforced by his readings in Schopenhauer, was decisive. He must, he wrote, find his destiny in the stars not in human companionship. "Strenuous intellectual work and the study of God's Nature are the reconciling, fortifying yet relentlessy demanding angels that will lead me through all of life's troubles."

In Aarau he declared his intention of pursuing the study of theoretical physics, a daring proposition as this discipline was still in its infancy at the end of the nineteenth century. One of its main attractions for Einstein was the independence that such study would afford him. Reveling in this freedom from attachment was underscored when he declared on his Swiss citizenship application in October 1900 that he was without religious affiliation.

On entering the ETH in the autumn of 1896, he was disappointed by the traditional physics coursework, which neglected such recent masterful work as that by James Clerk Maxwell.

He cut classes with abandon in order to study physics texts of his own choosing. Yet he was careful to pore over a friend's class notes in order to pass his examinations. On writing his dissertation for the University of Zurich he abandoned an earlier topic on the electrodynamics of moving bodies, which was considered overly speculative, and submitted in its stead a thesis which gave free play to theory within the constraints imposed by an experimentally oriented academic physics environment. In obtaining a post as technical expert at the Swiss Federal Patent Office in Bern he

secured a well-paying temporary position that allowed him to bide his time productively while an academic position was created for him at the University of Zurich. The Patent Office was not a backwater to which Einstein was relegated; he actively sought it as a marshaling ground for writing scientific papers that would keep alive his eligibility for the Zurich position. His Berne posting also

Bern, Kramgasse und

afforded him creative isolation and shielded him from excessive bureaucratic demands. An unorthodox schedule allowed him to pursue his own interests half the day after examining the practicability of electrical devices submitted for patent approval the other half. The schedule seems to have suited him well. In 1905, he managed within the space of eight weeks to produce three papers that revolutionized the world of theoretical physics: an extension of his doctoral dissertation on Brownian motion, an article on the electrodynamics of moving bodies, more commonly known as the special theory of relativity, and a paper on the light quantum hypothesis.

The *Zytglogge* in Berne, seen from the *Kramgasse*

Matching the unconventionality of his professional life in Berne at the Patent Office, Einstein founded and participated in the "Academy Olympia," a casual group of bohemians that mirrored more formal societies dedicated to inquiry into the natural sciences and philosophy. Much as he was to hew a solitary path to academic respectability, so he carved out an iconoclastic circle of friends, his fellow Olympians, Maurice Solovine and Conrad Habicht. A similar sense of kinship bound him to the two outsiders he had met at the ETH: his future wife, Mileva Marić and his best friend, Michele Besso. Mileva, a Serbian classmate at the ETH and fellow bohemian was a privileged comrade in his closed world, the only woman in the mathematical-physics division of the ETH. She shared, at least at this early stage, Einstein's pride in pitting himself against the conventions of bourgeois society. Besso, the "eternal student" and perfect foil, exasperated Einstein with his impracticality and was loved by him for his lack of guile and indifference to worldly ambition.

In 1907, Einstein submitted seventeen published articles

Here, at the *Äusseres Bollwerk* and opposite the Berne train station, the Federal Patent Office *(Eidgenössisches Patentamt)*, was situated where Einstein worked between 1902 und 1909

Einstein's workplace (reconstructed) at the Patent Office in Berne

The technical expert 2nd class © The Hebrew University of Jerusalem, Albert Einstein Archives

in theoretical physics for the habilitation, a post-doctoral degree that confers the right to teach at the university level. Included in this number were the groundbreaking papers of the *annus mirabilis*, 1905. But the University of Berne insisted on one customized article; Einstein cut his losses. He resubmitted the following year with "a specialized

tors. On hearing of his announced departure from Zurich to Prague, students circulated a petition: "Prof. Einstein has an amazing talent for presenting the most difficult problems of theoretical physics so clearly & so comprehensively that it is a great delight for us to follow his lectures." In the end he spent only three semesters at the Universi-

"Academy Olympia": Albert Einstein, Maurice Solovine and Conrad Habicht (from right to left)

investigation of a scientific topic" and was awarded the degree. His single-minded, if unorthodox, pursuit of an academic career began to bear fruit. The new position that he had anticipated and coveted at the University of Zurich was finally funded. His triumphal return to Zurich as an extraordinary professor in October 1909 began a new phase. Einstein enjoyed teaching, though he found it demanding work. He wrote a friend in December of that year "I like my new profession *very much*. I get much pleasure from teaching, even if it requires much work the first time around." Certainly he seemed to have made an excellent impression on a significant number of his audi-

ty. Though he was to sorely miss the liberal social and political atmosphere of Switzerland, he could not resist the Prague offer of a full professorship. A clear indication of the unfavorable light in which Einstein viewed his Prague surroundings in comparison to his beloved Switzerland is evident in one of his pungent characterizations: "Those who spent their developing years in a democratic society cannot really get used to such a caste system – as it is here. One does not cease to ridicule the affection and pomposity."

After the Prague interlude, which lasted from the spring of 1911 until autumn 1912, Einstein was called to a full professorship at his alma mater, the ETH in Zurich. The competition was fierce: both the universities of Utrecht and Leyden in the Netherlands vied for his favor. The lure of Switzerland proved too strong in the end. It is in this period that he began his collaboration with Marcel Grossmann in what would become a three-year struggle to generalize his relativity theory of 1905. In May 1913, Einstein and Grossmann published a preliminary but comprehensive version of a generalized theory of relativity and gravitation. The results remained unsatisfactory, and the final breakthroughs would come only after he took up an appointment at the Prussian Academy of Sciences in spring 1914.

Einstein's departure for his long-term appointment in Berlin (1914–1933) did not break the hold that Switzerland exercised over him. He returned from time to time to offer courses in Zurich, to visit his estranged family, and to maintain contacts with his closest friends and confidants. Until the end of his life, it remained for him the place where he found his bearings as a creative individual and which provided the crucible of his most important scientific achievements.

Scott Walter

Henri Poincaré and the Theory of Relativity

In the month of June, 1905, the theory of relativity came to light in the form of two scientific papers, one by the French mathematician Henri Poincaré (1854–1912), the other by a young patent examiner in Bern, Albert Einstein (1879–1955). The two scientists had never met, although Einstein was

Henri Poincaré, 1912

familiar with elements of both the science and the philosophy of Poincaré, whose name was celebrated in scientific circles.

Much like Einstein, Poincaré established his scientific credentials at an early age. In 1880 he proved the existence of a large class of automorphic functions he named Fuchsian functions. In doing so, he made an innovative employment of non-Euclidean geometry, as he noticed that the same relation exists between Fuchsian functions and non-Euclidean (hyperbolic) geometry, on one hand, and between certain elliptic functions and Euclidean geometry, on the other hand.

The following year, Poincaré was named assistant professor of analysis at the University of Paris, becoming professor of mathematical physics in 1886, and a member of the French Academy of Science in 1887. His scientific contributions remained little known outside of mathematics until 1889, when he was awarded the Grand Prize of King Oscar II of Sweden for his study of a thorny question in celestial mechanics known as the

"three-body problem": how do three masses behave under the influence of gravitation? The revised version of his study is a milestone in the history of both celestial mechanics and dynamics, although some of its more profound insights lay fallow for decades. For instance, Poincaré provided the first mathematical description of what is now known as chaotic motion.

Other results in the prize paper were rapidly assimilated, including Poincaré's Recurrence Theorem, which states (roughly) that a closed mechanical system with finite energy (like that of three planets gravitating in empty space according to Newton's law) will return periodically to a state very close to its initial state. The consequences of this theorem were significant for the molecular foundation of the second law of thermodynamics (according to which entropy increases over time for any closed system). Under the reading argued for by Ludwig Boltzmann (1844–1906), the second law is consistent with Poincaré's theorem, but must be understood as a probabilistic truth, i. e., one allowing for the occasional period of decreasing entropy.

Poincaré lectured on all aspects of physics, in an elegant, abstract style quite different from that practiced elsewhere. With the aid of student note-takers, he published fifteen volumes of lectures, four of which were translated into German, including his course on the theory of electromagnetism by James Clerk Maxwell (1831–1879). After ten years, Poincaré relinquished his chair in mathematical physics for another in mathematical astronomy and celestial mechanics, although on occasion he would still lecture on questions of physics.

Poincaré's lectures of 1899 provide an example of this, as they took up recent theories that promised to address certain lacunae of Maxwell's theory, including an adequate explanation of the electrodynamics of moving bodies. Whereas Maxwell's theory dealt with continuous macroscopic fields, the theory of Hendrik A. Lorentz (1853–1928) was based on the notion of elementary charged particles called electrons, the existence of which had been experimentally confirmed two years earlier.

Impressed with the ability of Lorentz's theory to explain a curious splitting of spectral lines in a strong magnetic field (the Zeeman effect), Poincaré considered this theory to be the best one avail-

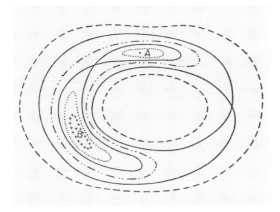

able, and in 1900 took it upon himself to eliminate what he saw as its major defect: the contradiction of Newton's third law (for every action there is an opposite and equal reaction). In doing so, Poincaré noted that in order for the principle of relative motion to hold, it was necessary to refer time measurements not to the "true time" of an observer at rest with respect to a universal, motionless carrier of electromagnetic waves known as the ether, but to a "local time" devised by Lorentz as a technical shortcut. For Poincaré, local time had a real, operational meaning: it was the time read by the light-synchronized clocks of observers in common motion with respect to the ether, corrected by the light signal's time of flight, but ignoring the effect of motion on light propagation.

There was more to this exchange of light signals than the utilitarian synchronization of clocks, as Poincaré noted in a philosophical essay entitled *The measurement of time* (1898). Passing over in silence twenty-five centuries of metaphysical de-

bates, Poincaré proposed that his readers examine how working scientists seek to establish the simultaneity of two events. The definition of time adopted by astronomers, for example, was merely the most convenient one for their purposes, and in general, he wrote, "no given way of measuring time is more *true* than another." The notion of the simultaneity of two events, Poincaré concluded, is not determined by objective considerations, but is a matter of definition.

Likewise, Poincaré wrote on a different occasion, there is no absolute space, and it is senseless to speak of the actual geometry of physical space.

Contrary to Poincaré's doctrine, astronomers thought they could measure the curvature of space, at least in principle, and thereby determine the geometry of physical space. For Poincaré, any physical measurement necessarily entailed *both* geometry and physics, because the objects of geometry – points, lines, planes – are abstract, and any external use of geometry requires a more-or-less arbitrary identification of these objects with physical phenomena, for example, that of equating straight lines with light rays. According to Poincaré's conventionalist philosophy of science,

Poincaré's representation of unstable trajectories in the three-body problem

The Poincaré disk model of hyperbolic geometry (cover J. Stillwell, *Sources of Hyperbolic Geometry*, Providence 1996)

Hermann Minkowski

scientists are often confronted with open-ended situations requiring them to choose between alternative definitions of their objects of study. In virtue of this freedom of choice, which marks the linguistic turn in philosophy of science, Poincaré was often thought to be upholding a variety of nominalism, an error he vigorously denounced. The choice scientists have to make, Poincaré explained, is not entirely free: in establishing a convention, scientists are "guided by experience."

The conventionalist philosophy of science gained greater recognition upon publication of a collection of Poincaré's essays, *La science et l'hypothèse* (1902), a work promptly translated into several languages. Among its early readers were the members of the "Olympia Academy" in Bern, made up of Einstein and his friends Conrad Habicht and Maurice Solovine. Einstein also read Poincaré's memoir of 1900 on Lorentz's electron theory, with the operational definition of local time via clock synchronization, although he may well have read this only after writing his first relativity paper of 1905. Einstein's paper features a penetrating analysis of the notion of simultaneity, wholly consistent with that of Poincaré, up to and including the procedure for synchronizing clocks in different locations by exchanging light signals. This particular analysis effectively reinforced the idea that Einstein's was a *kinematic* theory, and thereby more fundamental than any given theory of mechanics or electrodynamics.

Einstein's approach to relativity contrasts sharply in this respect with that of Poincaré, who neglected any mention of clock synchronization in his own relativity paper of 1905 (although he reviewed the topic a few years later). Nonetheless, Einstein's mathematical results agree precisely

Spacetime diagram representing Poincaré's and Minkowski's laws of gravitation

with those of Poincaré; in particular, both scientists derived the relativistic velocity composition law from the coordinate transformations connecting two inertial systems (christened Lorentz transformations by Poincaré). Likewise for the empirical consequences of the two papers, which are identical, with one exception.

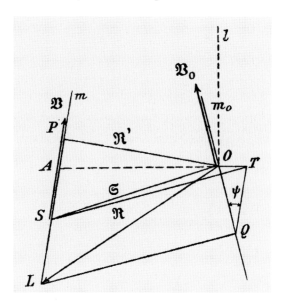

A significant difference between the relativity papers of Einstein and Poincaré concerns the status of gravitational phenomena. This subject was neglected by Einstein, although it had been identified by Poincaré in his 24 September 1904 address to the scientific congress of the World's Fair in Saint Louis as a potential spoiler for the principle of relativity. As professor of mathematical astronomy and celestial mechanics, Poincaré could hardly pretend that gravitation did not exist, and instead formulated a pair of relativistic laws of gravitation, the first of their kind. The laws themselves were observationally on a par with that of Newton, but hardly compelling on physical grounds; they retain interest, nonetheless, in virtue of their form, and the way in which Poincaré derived them. Poincaré introduced a four-dimensional space, in which three (spatial) dimensions are real, and the fourth (temporal) dimension is imaginary, such

From 30 October to 3 November 1911, the leading European physicists met in Brussels for the first Solvay Conference. Among the participants were Poincaré and Madame Curie (seated: first from right and second from right) as well as Planck and Einstein (standing second from left and second from right)

that coordinate rotations about the origin correspond to Lorentz transformations. Three years later, the mathematician Hermann Minkowski (1864–1909) extended Poincaré's geometrical approach in his theory of spacetime, and used this to express two relativistic laws of gravitation of his own.

Neither Einstein nor Poincaré let himself be impressed by Minkowski's sophisticated spacetime theory, but shortly after Poincaré died, Einstein changed his mind, and with the help of his friend, the mathematician Marcel Grossmann (1878–1936), like himself a former student of Minkowski's,

adopted it in view of a radically new approach to gravitational attraction. Three years later, in November 1915, Einstein discovered the field equations of general relativity, according to which the geometry of spacetime is curved in the presence of matter. Had Einstein finally refuted Poincaré's doctrine of space? Poincaré, that "astute and profound" thinker, Einstein wrote, was right, *sub specie aeterni* (*Geometrie und Erfahrung*, 1921). Nevertheless, Einstein continued, his own point of view was required by the current state of theoretical physics.

Horst Kant

Albert Einstein and the
Kaiser Wilhelm Institute for Physics in Berlin

On the basis a of memorandum worked out by Adolf von Harnack (1851–1930) on the occasion of the 100th anniversary of the University of Berlin, the *Kaiser Wilhelm Gesellschaft* (KWG) was founded in 1911 as a new, specialized research organization. The opening of its first two institutes, the Chemical Institute under Ernst Otto Beckmann (1853–1923) and the Institute for Physical Chemistry and Electrochemistry under Fritz Haber (1868–1934; Nobel Prize winner 1918),

Dedication of the Kaiser Wilhelm Institute for Physical Chemistry and Electrochemistry in Berlin-Dahlem, 23 October 1912. Behind Kaiser Wilhelm II, from right to left: Adolf Harnack, Emil Fischer and Fritz Haber

marked the beginning of its activity in Berlin-Dahlem in the fall of 1912. The beginning of the 20th century demanded the establishment of independent institutes that were dedicated exclusively to research. This was due both to the enormous increase in the need for research by industry and other practical fields, and to the development of universities into institutions of mass education. While non-academic research institutions had existed for some time – a prime example of which was the *Physikalisch-Technische Reichsanstalt* (PTR), which was founded in 1887 and enjoyed a strong international reputation – the KWG ultimately constituted an institutional innovation over such predecessors, merging a

multi-institutional association oriented toward theoretical research with a sophisticated combination of non-state and state scientific sponsoring, which allowed its own research program to proceed with a high degree of autonomy.

Even before founding the KWG, all participants agreed that the group of potential KWG institutes should include an institute for physics. However, because Berlin already had institutes in the field of physics, including not only the institutes of the university and the *Technische Hochschule,* but also the PTR and other institutions, an institute of physics for the KWG was not necessarily a high priority. Yet after the banker Leopold Koppel (1854–1933), who had provided decisive support for the realization of the KWG by providing significant foundation funds, hinted in late 1913 that he also intended to provide funds for research in physics, the Prussian Academy of Sciences seized the initiative. The physics among its members, Haber, Walther Nernst (1864–1941; NP 1920), Max Planck (1858–1947; NP 1918), Heinrich Rubens (1865–1922) and Emil Warburg (1846–1931; PTR president at the time) submitted a proposal for the "Founding of a Kaiser Wilhelm Institute for Physical Research." This proposal included the following text:

"We propose to set up a Kaiser Wilhelm Institute for physical research. The purpose of this institute is to construct alliances of qualified specialists in physics research, concurrently and subsequently, for the solution of important and urgent issues in the field, in order to guide the relevant questions in a planned manner, both through mathematical-physical observations and through experimental investigations, to be performed especially in the laboratories of the relevant researchers, toward as exhaustive a solution as possible.

We envision the headquarters of the institute in Dahlem in a small building which accommodates meetings as well as an archive, library and individual physical apparatus."

Albert Einstein and the Kaiser Wilhelm Institute for Physics in Berlin

This proposal contained an interesting innovation: not only was a new research institute to be created, but rather an institution of the kind that the organization of science would first begin seriously to focus on in the middle of the 20th century: researchers from different institutions were to come together for a certain period in order to solve a certain relevant problem, and thus in a specific manner that was altogether different from the already existing KWG institutes. Not only were cost savings anticipated from this proposal – as only a relatively small building, with a library, and a little laboratory capacity would be required – but in particular stimulating effects for the problem solving process.

Fritz Haber with Albert Einstein at the Kaiser Wilhelm Institute for Physical Chemistry and Electrochemistry, 1 July 1914

In accordance with the innovative concern of this institute, the proposed leadership structure adapted to this form: in addition to a board of trustees as the administrative body, an academic committee was to be formed, and presided over by a permanent honorary secretary. The first choice for this permanent honorary secretary was Albert Einstein, who had just been persuaded to come to Berlin. However, the day before World War I broke out, the finance minister rejected the third of funding that was to be supplied by the state; as a result the institute for the time being did not materialize.

Interestingly enough, the project was revived in the middle of the war, in this case without any direct relation to war-relevant research – but apparently with a view to the postwar period. A donation of 500,000 marks from the Berlin machine and tool manufacturer Franz Stock (†1939) made it possible for the KWG to pay from its own funds that third of the financing originally to be contributed by the state. On 6 July 1917, the KWG senate decided to begin operations at the Physics Institute in October 1917 – essentially on the basis of the memo from 1914. On significant change was that Einstein was appointed as director in the conventional sense. Besides Einstein, the board of directors included the five signatories of the 1914 proposal. The institute did not receive a building at this time, and no efforts were made to do so even after the war. Initially the board of directors explained that this meant that all available funds could be used to "completely benefit scientific research." Einstein, the institute's director, was quite content not to have to concern himself with the administration of an institute of larger scope. The institute's address was Einstein's private address, and the required meetings of the board of directors were held either at the homes of board members or in the rooms of the Prussian Academy of Sciences. Accordingly, in March 1919 the board of directors sent out a circular emphasizing the potential possibilities of the KWI for Physics:

"On the occasion of the recommencement of research work in physics, herewith we would like to bring to our esteemed colleagues' attention that the Kaiser Wilhelm Institute for Physics has considerable funds at its disposal, which are available to make possible or facilitate scientific research work for scientific institutes and for individual colleagues. Funding will be considered especially for the following:

Am 1. Oktober 1917 ist das
Kaiser-Wilhelm-Institut für physikal. Forschung
ins Leben getreten. Seine Aufgabe soll darin bestehen, die planmäßige Bearbeitung wichtiger und dringlicher physikalischer Probleme durch Gewinnung und materielle Unterstützung besonders geeigneter Forscher zu veranlassen und zu fördern.
Die Auswahl der Probleme, der Methoden sowie des Arbeitsplatzes liegt in der Hand des unterzeichneten Direktoriums. Doch sollen auch von anderen Physikern an das Direktorium gelangende Anregungen von diesem erwogen und die vorgeschlagenen Untersuchungen im Falle der Billigung gefördert werden.
Wenn das Institut auch naturgemäß erst nach Beendigung des Krieges seine volle Wirksamkeit wird entfalten können, so soll doch womöglich schon jetzt mit der Arbeit begonnen werden. Angaben über nähere Einzelheiten sind an den mitunterzeichneten Vorsitzenden des Direktoriums, Professor Einstein (Haberlandstr 5, Berlin-Schöneberg) zu richten.
Das Direktorium.
Einstein. Haber. Nernst. Rubens. Warburg.

Announcement of the Founding of the Kaiser Wilhelm Institute for Physics, *Vossische Zeitung*, 16 December 1917

I. the purchase of apparatus for specific studies.
II. scholarships grants to enable the execution of
certain scientific works."

In the following years, the actual work of the KWI
for Physics consisted in the board of directors'
considering proposals for
financing and presenting its
recommendations to the board
of trustees; the board of trustees
then decided on the allocation
of funds. If equipment was purchased, it generally remained
property of the KWI for Physics
and was then in later years
loaned out to other researchers
as needed. However, so much
skill was invested in this activity, especially in the difficult
years after the war, that a num-
ber of fundamental research topics in modern
physics could be further developed, or even just
continued, at different institutions, including the
other KGW institutes. In this respect the Institute
made an important contribution to physical
research in Germany in the 1920s. However, the
innovative idea from 1914, of bringing scientists
together for a time to collaborate on fundamental
research projects, was not realized and apparently not even tackled.

Although the emphasis of support was placed
on topics from the field of atomic physics in the
broader sense, it seems clear that no specific
research objective was actually pursued; rather,
works from a wide variety of fields of physics
were supported. From 1924 on, funds were made
available for three nominal assistants' salaries,
which were divided up and granted to other institutes for research scholarships.

Approximately twelve projects were supported
each year. Two examples may illustrate the way
the institute worked and functioned at the time.
The first is a proposal from 1918, which Peter

Erwin
Finlay-Freundlich,
before 1933

Debye (1884–1966; NP 1936), at the time a professor at the University of Göttingen, submitted to
the KWI. During this period Debye, together with
his assistant Paul Scherrer (1890–1969), was performing his famous work on the effects of X-ray
interference. For this work, however, he required
more powerful X-ray equipment than that which
was available to him in Göttingen. The KWI approved 16,030 marks for him to purchase such
a device from Siemens & Halske, under the condition that the apparatus would be made available
to the KWI after the study was concluded. Because of the postwar situation, however, the device could not be delivered until summer 1920,
when Debye had already assumed his professorial
chair at the ETH in Zurich. The KWI board of directors was still willing to lend him the apparatus
so that even there the planned work could finally
be performed. In the years 1922–1924, Werner
Kolhörster (1887–1946), working at the time as a
guest at the PTR (while Nernst was president),
received financial support for his research on the
origin of cosmic radiation. Because of inflation,
the sums granted are somewhat confusing: 10,000
marks in June 1922, 100,000 marks in March 1923
and 2 million marks in July 1923. The funds were
used to purchase measuring equipment and to
undertake trips to the Jungfraujoch in Switzerland,
where the attitude was suitable for taking measurements.

Also worth mentioning is that the astronomer
Erwin Finlay-Freundlich (1885–1964) was a member of the scientific staff between 1918 and 1920 –
the only one, hired to perform work testing the
general theory of relativity, a research topic
that was of direct interest to the director. He performed this research as a guest of the Astrophysics
Observatory in Potsdam. From 1929–1933 Karl
Weissenberg (1893–1976), one of the founders
of rheology, also worked as a scientific member
of the institute. There was no other staff.

In 1921 Max von Laue (1879–1960; NP 1914) was elected to the board of directors and soon thereafter appointed as its deputy director. Einstein increasingly left the management of the institute to him, although he remained nominal director of the institute until 1932. Initially, the reason was Einstein's extended lecture trip to the Far East starting in October 1922, but the political situation in Germany at the time, along with the well-known activities against Einstein, also had an effect. As director, Einstein received a salary of 5000 marks per year (after inflation, 278 marks per month), which was balanced with his Academy salary. As a full professor at the university, Laue apparently did not receive any separate salary for his role as deputy KWI director. In addition, a secretary was paid for three half-days per week. After the available funds dwindled, and over the course of the 1920s more tasks for research support were taken on by the *Notgemeinschaft der deutschen Wissenschaft* (Emergency Organization of German Science), in whose bodies some of the physicists of the KWI for Physics were also represented, the institute played a lesser role in research funding in the second half of the 1920s, although the tasks mentioned above continued to be fulfilled. Yet it became clear that the institute could no longer provide real advantages without any real research of its own. To do this, a thorough reorganization would be required.

More intensive considerations on this subject began in early 1929, with Laue taking the initiative. These efforts ultimately resulted in a kind of refounding, yet there were delays, due not least of all to the new political constellations after the Nazis took power. Money for the new building, erected in Berlin-Dahlem between 1935 and 1937, came from the Rockefeller Foundation. Peter Debye became the new director; Laue remained deputy director. The emphases of the research work were now set in the areas of nuclear physics and low temperature physics. For these purposes a high voltage system and a cryogenic laboratory were completed in 1938. After Otto Hahn (1879–1968; NP 1944) and Fritz Strassmann (1902–1980) had discovered nuclear fission at the KWI for Chemistry in late 1938 (which Lise Meitner and Otto Robert Frisch succeeded in explaining theoretically in Swedish exile shortly thereafter) at the end of 1939 the KWI for Physics was placed under the command of the *Heereswaffenamt* (Army Weapons Office), in order to participate in overall charge in the German uranium project. The objective of this project was at first to investigate possibilities for realizing nuclear energy and to explore potential military uses. Debye, a Dutch citizen, was suspended as director and went into exile in the U.S. When for various reasons the German military pulled out of the uranium project in mid-1942, Werner Heisenberg (1901–1976; NP 1932) became new director at the institute; from that on the only official goal of the uranium project was to create a nuclear reactor. In mid-1944, due to the effects of war in Berlin, the institute was relocated to Southern Germany. After the war it was recreated under Heisenberg's direction, initially in Göttingen. Ultimately it was moved to Munich as the Max Planck Institute for Physics and Astrophysics in the late 1950s.

Max Planck (left) and Max von Laue in Göttingen, around 1946

Christian Sichau

6 m² Wall Space and Two Misplaced Artifacts: The Theory of Relativity in the Deutsches Museum

A Physics Curator's Anecdotes

It was especially hot on this of all days in July 1923. Moreover, the topic of the lecture by Franz Fuchs (1881–1971), the curator for physics at the *Deutsches Museum*, was already complicated enough: The main features of the theory of relativity were to be elucidated to the "office and workshop personnel of the Deutsches Museum!"

was imminent and there was no time to lose in preparing various areas of the exhibition. While Einstein offered Miller his support, he was slow to offer concrete suggestions. Instead, he referred Miller to the professor for theoretical physics in Vienna, Hans Thirring (1888–1976), whom he deemed more likely to succeed in providing an "easily understandable explanation," having published a book about the theory of relativity in 1921.

Following Einstein's advice, on 17 March 1924 a letter with the Deutsches Museum's quite precisely formulated conceptions was sent to Thirring, with a request for his assistance: Einstein's works were to take their place in a room "in which the laws of energy are presented and explained"; by avoiding mathematical formulas, the visitors were to be able to "quickly get ... a general idea of the meaning of Einstein's theory of relativity." If possible, continued the museum's request, Thirring should elaborate about "the importance of the solution of these problems for natural science in general." A demanding agenda – for which, the letter continues, "wall space of approx. six square meters is to be provided." Thirring approached this "(in and of itself quite welcome) enterprise with a great deal of skepticism." His only suggestion was to take a summarizing diagram from his book – although it "must appear completely Greek to the layman."

In the face of these problems, the plans to exhibit the theory of relativity were dropped. Instead, a picture of Einstein was to be displayed in the "Energy Hall of Honor," but ultimately not even this project was realized.

Portrayal of the "Connections between the Basic Ideas of the Theory of Relativity" by Hans Thirring. From: Thirring, *Die Idee der Relativitätstheorie*, Berlin 1921

This was the wish of the museum's founder, Oskar von Miller (1855–1933). He wanted to give the future visitors of the Deutsches Museum "a general idea of the meaning of the theory of relativity," and for this purpose Fuchs had ordered several diagrams to be prepared "about the addition of velocities, about the Michelson experiment, about simultaneity, etc." Now their effectiveness was to be tested in the lecture. Fuchs labored for an hour and a half. Unsuccessfully. With these graphic aids it was not possible to explain the theory of relativity to a general audience. Therefore it was decided to dispense with these diagrams in the exhibition. In early 1924, Oskar von Miller spoke with Einstein personally about the matter. After all, the opening of the new museum building on the Isar island

History Adjusted

So much for Franz Fuchs' personal memories. The only vestige of these efforts was a model of an apparatus by Michelson, with whom he attempted in 1881 to trace a movement of the aether in relation

to the Earth. The accompanying text designates the Michelson experiment as the "pivotal point of Einstein's theory of relativity." Just like a number of other "popular" portrayals, the Michelson experiment was regarded as a decisive argument for the correctness of the theory of relativity, and was also used to instruct people about this theory. Apart from this single experiment, the theory of relativity was deemed an abstract creation in the realm of ideas and mathematics, without any close relation to experimental practice or material artifacts, as the Fuchs lecture and Thirring's proposals show. Yet in the course of setting up the "Electrical Rays" group, Franz Fuchs had just acquired two objects that could have provided a different perspective on the theory of relativity and its history: The apparatus by Walter Kaufmann (1871–1947) and Alfred Bucherer (1863–1927).

Both devices were used to measure the deflection of fast electrons in an electric and in a magnetic field, in order to determine the relationship between the electrical charge of the electron and its mass. This is why both devices were placed in the "Electricity" group: "Electrical Rays, X Rays." More precisely: As a development phase of discharge tubes, they were supposed to contribute to an impression of historical continuity in the exploration of the "Nature of Electricity," and thus also serve a didactic purpose. This corresponded to the interpretation of history firmly anchored in and imprinted on the conception of the museum, and ad-

Model of the Michelson apparatus. Manufactured by the *Deutsches Museum*, 1922

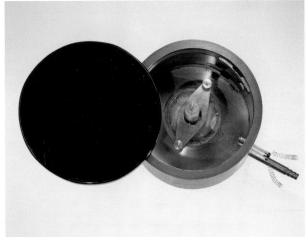

hered to the museum's goals of imparting information. Thus the devices by Kaufmann and Bucherer found their place here, set into a narrow scheme of explanation and interpretation, at the price of abandoning part of their historical identity – the part that related to the theory of relativity.

Apparatus by Alfred Bucherer for the measurement of the specific charge of electrons, 1908

Relativity in the Laboratory

This, despite the fact that Walter Kaufmann had pointed out this very aspect of his apparatus in his correspondence with the museum before handing it over: He had used the donated apparatus in his investigation of the "variability of mass in quick electrons" in 1906. In the case of Bucherer the "misconception" was even more apparent, for the experiments he had performed with the apparatus donated in 1922 had been published in 1908 under the title: *An Experimental Confirmation of the Principle of Relativity*.

At the time Bucherer had made explicit reference to the long-enduring debate about the reconcilability of Kaufmann's measurements and the spe-

Apparatus by Walter Kaufmann for the measurement of the "mass changeability of fast electrons", 1906

cial theory of relativity. While Einstein, in his first paper in 1905 had treated the question underlying the experiments of Kaufmann and of Bucherer merely theoretically and without any reference to real experiments. In 1907 he discussed the problem at length in the *Jahrbuch für Radioaktivität und Elektronik*. He appended to this discussion a schematic sketch of Kaufmann's apparatus (taken from Kaufmann's original treatise) and explained in detail how it worked: In the middle of the base of a brass cylinder was placed a grain of radium bromide, which emitted fast electrons during its

radioactive decay. This radiation, known as "β-radiation," passed through two electrically charged plates and was then registered in the upper part of the cylinder by means of a photographic plate. Not portrayed in the sketch was the magnet surrounding the cylinder. As regards the reconcilability of his own theory and Kaufmann's results from his deflection experiments, Einstein concluded: "Considering the difficulty of the investigation, one would be inclined to regard the correspondence as sufficient." Moreover, Einstein cleverly left unmentioned those rival theories which were more consistent with Kaufmann's results. Above all, he had turned Kaufmann's own interpretation on its head – a year earlier, Kaufmann had maintained that, "The measurement results are not reconcilable with the basic hypothesis of Lorentz and Einstein."

It was this original finding by Kaufmann that had motivated Alfred Bucherer to undertake his own experimental investigations. Bucherer adhered to the essential principle of the experiment, although the concrete realization changed. He had originally developed his own theory of the electron, but now he switched over to the Einstein camp: with his experiments he believed to have "furnished proof for the validity of the Lorentz-Einstein principle of relativity." Despite a few remaining uncertainties, the discussion among scientists now concentrated increasingly on the search for a new source of error in Kaufmann's experiments. Einstein had already tried to steer the discussion this way in 1907, when he conjectured a "source of error that had not yet been evaluated" in Kaufmann's work.

For the adherents of Einstein's theory of relativity, Bucherer's confirmation was grounds for celebration, as this appeared to have resolved a significant experimental contradiction. Hence Hermann Minkowski (1864–1909) hailed Bucherer's results at the close of the latter's lecture with the words: "From the theoretical standpoint, there was no doubt at all that this would be the case one day."

Desired Enrichments

This statement is an indication of the growing self-confidence of the "modern theoretical physics" emerging as an independent discipline at this time. The success of these efforts at demarcation from experimental physics was (not only) observable in the *Deutsches Museum* of the 1920s: the theory of relativity became an allegory for theoretical physics – although, as already mentioned, this perspective presented the museum with practically unsolvable problems in conveying the knowledge of the field. The experiments by Kaufmann and Bucherer were categorized at other locations; what remained were mere schematic diagrams and the model of the Michelson experiment. Although it was stylized as the "decisive experiment" and packed in a simple model, it was more of a "theoretical experiment" and thus at the same time expressed a certain understanding of scientific development: in this case theoretical physics had been able to solve a problem that had arisen in experimental physics, where it had not yet been understood. Arranging Kaufmann's and Bucherer's devices in the series of apparatus developed for "Electrical Rays" thus not only meant

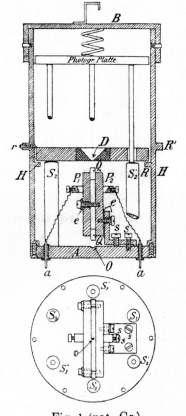

Fig. 1 (nat. Gr.).

Detailed sketch of Walter Kaufmann's apparatus
In: Albert Einstein, *Über das Relativitätsprinzip und die aus demselben gezogenen Folgerungen*, Jahrbuch der Radioaktivität und Elektronik 4 (1907)

a loss of historical accuracy, since it failed to elaborate on their role in gaining acceptance for the theory of relativity, but further prevented any progress in coming to terms with the predominant understanding of science. Nevertheless, the apparatus of Bucherer and Kaufmann were indeed preserved as a "valuable enrichment of our Electrical Rays group," as the Museum's letters of thanks formulated it at the time. And now, after so many years in the storeroom, they can finally tell a different story about the theory of relativity.

Domenico Giulini

What is Inertia?

The Concept of Inertia in Classical Mechanics

In physics, the concept of "inertia" is associated with the general property of bodies to resist a temporal change in motion. The body counters every change of the state of movement forced upon it with an inertial force known as *vis inertiae*. The force with which a broken down automobile pushes against the palms of the helpful passerby (often forcing him to give up his attempts at assistance!) and the centrifugal force that wrenches the hammer thrower's arms are just two of the countless everyday manifestations of inertial forces.

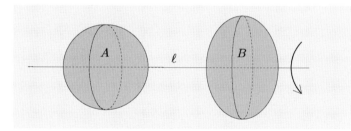

The term "change in motion" more accurately designates a change in velocity. In addition to its absolute value, which indicates "how fast" the body is moving, it also has a direction. Thus we understand that "change of velocity" always also means a change in at least one of these two determinant components – value and/or direction. In the case of the broken down car, this is primarily the value; in the case of the circling "hammer" (actually a sphere), the direction (at least in the final rotations shortly before release, when the athlete no longer significantly increases his frequency of rotation).

Thus changes in motion always cause inertial forces, which counter those external forces that cause the movement. According to Isaac Newton (1642–1727), the entire contents of the mechanical laws of motion can be summarized in a simple sentence: "With a given external force a body will move in such a way that the force of inertia cau-

sed by the movement is opposite and equal to the external force applied; the sum of all working forces thus disappears." It is correspondingly true that a body free of external forces always moves at constant velocity, that is, in a straight line (without any change to the direction of velocity) and homogeneously (without any change to the absolute value of velocity). This final assertion is known as the Galilean-Newtonian *principle of inertia*.

The rate at which the velocity changes temporally is called acceleration. Expressed precisely, the inertial force of an accelerated body is thus proportional to this *acceleration*, but is exerted in the op-

Figure left: An Einsteinian thought experiment: Consider two masses, *A* and *B*, in otherwise empty space. These masses consist of a fluid material, such that under the force of their own gravity they would each take on a spherical shape (disregarding the gravitational influence of the other mass). Relative to each other, the masses rotate rigidly around the line that connects them, *l*. If the space outside of the masses is truly empty, no reason can be derived from the directly observable facts for the masses to behave differently, for the only phenomenon that is directly observable is their rotation relative to each other. But assume that one of the masses (here *B*) is oblate. The Newtonian explanation for this would be that *B* rotates with reference to absolute space and is therefore oblate due to centrifugal forces, while *A* is at rest relative to absolute space. According to Einstein, this explanation is unacceptable, as it posits a "merely fictitious cause," while an explanation conform to the physical principle of causality may rest exclusively on "observable facts of experience." Indeed, without such facts no different behavior should occur.

posite direction. This constant of proportionality equals the mass of the body. However, in this context one should speak more accurately of the *inert mass*, as opposed to the physically different *heavy mass*. The latter is also a characteristic quantity possessed by the body, which indicates the force of attraction exerted upon a body by the gravitational field of another body. While it has

which it consists. And yet we still have in essence two characteristics, two main characteristics of the material before us, that could be regarded as completely independent of each other, and that through experience, and only through experience, prove to be fully equal. This agreement is much more than a marvelous mystery to be wondered at: it demands explanation. We may assume that

Christiaan Huygens does not declare the independence (invariance) of the laws of mechanics from the system of reference (here: ship and shore), but postulates it. Using the demand of invariance, he then derives laws. In the theory of relativity, too, it is customary to derive laws, such as the law of independent velocities of mass, from the demand for an equivalent description of the laws of collisions in different systems of reference. From: Christiaan Huygens, *De Motu Corporum*, *Oeuvres complètes*, Volume 16, 1659

been shown empirically that these two masses are always proportional to each other, such that they are even numerically equal if appropriate units of measure are selected, this must not belie the fact that they are logically defined independent of each other. Thus it is up to theorists to explain how this equality comes about. Back in 1884, this sentiment was expressed clearly by Heinrich Hertz (1857–1894), the physicist known for his direct proof of electromagnetic waves, who wrote the following passage in a lecture manuscript in that year (Hertz 1999):
"In textbooks, too, it is usually put forth as somewhat obvious, although not particularly emphasized, that weight, which is proportional to the mass of a body, is fully independent of the material of

a simple and understandable explanation is possible, and that this explanation will give us extensive insight into the constitution of matter."
This puzzle also perturbed Albert Einstein more than twenty years later, ultimately resulting in his formulation of the general theory of relativity.

Problems with the Newtonian Concept

According to the principle of inertia, a body free of forces thus moves homogeneously in a straight line. However, this assertion makes sense only if additional information is included about the measure of time relative to which the movement is homogeneous, and about the system of reference relative to which the movement proceeds in a straight line. At this juncture Newton introduced his metaphysical concepts of absolute space and absolute time. These are conceived of as autonomous units, although they cannot be identified using physical objects given in reality, and thus cannot be experienced with the senses directly. Nevertheless, Newton argued, their admittedly

Successive generations of physicists found this argumentation of Newton's unsatisfactory because of its unnecessary metaphysical baggage. The spokesman of this critique was the Viennese physicist, physiologist and philosopher Ernst Mach (1838–1916), who held the view that there was no room in a rational description of nature for ent-

Effect of centrifugal forces on a chain carousel

hidden existence becomes apparent through the fact that the law of inertia exists. In this context Newton cited the example of a bucket filled with water: When the water in the bucket is set rotating, the surface of the water becomes concave; this effect is independent of whether or not the walls of the bucket participate in the rotary motion. Because the centrifugal forces that cause the surface to change its shape do not result from a relative rotation of the water against the bucket, Newton postulates that the water's rotation relative to absolute space can provide the only explanation for this effect.

ities such as absolute space, which were supposed to produce physical effects like the forces of inertia without being able to receive any effects themselves, and which were not directly observable as a matter of principle. He deprived the supposed conclusiveness of the Newtonian bucket experiment of any foundation with the simple and relevant remark that the experiment did not exclude the possibility that the centrifugal forces might have their source in the relative rotation against the masses of the fixed stars in the sky. In this case all phenomena could be explained rationally without the presumption of absolute space.

Einstein later described a similar thought experiment; see Figure on page 174.

In fact, the law of inertia distinguishes certain systems of reference and measures of time itself, namely those against which bodies free of forces

move homogeneously in a straight line. Such inertial systems or inertial time scales can even be constructed operationally using bodies free of forces, as was demonstrated by Ludwig Lange (1863 – 1936), and also by James Thomson (1822 – 1892) and Peter Guthrie Tait (1831–1901). With this, there is no need for a concept of absolute space in a rational formulation of classical mechanics, at least under the presumption of the existence of bodies free of force. Strictly speaking, this is a fiction too, as all bodies without exception exert at least gravitational forces upon each other; yet given sufficiently large distances between bodies, these forces can be reduced to insignificant levels.

The General Theory of Relativity

In the general theory of relativity (GTR), gravitation and inertia are traced back to a common structure, connected with the geometry of the space-time continuum. Gravitation is no longer a force in the Newtonian sense, but is rather identical to that structure which actually determines which motions are free of force, and thus can be regarded as inertial. Here the concept of a global system of inertia is replaced by s concept of many inertial systems defined locally in space and time, which are determined by the gravitational field. In turn, Einstein's field equations bind the gravitational field to the local distribution of matter and its motion. In this manner matter influences the local inertial systems, just as Mach had considered hypothetically in his critique of Newton's conclusion from the bucket experiment. ("Mach's Principle" as formulated by Einstein went beyond this, however, arguing not only that local inertial systems are influenced by the distribution and motion of the material surrounding them, but actually claiming that these determined them completely. Although this intensified version guided Einstein in his formulation of the GTR, for it the GTR is not sufficient – with good reason.) Due to this influence, local inertial systems surrounded by rotating masses, such as our earth, even rotate around each other. This effect is currently being investigated in the satellite-based Gravity-Probe B experiment (cf. the experiment's homepage at www.einstein.stanford.edu).

The "frame-dragging" effect, represented here in the surroundings of a black hole by the distortions of the coordinate lines (graphic: Joe Bergeron, *Sky and Telescope*)

Daniel Kennefick

Astronomers Test General Relativity:
Light-bending and the Solar Redshift

The light-bending and gravitational redshift tests of relativity have their origins in Einstein's first realization that a generalized theory of relativity must also be a relativistic theory of gravity. Once he enunciated his equivalence principle in 1907 Einstein saw at once that light, which by special relativity must have mass even though it is a wave (because all energy has mass), must fall in a gravitational field. Its great speed means that the amount by which it falls near a large mass, even a mass as large as the Sun, will only be noticeable by a slight deflection of the light beam from its path. In addition the equivalence principle demands that if clocks run more slowly when traveling on board an accelerating spaceship, they must also run more slowly on the surface of a planet or star. Einstein claims that there is a complete equivalence between gravity and acceleration and that no experiment can tell the difference between being accelerated inside a spaceship and feeling the affect of being inside the same spaceship standing on a planet of the appropriate size. If so then special relativistic effects such as time dilation ought also to be characteristic of gravitational fields. Since the spectra emitted by atoms in the atmosphere of the Sun can be envisaged as a sort of clock, this suggests that the light from these atoms should be redshifted by a small but predictable amount, relative to the light from similar atoms on the Earth, with its much weaker gravitational field.

Albert Einstein in Princeton

At an early stage in the development of his thinking on a general theory Einstein was able to work out the amount of deflection, which a ray of light must undergo in grazing the limb of the Sun, based on the principle of equivalence. This calculation was simply based on the rate of fall of a body in the gravitational field of the Sun and gave the amount that an observer on Earth would observe a star near the Sun to be deflected from its true position relative to the other stars. Unfortunately stars near the Sun are exceptionally difficult to observe, so Einstein realized that an eclipse would be the only time during which the required measurement would be made. Similarly he was able to predict the amount by which lines in the spectrum of the Sun should be redshifted due to gravity. Although the amount here was also small, astronomers had already noticed a tendency of spectral lines in the Sun to be redshifted by about the right amount. However there were many competing hypotheses as to the origin of this solar redshift, and if Einstein's explanation was correct it would rule out most of these. In this respect the novelty of the light-bending hypothesis was an advantage, there were few competing theories for an affect no one had imagined even existed.

Even before his theory had obtained any complete form Einstein was urging astronomers to make a test of the light deflection prediction, even enquiring whether the test could be made with daytime photography of stars near the Sun or with stars passing near to Jupiter, the largest planet. In 1914 Einstein himself helped to obtain funding for an expedition to the Crimea, led by his astronomer colleague Erwin Finlay Freundlich (1885–1964), to observe an eclipse for the purpose of testing the theory, but the outbreak of World War I led to Freundlich and his team being interned by the Russians as enemy aliens. Perhaps this proved fortunate in the long run, for during the course of World War I Einstein finalized his theory, and in doing so discovered that in addition to the falling

of light near the Sun, the theory predicted actual curvature of the geometry of spacetime near the Sun which caused the path of the light beam to deviate further by an angular amount equal to that caused by falling alone. Thus Einstein's final prediction was for twice as great an angular deflection of light grazing the surface of the Sun as his original prediction. The curvature of spacetime near the Sun had no effect on the redshift prediction, but unfortunately opinion amongst solar astrophysicists had begun to run strongly against Einstein's theory. The data they had from measurements of solar lines were frequently inconsistent with Einstein's prediction. It was clear that some other effect or effects was operating, and many of the experts doubted that Einstein's theory could be accommodated with other hypotheses within the constraints of the data. During the war contact between scientists in belligerent countries was severely restricted. But neutral Holland provided an avenue for news of important results. The Dutch astronomer Willem de Sitter (1872–1934) was an early enthusiast for general relativity and wrote a highly influential series of articles in the leading British astronomy journal discussing the various experimental consequences of Einstein's theory, including the light-bending test. His papers provoked the interest of two of the leading men of British astronomy, Sir Frank Watson Dyson (1868–1939), Astronomer Royal and Director of the Greenwich observatory, and Arthur Stanley Eddington (1882–1944), director of the Cambridge University observatory. It so happened that the total solar eclipse predicted for 1919 would take place while the Sun was in the star field of the Hyades, the closest star cluster to the Earth. Unusually there would be a large number of bright stars near the Sun during the

eclipse, which would greatly increase the chances of good images being taken during the brief time of the eclipse. Motivated by the unique opportunity which was presented by this eclipse to test what seemed to be a very important theory for astronomy (here Einstein was undoubtedly greatly helped by the success his theory had in explaining the long-standing anomaly of the perihelion shift of Mercury), Dyson proposed to mount an expedition to test the theory, and placed the full resources of the joint permanent eclipse committee (JPEC) of the Royal Society and the Royal Astronomical Society behind the effort. Dyson was the chair of this committee, and his influence was such that he was able to obtain more than a thousand pounds in funding from the British government for the project, even in wartime. It was decided to send two separate expeditions. One, led by Eddington, would take up a station on the island of Principe of the coast of West Africa. The other expedition, mounted by Dyson's own Greenwich observatory, would observe the eclipse from Sobral in northern Brazil.

Even after the war suddenly ended in November 1918 it was not clear whether the expedition would be able to go forward. Certainly Freundlich's hopes of mounting a German expedition were dashed by the disastrous conclusion to the war, from the German viewpoint. Nevertheless the British persevered in the face of many obstacles, and the two expeditions departed together from Liverpool on 8 March 1919, arriving at their respective stations in plenty of time to prepare for the eclipse on 29 May. News that data had been successfully taken at both sites was communicated by telegram back to Dyson in Britain, who made public the news from the sites. After that there was to be a long wait for further information on the success or otherwise of the observations because poor weather at Principe and technical difficulties at Sobral greatly complicated the job of reducing and analyzing the data.

Arthur Stanley Eddington

Einstein's anxious wait for the results (he even began to ask his Dutch colleagues if they had any news) was alleviated when Hendrik Lorentz (1853–1928) reported to him by telegram that Eddington had told a meeting of the British Association for the Advancement of Science in Bournemouth that the result lay between the two theoretical predictions made by Einstein. But the final verdict, announced at a joint meeting of the Royal and Royal Astronomical Societies on 6 November went further, deciding firmly in favour of the higher result. Eddington and Dyson chose to frame the test as a decision between the gravitational theories of Newton and Einstein, since the lower test, though dependent on the special theory of relativity was consistent with Newton's law of gravity. In doing so he assured that the announcement of the result would be highly dramatic. In addition the whole story, when the result was finally announced, struck a cord with the public as an in-

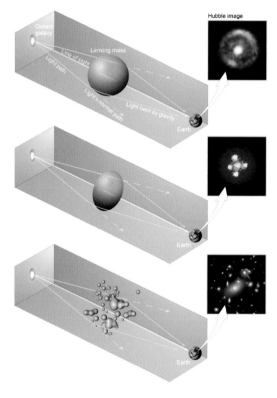

Galaxies as gravitation lens. Gravitation lenses generate images that are independent of the form of the lens body. If the lens is spherical, the image appears as an Einstein ring (or light ring), above; if the lens is elongated, the image will be an Einstein cross (it appears subdivided into four separate pictures), center; if the lens is a galaxy cluster, banana-shaped arcs of light appear, below

stance of scientific collaboration between hostile nations. In this respect it was an odd coincidence of history that both Eddington and Einstein had been vocal opponents of the war in their own country, and even more vocal proponents of the resumption of normal scientific relations after the war. In Britain the result made headlines at once, and a public fascination with Einstein quickly spread to many other countries. In 1919 the name Einstein would have been unfamiliar to most Americans. In 1922 when he visited the country for the first time he received a ticker tape parade down Broadway in New York.

Although very many contemporary physicists and astronomers were highly skeptical of general relativity, the authority of Dyson and Eddington, the care in analysis and presentation of the results and the fact that subsequent eclipse expeditions agreed with those of 1919, all combined to restrain criticism of the expedition's work, though it was subject to intense scrutiny. Until the 1970s eclipse expeditions remained the leading test of the solar light bending prediction of relativity, but remarkably no expedition ever succeeded in improving substantially on the accuracy obtained by the 1919 team. In the 1970s radio astronomy began to be used to measure light bending during occultations of quasars by the Sun, which does not require an eclipse. This method gives a far more accurate agreement with Einstein's prediction.

In the last month of 1919 Einstein received further good news. Two physicists based in Bonn, Leonhard Grebe (1883–1967) and Albert Bachem (1888–1957), made a careful reanalysis of spectral lines used in the solar redshift problem. Among other things, they noted that in comparing spectral lines seen in the Sun with those observed in the laboratory, early researchers apparently misidentified some lines in parts of the spectrum with many lines close together because they had assumed that the lines would be unshifted. Since spectral lines can generally only be identified by

physicists to presume the existence of the gravitational redshift in interpreting their data, in which the general relativistic effect is mixed in with other kinds of redshift, such as ordinary Doppler shifting. Within a few years evidence emerged of a redshift of gravitational origin in dense stars, such as White Dwarfs. Really precise measurements of the gravitational redshift had to wait until after the discovery of the Mossbauer effect, permitting the measurement of the redshift generated by the Earth's gravitational field in the late 1950s. Although the testing of general relativity has spanned many decades in a search for every greater precision in measurement, it is remarkable that in a few short months at the end of 1919 the experimental questions whose answers Einstein had sought since 1911 were answered largely to his own satisfaction, and apparently the world's.

their position in the spectrum, shifts in position have to be taken into account in identifying lines between two sources. In the wake of the eclipse results it became increasingly common for solar

Photo by the Hubble telescope from the year 1999: The galaxy cluster Abell 2218 functions as a gravitation lens and makes objects 13.4 billion light years away visible for the first time. The full-scale galaxies around us today probably emerged from comparable systems of stars. The left photo shows the entire galaxy cluster Abell 2218

Gerhard Hartl

The Confirmation of the General Theory of Relativity by the British Eclipse Expedition of 1919

In his works between 1907 and 1915, Albert Einstein listed three possibilities for testing his theories against experimental results:

1. the advance of Mercury's perihelion
2. the gravitational redshift of the solar spectral lines
3. the diffraction of starlight in the gravitational field of the Sun.

As early as 1859, the French astronomer Urbain-Jean-Joseph Le Verrier (1811–1877) had discovered that the slow movement in the orbit of the planet Mercury, the advance of Mercury's perihelion, could not be described precisely by means of Newtonian mechanics.

The perihelion is the point of the elliptical orbit closest to the Sun (the point furthest from the Sun is called the aphelion). However, this point is not fixed in space, especially for Mercury, the planet closest to the Sun, but shifts in position, in a movement caused in part by the other planets.

Le Verrier established that the movement in the Mercury's orbit is greater than it should be as a result of the gravitational influences of the other planets. Newtonian mechanics provided for a movement of 530 arcseconds over 100 years. What was observed was an advance 43 arcseconds greater, which could not be explained until the general theory of relativity. Einstein writes on 9 December 1915: "The result of the advance of Mercury's perihelion fills me with great satisfaction. How helpful was the pedantic accuracy of astronomy here, which I often used to make fun of on the sly."

Due to the measurement technology available in his day, not even Einstein himself thought it possible to apply the second method, that is, the proof that the solar spectral lines shift toward greater wavelengths in the gravitational field of the sun by the tiny amount of 0.01×10^{-10} m.

Thus only the diffraction of starlight by the Sun remained as a final experimental confirmation of the correctness of his theory.

Einstein challenged astronomers to perform this measurement as early as June 1911:

"Hence it follows that, according to the proposed theory, light rays which pass by in the vicinity of the Sun experience a diffraction through the gravitational field of the latter, such that the angular distance of a fixed star appearing near the Sun is apparently magnified. A light ray passing by the Sun would accordingly sustain a diffraction of the amount $4 \times 10{-}6 = 0.83$ arcseconds. The angular distance of the star from the middle of the Sun appears magnified by the curvature of the ray. Because the fixed stars in the sector of the sky directed toward the Sun become visible during total eclipses, this consequence of the theory can be compared with experience. [...]. My urgent hope is that astronomers take on the issue broached here, although the deliberations given above might appear insufficiently grounded or even fantastic. For aside from every theory one must ask whether an influence of the gravitational fields on the diffusion of light can be confirmed with contemporary means."

In 1915 Einstein had to correct the value of 0.83 arcseconds for the diffraction on the edge of the Sun, which he had predicted four years before the final formulation of the general theory of relativity, to 1.76 arcseconds: in his preliminary version of the theory he had not taken into consideration the curvature of space by the Sun.

"Full of optimism, on June 25, 1913 Einstein wrote: In the solar eclipse next year it should become apparent whether the rays of light are curved at the Sun, and whether, in other words, the fundamental assumption upon which it is based, the equivalence of the acceleration of the reference system on the one hand, and the gravitational field on the other hand, is actually correct."

For this Einstein set his hopes on the total solar eclipse of 21 August 1914, to which a German expedition to Russia led by Erwin Freundlich (1885–1964) was planned. However, the expedition was frustrated by the outbreak of World War I, and Einstein had to observe anxiously on 19 August 1914: "My good astronomer Freundlich will experience captivity in Russia rather than the solar eclipse."

comes visible. During the eclipse the brightest stars appear in the sky.

The astronomers of the English expedition had to record this phenomenon on photographic plates, in order to compare it with photographs taken of

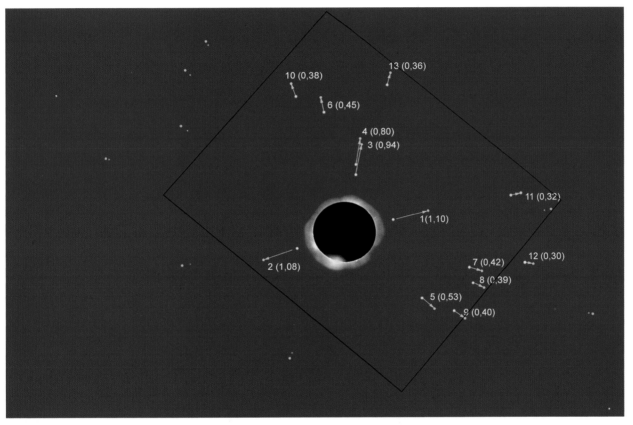

The diagram shows the changes in position of the thirteen stars near the Sun during the total solar eclipse of 1919. The black frame marks the section of the sky that was captured on the photographic plates, the numbers identify the stars according to Eddington's list, and the expressions in parentheses are the theoretical deviations in arcseconds. In the diagram, the changes in the positions are magnified by a factor of 1000

The next opportunity to observe a total solar eclipse presented itself on 29 May 1919. In this case the narrow path of totality led from South America over the Atlantic to Africa. This meant an expedition involving great expense to tropical areas barely explored by civilization. The English Royal Astronomical Society resolved to observe this eclipse and began planning an expedition in March 1917.

A total solar eclipse offers the only possibility of observing stars in the vicinity of the Sun. For the few minutes of total darkness, the Moon covers the bright disc of the Sun, the sky darkens and the wreath of rays around the Sun, the corona, be-

the same position of the sky at night in the months before or after the eclipse.

The difficulty of this task consisted in measuring on the photographic plates the shift in the positions of the stars close to the Sun as related to the positions of stars distant from the Sun, which did not shift. Required for this were not only bright stars in the vicinity of the Sun that were not covered by the corona, but also stars of sufficient brightness somewhat distant from these that could serve as reference stars during the measurement. Fortunately, these conditions were met in the eclipse of 1919. The eclipse took place near the open star cluster of the Hyades in the

constellation Taurus, a region of the heavens rich
in stars.

According to Einstein's theory, the positions of
the same stars had to deviate from each other,
and this deviation would be greater for stars located nearer to the Sun. The maximum theoretical

In Leiden
on 26 September
1923. From left to
right: back row,
Albert Einstein,
Paul Ehrenfest,
Willem de Sitter.
Front row,
Arthur Stanley
Eddington,
Hendrik Antoon
Lorentz

diffraction according to the general theory of relativity should amount to 1.75 arcseconds directly
at the edge of the Sun. This is approximately the
angle at which a 1 Euro coin appears at a distance
of 3 km.

In order to minimize the risk of bad weather, or
even of a cloudy sky, at the location of the observatory, the plan involved sending two observation
teams to perform measurements at different locations along the path of the eclipse.

The locations were selected on 10 November
1917, according to such aspects as the probability
of good weather and the sufficient observation
height of the Sun above the horizon. The latter
was important to obtain minimum refraction, that
is, to minimize the influence on the path of light
exerted by the Earth's atmosphere.

The location selected for the first part of the path
of the eclipse was Sobral in Northern Brazil. Dr.
Andrew Claude D. Crommelin (1865–1939) and
Charles Davidson (1875–1970) were selected as
this team's observers.

The location selected along the second section of
the path of the eclipse was Principe Island in the
Gulf of Guinea, at the time a Portuguese colony
located 120 km off the coast of western Africa.
Arthur Stanley Eddington (1882–1944) and Edwin
Turner Cottingham (1869–1940) were to be the observers on this team.

The telescope available for the Principe team was
an astrograph with a focal length of 3.45 m from
the Oxford Observatory. This telescope had already proven effective in the observation of the
eclipse of 1905. Thirteen stars would appear on
the photographic plates measuring 16 x 16 cm
with this telescope. Eddington gives the objective
opening as 32.5 cm, but it was reduced to 20 cm
in order to obtain sharper images. The individual
plates were to be exposed for just a few seconds.
Even so, the movement of the stars during the photograph had to be taken into consideration. Mounting the instrument as a parallactic telescope and
tracking the movement would have been too expensive in the improvised arrangement for measurement and observation. Therefore it was decided to set up the telescope in a stationary, horizontal position and allow the starlight to enter it by
means of a flat mirror, which swiveled clockwise
at the right speed. As to the evaluation of the
photographic plates, Eddington writes: "all the
measurements were made [...] with the Cambridge
measuring machine."

This measurement instrument for star plates is described in a publication by Arthur Robert Hinks (1873–1945). Although the observation of the eclipse and the photographic documentation of the stars are of central importance, the quantitative proof of diffraction is based not least of all on the quality of this measurement instrument. Both teams set out from Liverpool on 8 March 1919. On 23 April 1919 Eddington's team arrived on Principe Island, a small island just 10 by 6 miles in size.

They had over a month to get established and prepare the scheduled measurements. Principe had bad weather on the day of the eclipse. It was overcast and a severe storm raged between 10.00 and 11.30 a.m. local time (= UT). Afterward the sun was visible only for short periods, through passing clouds.

The total eclipse was expected to last from 2.13 until 2.18 p.m. local time. As Eddington writes, at the beginning of the total eclipse, the darkened Sun with its corona could be recognized only as a shadow behind the clouds, and no stars were visible.

Nevertheless, the researchers reeled off their prescribed measurement program. Eddington changed the photographic plates, Cottingham attended to the correct exposure of the plates by holding a shutter in front of the objective to prevent the telescope from vibrating mechanically. The observers themselves had no time to view the spectacle of the eclipse more closely; they were far too busy with the exactly planned manipulations during the 302 seconds of darkness. Eddington and Cottingham exposed sixteen photographic plates in those five minutes. Due to thick clouds, only one or two individual stars were visible on each of the plates – too few for a measurement.

Only one single plate showed five stars, making this the only one suitable for evaluation in the field.

The necessary reference photographs of the same position of the sky had been photographed in England in January, using the same instrument. According to Einstein's calculations, the greatest shift in the positions of the stars directly at the

edge of the Sun should amount to 1.75 arcseconds. This corresponds to a shift of 0.029 mm on the photographic plate. The diffraction of the star closest to the Sun (No. 1 on Eddington's list) amounted to 1.10", and that of the star furthest from the Sun (No. 12) only 0.30". This required accuracy of at least 0.004 mm from the plate measuring machine!

Measuring out this one plate (the plate labeled X, recording time of 2h 17m 33s, length of exposure 3 seconds) showed a shift in the positions that was quite consistent with Einstein's theory. Although they had obtained only one plate that could be evaluated, the departure from Principe

The plate measuring machine from Cambridge Eddington used to measure out the photographic plates

Original photograph (negative) of the total solar eclipse of 1919. The visible fixed stars can be recognized as small black points between the pairs of markings

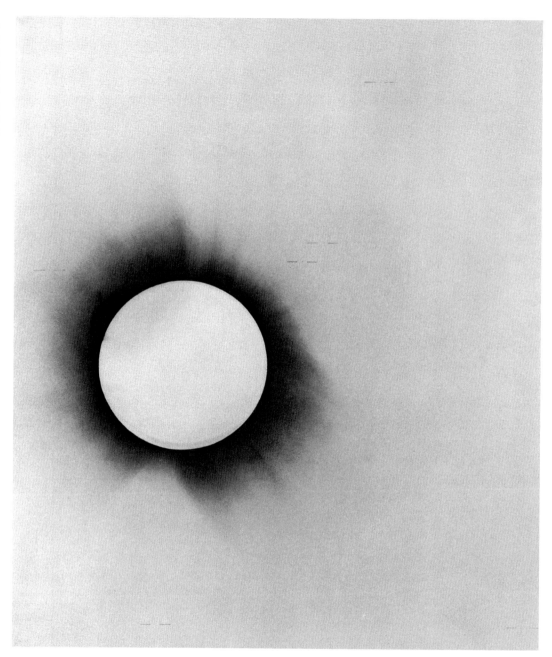

Island on June 12, 1919 was marked by careful optimism and sanguine expectations for the Sobral measurements. The team arrived in Liverpool on 14 July 1919.

As a precaution because of the problematic climatic conditions in Principe, four of Eddington's plates had been brought back to England undeveloped. One of these (the one labeled W, recording time 2h 17m 15 s, length of exposure 10 seconds) showed a sufficient number of stars. Its evaluation on 22/23 August confirmed the result of the plate measured in the field!

On 29 May 1919 there was good weather in Brazil, allowing practically optimal observation conditions. The Sobral team stayed on location longer than the Principe team, so that they were able to photograph the area of the sky around the eclipse at night before sunrise until the end of July. The idea was to eliminate any influence on the photographs exerted by different instrumental and external conditions. However, the development of the voluminous photographic material in England brought great disappointment: the stars shown in the photographs were blurry. Crommelin and Davidson commented:

"The images were diffused and apparently out of the focus, although on the night of May 27 the focus was good. Worse still, this change was temporary, for without any change in the adjustments, the instrument had returned to focus when the comparison plates were taken in July. These changes must be attributed to the effect of the Sun's heat on the mirror, but it is difficult to say whether this caused a real change of scale in the resulting photographs or merely blurred the images."

As much as the team in Principe had wrangled with the weather conditions, the bad weather there turned out to be a stroke of luck: the clouds before and, in part, during the total eclipse had prevented a thermal distortion of the heliostat mirror.

Only seven plates from Sobral, which had been photographed with a smaller telescope and heliostat, fulfilled the quality requirements for evaluation. They confirmed the Principe team's result. Not only a summary of the observations and evaluation, but also a critical assessment of the results of both teams was published in 1920 by Sir Frank Watson Dyson, Royal Astronomer (1868–1939), Eddington and Davidson.

The paper in the *Philosophical Transactions of the Royal Society of London* is entitled *A Determination of the Deflection of Light by the Sun's Gravitational Field, from Observations made at the Total Eclipse of 29 May 1919.*

As final conclusions of the English solar eclipse expeditions, they give the following diffractions of starlight, as related to the radius of the Sun:

Sobral 1,98" ± 0,12"

Principe Island 1,61" ± 0,30"

From this the authors conclude:

"Thus the results of the expeditions to Sobral and Principe can leave little doubt that a deflection of light takes place in the neighborhood of the Sun and that it is of the amount demanded by EINSTEIN'S generalised theory of relativity, as attributable to the Sun's gravitational field".

A few days after the Royal Society and the Royal Astronomical Society announced the results of the English solar eclipse expedition in November 1919, Albert Einstein had become the most famous natural scientist of his day.

Hans Wilderotter

The Einstein Tower:
Its Genesis and Function

"Friends of Einstein," according to the *Wochenpost* article *The Einstein Tower in Potsdam* in September 1930, "in their initial enthusiasm for his theory of relativity, erected at great expense a peculiar observation tower whose primary purpose is to confirm certain conclusions derived from the relativity hypothesis." This sentence contains all of

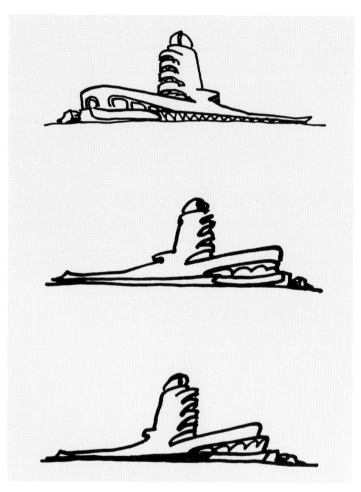

Erich Mendelsohn: Three sketches of the Einstein Tower, 1920

the central motifs important for the history and the function of the Einstein Tower. The building was erected in order to prove empirically one of the predictions Albert Einstein had formulated as part of the general theory of relativity; its construction was possible only because a number of donors were willing to provide considerable funding; the "peculiar" architecture of the tower ultimately ensured the great public attention the Ein-

stein Tower enjoys even today, of which the mere fact that an article in the mass media attended to it – and not for the first time – testifies at least indirectly.

One of the predictions Einstein had already advanced as part of his Prague theory in 1911, under the title *On the Influence of Gravity on the Propagation of Light*, (Über den Einfluss der Schwerkraft auf die Ausbreitung des Lichtes) postulated that light is deflected in the gravitational field of bodies, an effect which should be observable and measurable in the gravitational field of a body with great mass such as the sun. In May 1919, a British research expedition to Brazil led by Arthur Stanley Eddington succeeded in proving this deflection of light during a total solar eclipse. The publication of British expedition's results led to a press campaign starting in November 1919, which made Albert Einstein, who until that time had enjoyed great respect in a broad circle of specialists, but was largely unknown in the broader public, a media star. His popularity in the following years and decades grew constantly and has hardly abated even today. Without this "media hype" the magnitude of "Einstein's friends" enthusiasm would be incomprehensible and would hardly have resulted in the conceptual and material commitment which ultimately made the erection of the tower possible. The driving force and spiritus rector in the planning and construction of the tower was the mathematician and astronomer Erwin Finlay Freundlich (1885–1964), who had worked at the Berlin Observatory since 1910, and who received a letter from the astronomer Leo Wenzel Pollak in Prague in 1911, informing Freundlich (and probably a whole slew of other astronomers at the same time) about Einstein's predictions and requesting advice and assistance in their empirical proof. Freundlich, who appears to have been the only one to respond, accepted this challenge with great "enthusiasm" and went straight to work. Initially, his efforts focused on the proof of the deflection of light in the gra-

The Einstein Tower
on the Telegraphen-
berg, Potsdam,
1927

vitational field of the sun. He hoped to be able to
produce this by reanalyzing already existing photographic plates that had been taken during earlier solar eclipse expeditions. This attempt failed,
as did a number of additional attempts, including
the ill-fated expedition to the Crimea to observe
a solar eclipse there in August 1914. Freundlich
succeeded in securing the financing of the expedition, with the support of Einstein, who had arrived

in Berlin in the spring of 1914 and was now in contact with Freundlich not only by letter, but also
personally; however, the start of World War I resulted in the confiscation of the expedition's equipment in Russia and the internment of Freundlich
and his staff for several weeks.

Freundlich had since begun to strive for the proof
of the second prediction Pollak had related: the
assumption that a characteristic line shift toward

the red range must result in the spectrum of sunlight as compared to the spectra of terrestrial light sources, which could be attributed to the gravitation potential of the Sun. As in the investigation of the deflection of light, the evaluation of already existing data did not produce any convincing result. The fact that not only the existing data were useless, but that the existing instruments, too, proved insufficient for the collection of new data,

Einstein himself, having expressed repeatedly that his entire theory stands and falls with these proofs.

For his part, Erwin Freundlich had not been idle either. Since 1918 – by now the first and only scientific employee of the Kaiser Wilhelm Institute for Physical Research, founded in 1917 expressly for Albert Einstein – he developed plans for the construction of a large-scale instrument to prove

Albert Einstein visiting the Einstein Tower during the "Astronomer Days" in Potsdam, 1921

was clear by 1916 at the latest. In that year Karl Schwarzschild, the director of the Astrophysical Observatory in Potsdam, a renowned physicist and experienced astrophysicist who had published the first exact solution of the equations of the general theory of relativity shortly before, failed in the attempt to furnish proof of the gravitational red shift using existing instruments from the roof of the "official residence" of the Astrophysical Observatory. Only the development of new instruments with considerably improved measurement accuracy offered the prospect of success, a success nobody could be more interested in than

the gravitational red shift. After a failed attempt in summer 1918, the prospect of realizing this plan appeared a long way off after the end of the war, as the economic situation hardly allowed the financing of a project that was not only costly, but seemingly esoteric.

Ultimately the plans benefited decisively from the overwhelming publicity granted to the British solar eclipse expedition. Einstein's sudden fame created at least the impression that the theory of relativity was less an esoteric matter than one of interest and importance to the public, worthy of great attention. Besides, after national prestige had been gambled away in the military defeat and its consequences, many *opinion leaders* in science and politics held the fact that a theory that had been formulated by a scientist born in and working in Germany had been proven by scientists from one of the victorious powers to be a further defeat in the only field left as a source of national prestige.

A strategy of cultural propaganda was pursued, which was supposed to not only stabilize, but actually increase national prestige, at least in the field of science. Part of the strategy involved an attempt to make use of Albert Einstein's world renown, which took little notice of the scholar's he-

sitant – to put it mildly – understanding of himself as a German national and took the political form of support for the construction of the planned tower telescope, so that the second proof of the supposedly German theory could be provided by German scientists. The Prussian Parliament, (probably not without encouragement by Erwin Freundlich) and probably also still under the impression of the light-deflection proof, in that same month was willing to provide 150,000 marks, an exorbitant amount considering the financial situation of the public treasury, even though it was far from sufficient to secure the financing of the project. In this situation, again it was Erwin Freundlich who seized the initiative. At the end of the year he drafted an "Appeal for the Albert Einstein Donation," which, signed by "mandarins" of the scientific community in Berlin – including Fritz Haber, Walter Nernst and Max Planck – was sent to prosperous representatives of the German economy. The appeal, which appealed to patriotism

Left:
Erwin Finlay-Freundlich
(1885–1964),
woodcut from
Hermann Max
Pechstein, 1918

Below left:
Erich Mendelsohn
sketching, 1921

by referring to the fact that the "academies of England, America and France, in exclusion of Germany, recently [had established] a commission to energetically lay the experimental foundation of the general theory of relativity," asked for donations "to make the examination of the theory in direct collaboration with its creator possible for at least one German observatory." The success was overwhelming. Thanks to the generosity of large companies from the chemical and electric industries, but also to the willingness of smaller companies, business associations and private persons to donate, over the course of a few months more than 300,000 marks were collected. It was thus possible to begin construction as early as 1920; the building was completed in December 1921 and presented to the board of trustees of the Einstein Foundation. This foundation, chaired by Albert Einstein, had been founded to administer the donations, which by then had grown to over one million marks.

The new building, which became known to the public through press reports, aroused great attention. It appeared to be comparable with none of the models of European architecture, seeming to "the average citizen," as Phillip Frank remembered, "like a cross between a New York skyscraper and an Egyptian pyramid." Even today, in the world of architectural historians, the question as to whether the Einstein Tower is attributed more to expressionism or futurism, or whether perhaps motifs of Art Nouveau were most prominent, has not been answered conclusively; attempts to interpret it as the architectural manifestation of the theory of relativity have failed. The tower stands as a solitaire, not only the history of architecture, but also in the work of the architect Erich Mendelsohn (1887–1953).

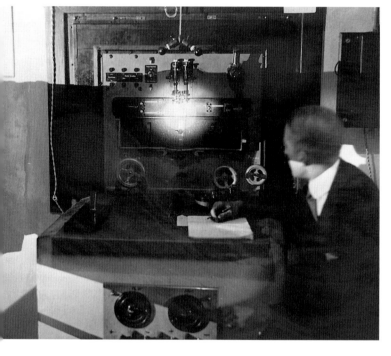

In the underground laboratory, an image of the Sun projected onto a wall with a slit, 1928

Years before construction began, Mendelsohn had been in close personal contact with Erwin Freundlich, and this contact was maintained through lively correspondence after the young architect was called up in 1915. In a letter to Mendelsohn of July 1918, Freundlich developed a very concrete plan for the construction of a tower telescope, oriented on the model of the tower telescope that had been erected on Mount Wilson in California in 1912. At this facility the telescope had been set up vertically and installed permanently, whereby a system of two mirrors – known as the "coelostat" – directs the light of the Sun or the stars into

the telescope, at the end of which the spectral analysis tests can be performed in a laboratory. The architect started immediately, producing a number of sketches to record his first design ideas, which are already remarkably similar to the building that was ultimately realized. Changes resulted primarily from the necessity to take consideration of the large-scale instrument in the core of the building. On the one hand, the building has the function of protecting this sensitive instrument – a wooden tower, at the tip of which the two mirrors of the coelostat are attached under a dome with a diameter of 4.5 meters, and which supports the lens, whose focal length amounts to a height of around 14 meters – from disturbing environmental influences. On the other hand, it was not allowed to have any contact with this inner tower, which stood on its own foundation, in order to prevent any tremors that might affect the building from being transmitted to the technical equipment.

However not only because of its striking architectural casing, the Einstein tower was different from the Mount Wilson Observatory, in which the device stands isolated. Also, in Potsdam, the light at the foot of the telescope is bent 90° into a horizontal position using a third mirror attached at a 45° angle. In the focus it generates an image of the Sun around 13 cm in diameter on a wall with a slit, which allows the given section of the image of the Sun to be directed with great precision into the dark spectrograph room located behind it, where, 12 meters further, it is broken down into a spectrum by the scientist's choice of either a prism spectrograph or a grid spectrograph. Then it this spectrum is reflected, and, inclined at a small angle with respect to the incident light, directed back to the back of the wall with a slit, where it is recorded on a photographic plate installed there. The equipment in the spectrograph room, whose subterranean location isolates it for the most part from variations in temperature and humidity, are operated from the room in front of the wall with a slit, in which an oven and an arc lamp are located to generate terrestrial spectra for comparison.

effect predicted by Einstein was overlaid by numerous other effects resulting from the complicated solar activities to which research had to dedicate itself first. In the following years and into the

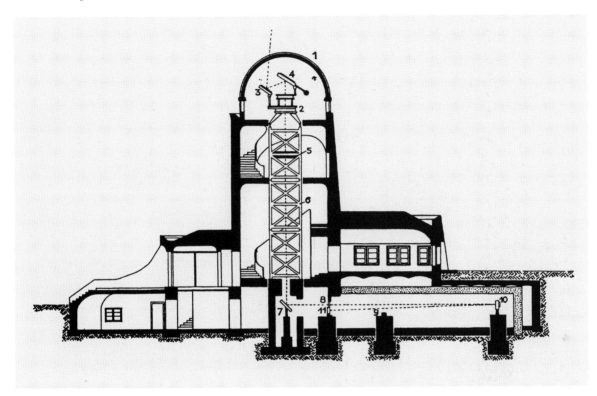

Construction of the telescope with mirrors and spectrograph:
1. Dome
2. Swivel for mirror
3. Coelostat
4. Secondary mirror
5. Objective
6. Wooden framework
7. Deflector
8. Slit
9. Prism apparatus
10. Diffraction grating
11. Photographic camera

Not until late 1924, three years after the building was completed, could the installation of the complicated and innovative instruments be concluded and research operation begun. However, it soon became apparent that the proof of the gravitational red shift would have to wait a while, as the

present, this research was not only performed, but also at the Einstein tower, a success that was no more predictable than the success of the architectural casing, which makes the "peculiar" Einstein Tower a building of world renown even today.

Christian Sichau

The Gradual Disappearance of Einstein:
Georg Joos' Experiments on the Theory of Relativity

Newspaper Headlines

Physicists were already familiar with the "nasty surprise" from America, even before the first headlines appeared in the press in early 1926: "Theory of Relativity Shaken!" The occasion was the publication of experimental results by the American physicist Dayton C. Miller (1866–1941). Contrary to the basic propositions of the special theory of relativity, he believed he had found an indication of the existence of the aether. As the Munich physicist Jonathan Zenneck (1871–1959) wrote to his colleague Georg Joos (1894–1959) in Jena, Miller's findings would bring "the entire

Explanation of striped pattern observed in Joos' experiment, Die Umschau, 1931

A Speedometer for the Earth

Yet where did the error in the experiments of the renowned physicist Dayton Miller lie? Not only did Georg Joos discuss this question with various specialist colleagues, he also established contact with the company Carl Zeiss to request that it construct a new experimental apparatus with increased precision in order to refute Miller's assertions.

In this apparatus he maintained the "classical" experimental principle used in Miller's experiments as well, which had been developed by Albert A. Michelson (1852–1931) and Edward W.

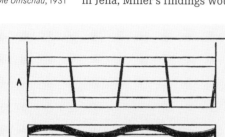

Fig. 5. So müßte ein Interferenzbild (Registrierstreifen) aussehen, wenn es einen Aetherwind gäbe

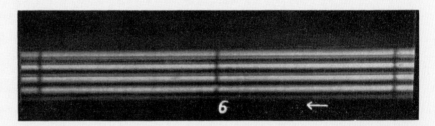

Fig. 6. Einer der 384 Registrierstreifen mit den Photographien der interferierenden Lichtstrahlen. — Beweist das Fehlen eines Aetherwindes

question of ether into a completely new stage," although "until now [...] aether [had been] a mere expression of physicists' utmost embarrassment." Einstein reacted to the debate publicly with a newspaper notice and conceded right from the beginning: "If the results of Miller's experiments should be confirmed, then the theory of relativity could not be upheld." Yet he doubted that Miller's experiments would stand up to scrutiny, and proffered a bet with a recommendation to the readers: "[...] it is a better bet that Miller's experiments prove faulty, or that his results have nothing to do with an 'aether wind'."

Morley (1838–1923) at the end of the 19th century. In this setup a beam of light is divided, the two partial beams propagate toward each other vertically and are then, after (in some experiments multiple) reflection, reunited. In this superimposition, a characteristic pattern of stripes is generated, on the basis of which it can be recognized whether the two partial beams propagate at the same speed. If there were an all-encompassing aether that conveyed beams of light, and through which the Earth moved during its orbit around the Sun, then the speed of light would depend on the direction in which light was propagated. This would have to be visible in the pattern of stripes

Discussion of an experiment by Dayton Miller in a German popular science journal, 1926

DIE UMSCHAU

VEREINIGT MIT

NATURWISSENSCHAFTL. WOCHENSCHRIFT U. PROMETHEUS

ILLUSTRIERTE WOCHENSCHRIFT ÜBER DIE FORTSCHRITTE IN WISSENSCHAFT U. TECHNIK

Bezug durch Buch-
handl. u. Postämter

HERAUSGEGEBEN VON
PROF. DR. J. H. BECHHOLD

Erscheint einmal
wöchentlich

Schriftleitung: Frankfurt M.-Niederrad, Niederräder Landstr. 28 | Verlagsgeschäftsstelle: Frankfurt-M., Niddastr. 81/83, Tel. Main-
zuständig für alle redaktionellen Angelegenheiten | gau 5024, 5025, zuständig f. Bezug, Anzeigenteil, Auskünfte usw.
Rücksendung v. Manuskripten, Beantwortung v. Anfragen u. ä. erfolgt nur gegen Beifügung v. dopp. Postgeld für unsere Auslagen
Bestätigung des Eingangs oder der Annahme eines Manuskripts erfolgt gegen Beifügung von einfachem Postgeld.

HEFT 17 / FRANKFURT A. M., 24. APRIL 1926 / 30. JAHRG.

Ist die Relativitätstheorie widerlegt?

VON DR. HANS REICHENBACH

Nachdem der Streit um die Relativitätstheorie schon seit einigen Jahren abgeklungen ist und die neue Theorie sich immer mehr durchgesetzt hat, sind in letzter Zeit Angriffe von einer Seite her erfolgt, von der aus man sie am wenigsten erwartet hatte. Denn es sind nicht Angriffe aus philosophischen Motiven, also nicht die bekannten Vorwürfe der „Unvorstellbarkeit" oder „Widersinnigkeit", die jetzt von neuem erhoben werden, nachdem sie gründlich genug widerlegt sind, sondern es ist ein physikalisches Experiment, das in offenem Widerspruch zu einer Aussage der Relativitätstheorie steht. Dieses Experiment ist von dem Amerikaner D. C. Miller auf dem Mount Wilson ausgeführt und in den Proceedings of the National Academy, Washington (11,382, 1925), publiziert worden.

Es handelt sich dabei um den sogenannten Michelson-Versuch, der einen der wichtigsten Stützpunkte der speziellen Relativitätstheorie bildet. Dieses Experiment geht auf einen Gedanken Maxwells zurück und ist von Michelson, einem durch seine optischen Präzisionsmessungen berühmten Gelehrten, zuerst durchgeführt worden. Michelson begann seine Versuche bereits in den siebziger Jahren in Berlin, als er noch Assistent von Helmholtz war; durchgeführt hat er sie in den achtziger Jahren in Amerika. Wir können den Versuch in schematischer Form folgendermaßen beschreiben (Abb. 1): Zwei starre Arme bilden einen wagrecht liegenden rechten Winkel, an ihren Endpunkten sind in der Querrichtung Spiegel S_1 und S_2 befestigt. Längs eines jeden Armes wird ein Lichtstrahl vom Scheitel O bis zum Spiegel geschickt und dort reflektiert; es steht zur Untersuchung, ob die beiden Strahlen, die zur gleichen Zeit O verlassen haben, auch zur gleichen Zeit an den Ausgangspunkt O zurückkehren. Nach der alten Aethertheorie kann dies nur gelten, wenn der Apparat in bezug auf den Weltenäther

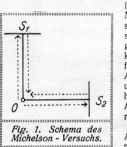

Fig. 1. Schema des Michelson - Versuchs.

ruht; da aber die Apparatur mitsamt dem Erdball durch den Weltraum bewegt wird, ist nach dieser Theorie eine Abweichung zu verlangen: der nach S_2 gehende Strahl muß etwas später zurückkommen. Mit Hilfe dieses Versuches, der eine sehr genaue Messung durch Benutzung von Interferenzstreifen gestattet, läßt sich also die Bewegung der Erde gegen den Weltenäther bestimmen. Das im Jahre 1883 und später 1887 in Verbindung mit Morley von Michelson gefundene Resultat setzte jedoch die Welt der Wissenschaftler in Staunen; trotz allergrößter Meßgenauigkeit zeigte sich keine Differenz für die Durchlaufungszeiten auf beiden Armen der Apparatur. Eine 1904/05 von Morley und Miller unternommene Wiederholung hatte denselben negativen Ausfall, trotz abermaliger Steigerung der Meßgenauigkeit.

Diese Tatsache war für die Aethertheorie nur zu erklären, wenn man annahm, daß der Aether von der Erde auf ihrer Bahn innerhalb einer gewissen Schicht mitgeführt wird, ähnlich wie ein fahrendes Schiff im Wasser eine dünne Wasserhaut mitreißt. Dies aber widersprach wieder anderen Tatsachen der Lichtbeobachtung; so wäre unter solchen Umständen nicht erklärlich, weshalb das von den Sternen kommende Licht keine Abbiegung erfährt, wenn es diese Schicht passiert. Zwar ist die Tatsache der Aberration der Sterne bekannt, wonach der scheinbare Sternort mit der Relativgeschwindigkeit der Erde zum Stern wechselt; aber gerade den Betrag der Aberration konnte die Theorie des mitgeführten Aethers nicht erklären. So befand sich die theoretische Optik in einem unlösbar scheinenden Dilemma, als H. A. Lorentz in Leyden eine Erklärung gab, die eine Verkürzung aller starren Körper bei Bewegung gegen den Aether annahm. Die Verkürzung betrifft nur den Arm OS_2, weil dieser in Richtung der Geschwindigkeit liegt, und sie ist gerade

that emerged. Originally, Michelson and Morley aspired to determine this speed of movement, yet they were not able to find any effect in the expected order of magnitude. This negative finding was one argument, although not the central one, in Einstein's development the special theory of relativity. According to this theory, light is propagated at constant speed (regardless of the movement of the light source). Further, there is no aether in which light propagates, and thus, obviously, no static aether that could have served as a special point of reference for physics – as "absolute space."

An Experimental Record in Precision

Yet how can it be shown experimentally that a presumed effect, an "aether," does not exist? This very question was a great problem, especially in a complex experiment saddled with so many exper-

stood? Such questions plagued Dayton Miller until he finally, in 1925 – more than forty years after the first experiments by Michelson – found that effect which apparently shook the basic assumptions of the theory of relativity.

Georg Joos wanted to check this result with an even more precise experiment. But exactness and precision cannot be established easily in such complex experiments; scientists must convince their colleagues – and in this case, the general public as well – with sophisticated argumentation. When the apparatus was completed in 1930, after five years of preparation, Joos thus made every effort to state in a publication the reasons why his experiments were more precise than Miller's.

In order to make the sensitivity of the equipment more "obvious," he demonstrated the enormous dimensions of his apparatus, on many occasions including a photograph showing a member of his staff next to the machine – and thus making its

large dimensions apparent. Joos had painstakingly selected the material used in its construction. Assisted by the company Schott, he searched for a suitable material that exhibits no appreciable expansion to reduce the apparatus' sensitivity to temperature. Temperature variations during measurement thus would not lead to a change in the distances between the optical components such as mirrors. These were then mounted on quartz plates produced especially for this purpose. In order to reduce disturbances caused by shocks, Joos suspended these plates from hundreds of springs – and only experimented on weekends, when he was free to perform his measurements in peace in a basement of the otherwise empty Zeiss factory.

Apparatus by a Georg Joos in the cellar rooms of the Carl Zeiss corporation in Jena

imental difficulties. Was the apparatus used sensitive enough? Was the apparatus perhaps shielded from the aether wind by the room in which it

Resting on a specially constructed support, the apparatus could be moved slowly by means of an electrical motor (a complete revolution took ten minutes), so that the apparatus took on various positions with reference to the direction of the Earth's presumed movement through the aether. At the end of the description of these technical details – and a multiplicity of others – Joos pointed out what was probably the most important innovation: he had installed a photographic camera with automatic registration. As Joos emphasized, he would thus have "documents," which could be

scrutinized by anyone. The photographic registration was praised as an "absolutely objective" procedure, and criticized the "subjective" character of the otherwise standard direct observation with the eye was critized.

To make an overall assessment of the achieved precision easier for the reader, Joos added yet another comparison: "The precision is as great as if the distance of the Earth from the Moon could be determined within one centimeter." The physicist Arnold Sommerfeld (1868 –1951) was convinced:

"These trials represented an experimental record in precision!" Joos hoped that this would lead to an end of all debates about aether.

A Matter that Imperiled the State

In any case, the findings of his experiments were clear: no aether effect could be established. Einstein's theory of relativity was thus again confirmed experimentally. Yet one searches in vain for the words "Einstein" and "theory of relativity" in Joos' publication in the journal *Annalen der Physik* in 1930. For Joos was aware of the explosive political power before he even began planning his experiments. In a letter to Arnold Sommerfeld in 1925, he wrote: "But we want to treat the matter here rather discretely, since intrigues are possible if it becomes known. Among non–physicists here, some regard the theory of relativity to be a peril to the state." Because of the delicate political situation, Joos found himself compelled to caution. In the *Annalen der Physik*, at least, he abandoned the close connection between his experiments and the theory of relativity. The result of his measurements was summarized in the form: "The amount of the aether wind would have to be less than 1,5 km/sec."

In the first edition of his textbook on theoretical physics of 1932, as well, Joos rarely mentioned the name "Einstein" in his explanation of the theory of relativity; in later editions it disappeared completely – although the theory of relativity continued to constitute an important chapter of his book. By contrast, in the journal *Die Naturwissenschaften*, in 1931 Joos emphasized how his

Detail of Joos' apparatus: mirror, mounted on quartz plates, suspended on springs

Report on Georg
Joos' experiments
in the journal, *Die
Umschau*,
1931

894   DR. SALLER, GIBT ES EINEN ÄTHERWIND?   35. Jahrg. 1931. Heft 45

Gibt es einen Aetherwind? — Nein / Von Dr. Saller

Die Jenaer Wiederholung des Michelsonversuchs

In der speziellen Einsteinschen Relativitätstheorie spielt auch die Frage der Existenz eines Weltäthers als Trägers der Lichtbewegung zwischen den das All bevölkernden Gestirnen eine große Rolle. Durch die Relativitätstheorie wird Weltäther in Frage gestellt. Die Wahrheit kann sich nur aus Versuchen ergeben. Zwei Grundversuche sind maßgebend, der von Fizeau (1851) und der berühmte Versuch von Michelson-Morley (1886/87). Die Ergebnisse beider widersprechen sich. Nach dem Versuch von Fizeau, der Licht durch eine strömende Flüssigkeit schickte, war anzunehmen, daß der Aether von der die Erde umgebenden Luft bei ihrer Bewegung durch den Weltraum nicht mitgenommen wird, daß ein „Aetherwind" entstehen müsse. Bei dem Versuch mit dem Michelsonschen Interferometer wird ein in der Bewegungsrichtung der Erde hin- und hergehender Lichtstrahl mit einem senkrecht dazu lau-

und Zeitbegriffe umstürzte. Seine Theorie ist aus rein mathematischen Erwägungen entstanden; die späteren Folgerungen und Erfolge auf astronomischem Gebiet waren ursprünglich wohl nicht vor-

Fig. 1.   Der Strahlengang im Interferometer

Fig. 2. Schnitt durch das Interferometer, mit dem der Michelson-Versuch von Joos in Jena wiederholt wurde

Die Pfeilstriche bezeichnen den Strahlengang

fenden zur Interferenz gebracht. Auf Grund dieses Versuchs schien der Aether vollständig mitgenommen zu werden. Die aus den beiden Beobachtungen abgeleiteten Anschauungen und Gesetze schlossen sich gegenseitig aus.

Die beiden Versuche unterscheiden sich darin, daß bei dem einen der Beobachter sich außerhalb des bewegten Mediums befindet, daß er sich beim anderen, dem Michelsonschen, mit dem bewegten Medium, der Erdatmosphäre, mitbewegt. Diesen Unterschied und die Bedeutung des Beobachterstandpunktes aufgegriffen und durch sein Gesetz von der allgemeinen Relativität der Bewegungen einer Lösung zugeführt zu haben, ist die Tat Einsteins, der bisherige Raum-

ausgesehen. — Nun blieb aber der Michelsonsche Versuch nicht unwidersprochen von solchen, die den Versuch wiederholten. Insbesondere erregte eine Versuchswiederholung von D. Miller, der dabei eine Aetherbewegung, den sog. Aetherwindeffekt, gefunden zu haben behauptete, Aufsehen. Sie war die Veranlassung, daß der Michelsonsche Versuch an einer Reihe von Orten nachgeprüft wurde. Die von G. Joos in Jena unter Benutzung der großen Hilfsmittel der Firma Zeiss angestellte Wiederholung des Michelsonversuches ist der Gegenstand einer Abhandlung in den „Annalen der Physik" (5. Folge, Band 7, 1930, Nr. 4). — Die Jenaer Wiederholung bedient sich, ebenso wie der ursprüngliche Michelsonsche Versuch, eines

experiments related to the theory of relativity – yet he avoided naming Einstein. Reports about Joos' experiments also appeared in various popular magazines. Here, too, it is apparent that Einstein and the theory of relativity were a delicate topic in 1931. The Zeiss corporate newspaper generally confined itself to speaking of the "basic conceptions of modern research" in 1931, while *Die Umschau* still openly established the connection between Joos' work and the "special theory of relativity by Einstein."

When the apparatus was added to the collection of the *Deutsches Museum* in 1935, the suppression of Einstein's name continued. In accordance with the founding idea of the museum, the apparatus was designated as an "outstanding masterpiece of modern precision mechanics and optics" and thus presented as a result of a presumably "German" tradition. Einstein and the theory of relativity were not mentioned at all – neither by Zeiss, nor by Joos, nor by the museum. A museum guide of 1938 did refer to the apparatus explicitly, but again, only the precision of the measurements was emphasized; their result was summarized as

follows: "that the movement of the earth has no influence on the propagation of light." Not until 1956 was another clause appended to a similar formulation in a museum guide: "[...] which is of importance for the theory of relativity."

By this time, the dispute about the existence of the aether and about the theory of relativity had blown over for the most part. For the majority of physicists, the search for an ether increasingly resembled the search for a perpetuum mobile. Investigation of Miller's measurement values in 1954 confirmed the presumption that his results could be traced back to variations in temperature. Accordingly, when he was asked in 1955 to make his apparatus available for new experiments, Joos refused with the words: "I believe we have sufficient touchstones, such that we can utilize our time more effectively than with the physics of the turn of the century."

Yet experiments of this kind are not only history; today they are performed with even greater precision, in the interest of further establishing the limits of their validity.

Tilman Sauer

Gravitational Lensing

The basic idea of gravitational lensing is simple. Consider a beam of light passing through the spherically symmetric gravitational field of a spherical massive object. According to Einstein's theory of general relativity (in Newtonian appro-

The basic idea of a gravitational lens: Light rays emitted from star *S* are deflected in the gravitational field of a massive astrophysical object *L* by an angle α that depends on the distance *r*. The observer *O* sees two images of S at either side of *L*

ximation) it would be bent inwards by an angle that is proportional to $1/r$. Here r is the distance from the point of closest approach of the light ray to the center of the massive object. Given this law of light bending, it is easy to study the imaging properties of the massive object as an optical system, i.e. as a lens. In this case, light rays passing closer to the center of the lens are bent more strongly than those passing at a greater distance. Hence, a gravitational lens does not work like a common convex shaped lens as we would have in a looking glass. If we were to simulate the effect of a gravitational lens by a glassy object, it would be shaped rather like the foot of a wine-glass. Due to the spherical symmetry of the gravitational field, the simple problem is characterized by only three points, the light source, the center of the massive object, and the observer. These three points uniquely characterize a plane. Corresponding to the two solutions of a quadratic equation, there are two light rays in the plane that are emitted from the source and may be seen by the observer, one passing at each side of the gravitational lens. Only for perfect alignment of source, lens, and observer, will the observer see a ring-shaped image of the source. In contrast to a convex looking glass where many light rays are collected in a common focal point, an untilted foot of a wine-

Imaging properties of a gravitational lens differ from those of an ordinary convex looking-glass lens. A convex lens collects incoming light rays into a common focal point. A simple gravitational lens only has two distinct light rays intersect for an observer

glass only lets two light rays intersect for any given position of the observer behind the lens. Nevertheless, one can derive an expression for light magnification by comparing the solid angles under which a light source of small but finite extension is perceived by the observer to the solid angle as it would be perceived without a lens.

As early as 1907, Einstein had first formulated his equivalence hypothesis. According to this hypothesis a static and homogeneous gravitational field is completely indistinguishable from a gravitation-free frame of reference that is moving with uniform and rectilinear acceleration. From this hypothesis alone it follows, as Einstein argued, that a ray of light passing through a gravitational field would be bent. In 1911, he even calculated quantitatively that the deflection of a light ray passing near the rim of the sun would amount to an angle given by $\alpha = 2kM/c^2\,\Delta$ where M and Δ

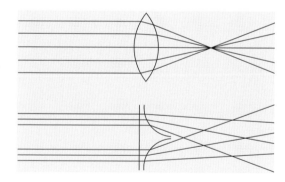

denote the mass and radius of the sun, k the gravitational constant, and c the speed of light. In the final theory of general relativity of 1916, the angle of deflection would in fact be twice that amount, since in that theory spatial curvature around the massive object contributes to the deflection. More generally, the theory was predicting a deflection $\alpha \propto 1/r$ where r is the distance from the center of the sun or star. The observational confirmation of this predicted light bending in 1919 during a solar eclipse by a British eclipse expedition led by

Arthur Stanley Eddington (1882–1944) made Einstein world-famous.

Eddington's photographic plates only showed single images of stars in the vicinity of the sun that were slightly displaced from their usual position. But as early as 1912, three years before the final breakthrough to general relativity, Einstein had already sketched the basic idea of geometric gravitational lensing, i.e. the double image configuration, into a little scratch notebook.

Einstein probably scribbled down his notes during a visit to Berlin in April 1912, when he met with

the astronomer Erwin Freundlich (1885–1964) and discussed possible ways of testing his ideas by astronomical observation. At the time, he must have concluded that the double-image effect would never be observable and did not publish his idea.

In the following years, the idea of gravitational lensing was discussed time and again. Einstein himself mentions it in a letter to his friend Heinrich Zangger (1874–1957) from 15 December 1915; Oliver Lodge (1851–1940) published a short note about it in Nature in 1919; A. S. Eddington discus-

sed the idea in his 1920 textbook on relativity *Space, Time, and Gravitation*; the Russian physicist Orest Chwolson (1852–1934) published the idea in the prestigious journal *Astronomische Nachrichten* in 1924. But the prospects of actually finding a gravitational lens in the observable sky remained hopeless.

In 1936, a Czech emigrant and amateur scientist, Rudi W. Mandl, visited Einstein in Princeton to discuss some far-ranging ideas of his. He, too, had the idea that a massive star might act as a gravitational lens. Building on this idea, he conjectured that biological evolution had been accelerated by lensed starlight causing strongly increased rates of genetic mutation in living organisms. Einstein received the amateur scientist friendly. Having forgotten about his own earlier investigations and apparently unaware of all published work, he conceded that the idea of gravitational lensing was a valid one but advised Mandl to refrain from any further speculations. Asked to publish a note on the idea of gravitational lensing, Einstein at first refused saying that the phenomenon would not be observable. Only after insistent prodding by Mandl, did Einstein eventually agree to publishing a short letter to the Editor of *Science*, entitled *Lens-like action of a star by the deviation of light in the gravitational field*. In this short paper, Einstein gave exactly the same expressions for the geometric optics of a gravitational lens, including the magnification factor, that he had derived in his scratch notebook some twenty-four years

This page (Einstein Archives, Call-No. 3-013, pp. 22f.) shows Einstein's earliest calculations on gravitational lensing together with mathematical riddles and an address of an aquaintance. On the left-hand side, Einstein sketches the geometrical setup of a gravitational lens and derives two algebraic equations for two unknowns. On the right-hand side, he identifies the two solutions ("*gibt zwei Wurzeln*") and also writes down an expression for the relative luminosity ("*gibt relative Helligkeit*").
© The Hebrew University of Jerusalem, Albert Einstein Archives

The manuscript of Einstein's 1936 paper (Einstein Archives, Call-No. 1-131) on gravitational lensing. The formulae given here are equivalent to the ones Einstein had derived already in 1912. ©The Hebrew University of Jerusalem, Albert Einstein Archives

earlier. He also acknowledged Mandl's role of prodding him to derive and publish these results. Immediately after the publication of Einstein's little note, several papers appeared by notable scholars that further elaborated on the idea. Thus, Fritz Zwicky (1898–1974) from the California Institute of Technology discussed the possibility of observing the phenomenon with the newly discovered extragalactic nebulae, i. e. with galaxies, and the Princeton astronomer H. N. Russell published a paper speculating how an eclipse "might be seen from a planet conveniently placed near the companion of Sirius" as an example of

"perfect tests of general relativity that are unavailable."

With the credibility that Einstein's little publication had lent to the idea, gravitational lensing had become part of the body of theoretical astronomical knowledge. A renewed interest in the theory of gravitational lensing arose in the early 1960s with the discovery of quasars. By then the idea was discussed on a much more sophisticated level. Gravitational lensing is, in fact, simple only when it is seen as a problem of geometrical optics for the deflection of light rays emitted from point like light sources and passing through perfectly spherically symmetric gravitational fields. In any real scenario, neither the light source nor the lensing gravitational object would be point-like. In these more general situations, gravitational lensing then produces more complex images: quadruple images arranged like a cloverleaf around a center ("Einstein cross"), or other multiple image structures, rings, ring-like shapes, or arc-like distortions.

The theory of gravitational lensing then becomes a science of very sophisticated modeling in general relativity. Realistic but tractable models of extended, inhomogeneous mass distributions as candidate gravitational lenses give rise to curved space-times and the passage of light rays through them generates mappings between the light source and an observer. Other effects must also be taken into account such as time delays arising from the different paths of the light rays. Observable light patterns computed on the basis of these model lensing systems have to be compared with actual astronomical data, and the probabilities of a match have to be assessed.

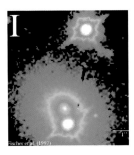 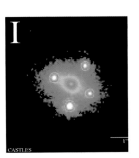 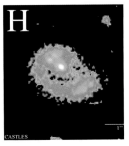

Survey (CASTLES) listed sixty-four undoubtedly identified multiple image lensing systems and another eighteen objects as candidates for such systems.

From left to right: Images of a) the double quasar Q 0957+561A, B, found to be a double-image of the same source and the first observationally confirmed example of a gravitational lens, b) the QSO 2237+0305 which is a quadruple image of the same source, i. e. an "Einstein cross," and c) the extended radio source MG1131+0456, the first example of an "Einstein ring." Images are taken from the CfA-Arizona Space Telescope LEns Survey (CASTLES)

The first gravitational lens was eventually observed in 1979 by D. Walsh, R. F. Carswell, and R. J. Weymann. They identified the double quasar Q0957+561 as a lensed double image, separated ca. 6" on the sky, of one and the same single background quasar (above left). A beautiful example of an Einstein cross, i.e. a quadruple-imaged lensing systems is the QSO 2237+0305 which was discovered in 1985. In 1988, the extended radio source MG1131+0456, was discovered and turned out to be the first example of an "Einstein ring" with a diameter of about 1.75". By now many dozens of multiple-imaged systems and a few Einstein rings have been identified. At the end of 2004, the CfA-Arizona Space Telescope LEns

In the twenty-first century gravitational lensing is a highly active field of astrophysical research. The first conference devoted exclusively to the topic was held at Liège, France, in 1983, and by now international conferences are regularly held on the topic every year. With an appreciable number of lensing systems undoubtedly identified, the lensing phenomenon is now more and more used as an observational tool that may help to answer other astrophysical and cosmological questions. Among these are assessments of the amount of dark matter in the lens, and the determination of the values of Hubble's constant or the cosmological constant.

Hubert Goenner

Einstein in Berlin: Unified Field Theory

In the science of physics, integrating previously separate fields into a shared conceptual framework and into a common mathematic notation brought decisive progress. Such successes include the merging of electricity with magnetism, optics with electromagnetism, and thermodynamics with statistical mechanics. In the second half of the 20th century, electromagnetic reciprocal action was successfully merged with the weak reciprocal action of radioactive decay, and even with strong nuclear forces, to construct a standard model of elementary particles. Einstein's relativistic field theory of gravity, his general theory of relativity of 1915, even succeeded in joining inertia with gravity!

At the end of World War I, gravity and electromagnetic forces were the only fundamental reciprocal actions of matter known. After his great success with the general theory of relativity, in 1918 Einstein set the goal of generalizing this theory of gravitation: gravitational field and electromagnetic field were to be unified in a conceptual model. The general theory of relativity encodes the six potential gravities – instead of the *one* in the Newtonian theory of gravity – in the geometry of space and time, more precisely in the space-time interval that defined Riemann's geometry. This raised the question of how and where the additional six variables for the electric and magnetic fields (one vector each), or at least the four involved in the *electromagnetic* four-vector potential, could be accommodated. Different approaches had already been proposed by other researchers: a generalization of the definition of the space-time interval by Rudolf Bach (1917/18), different extensions to the geometric framework by Hermann Weyl (1918) and Arthur S. Eddington (1921), and the addition of a

Erwin Schrödinger (1887–1961), Einstein's colleague in Berlin from 1927 to 1932

Right: Leibniz Session of the Prussian Academy, July 1931. In the back row, second from left: Albert Einstein; in the first row, second from left: Max Planck

further spatial dimension by Theodor Kaluza (1919). During his time in Berlin, Einstein had concerned himself with all of these potential solutions for a geometric basis shared by gravitational and electromagnetic fields. He also developed existing solutions further, especially Kaluza's five-dimensional field theory and what was known as

Eddington's "affine" theory. From 1928 until 1931, Einstein favored a subsidiary case of the latter, the "distant parallelism theory" introduced by the mathematician Cartan.

In addition to the unification of physical fields, Einstein's agenda for a unified field theory included the incorporation of field sources, i. e. the charged and heavy elementary components of *matter*. The tension between the model ideas of "particle" and "wave" in the description of matter proved to be a decisive factor in the development of concepts. As early as 1905 Einstein had already introduced the idea of light as a kind of gas composed of "light particles," or "photons," although it should have been a wave phenomenon according to Maxwell's theory. With the conceptualization of photons, which convey an energy quant, a "quantum physics" also came into play. After an intermezzo during which he described the elementary particles of matter as elongated objects, Einstein conceived of them as punctuate "field compressions," i.e., as the points of the "unified field" at which this field reaches infinitely large values.

Einstein's manuscript on *Unified Field Theory of Gravitation and Electricity*, 1925 © The Hebrew University of Jerusalem, Albert Elnstein Archives

A layman's commentary on Einstein's theory of distant parallelism

Einheitliche Feldtheorie von Gravitation und Elektrizität.

von A. Einstein.

§1. Die allgemeine Theorie.

$$dA^{\mu} = -\Gamma^{\mu}_{\alpha\beta}\,A^{\alpha}\,dx^{\beta} \quad \ldots (1)$$

$$R^{\alpha}_{\mu,\nu\beta} = -\frac{\partial\Gamma^{\alpha}_{\mu\nu}}{\partial x_{\beta}} + \Gamma^{\alpha}_{\sigma\nu}\Gamma^{\sigma}_{\mu\beta} + \frac{\partial\Gamma^{\alpha}_{\mu\beta}}{\partial x_{\nu}} - \Gamma^{\alpha}_{\sigma\nu}\Gamma^{\sigma}_{\mu\beta}$$

$$R_{\mu\nu} = R^{\alpha}_{\mu,\nu\alpha} = -\frac{\partial\Gamma^{\alpha}_{\mu\nu}}{\partial x_{\alpha}} + \Gamma^{\alpha}_{\mu\beta}\Gamma^{\beta}_{\alpha\nu} + \frac{\partial\Gamma^{\alpha}_{\mu\alpha}}{\partial x_{\nu}} - \Gamma^{\alpha}_{\mu\nu}\Gamma^{\beta}_{\alpha\beta} \quad (2)$$

A. Einstein Archive
1-047

© The Hebrew University of Jerusalem   © האוניברסיטה העברית בירושלים

lecture in England in June 1933, Einstein was still convinced of the capacity of his approach to achieve a "unified field theory" in the context of his theory of "semi-vectors."

The interest of the German public in Einstein's "unified field theory" reached its peak in the year 1929. One thousand copies of Einstein's work *On Unified Field Theory*, in the proceedings of the Prussian Academy of Sciences, were sold within three days, necessitating the reprint of another thousand copies – generally only one hundred offprints were required. In his book *Geheimnisse des Weltalls* ("The Secrets of the Universe") (1930), the writer Maurice Maeterlinck commented: "Yet again, in his last publication Einstein [...] brings us mathematical formulas which are utilizable for both gravity and electricity, as if these two forces, which appear to govern the universe, were identical and subject to the same law." Albert Einstein, at great pains to impart the progress of his "unified field theory" to the general public, found more vivid words in an interview in early 1929: "Now we finally know that the force which moves electrons in their elliptical paths around the nuclei of atoms is the same force as that which makes the planets circle around the sun in their yearly orbits. It is the same force that brings us rays of light and warmth, which makes life on this planet possible." Unfortunately, his theory was not able to fulfill these optimistic expectations.

A further characteristic of the elementary components of matter known at this time, namely hydrogen nuclei (protons) and electrons, was to emerge from this theory: the possession of *different* rest mass at equal but opposite charges. Thus, long before Schrödinger's wave mechanics and Heisenberg's quantum mechanics even existed (1925/26), Einstein raised his expectations of "unified field theory" sky-high; with it he aspired to not only a unified description of the two classic fields (electric and gravitational fields), but also to bridge the gap to quantum physics (atomic and electron theory). At the end of his time in Berlin, in his Spencer

Karl von Meyenn

Pauli and Einstein

When the Vienna native Wolfgang Pauli took up his studies at the Ludwig Maximilian University in Munich during winter semester 1918, the chaotic conditions of the last months of the war still predominated, proceeding from the proclamation of a Soviet Republic to its abrupt termination through the intervention of government troops. Yet Pauli, already considered a mathematical genius by his secondary-school teachers, did not let these events distract him from his studies, for he was too impressed by the great scientific discoveries stirring the world public at the time. Rutherford and his staff had been successful in artificially breaking a nitrogen nucleus down into simpler components by bombarding it with radioactive

Passport photo of Wolfgang Pauli

α-rays; even more attention and a flood of writings about Einstein and his world of ideas attended the reported confirmation of his theory of gravitation through the diffraction of light in the gravitational field of the Sun in summer 1919.

In contrast to most of his contemporaries, who were informed about these new findings only through "more or less popular introductions," the young Pauli was able to participate in this research not only as a recipient, but also as an actor. Instructed by a Viennese physicist, as a highschool student he had already studied the form of the theory of relativity expanded by Hermann Weyl. Even today, in the Pauli estate preserved at CERN in Geneva, we can view a copy of the much-read work *Space-Time-Matter* with copious notes by Pauli, in which Weyl presented his lectures on the general theory of relativity of summer 1917, highly praised even by Einstein.

Right from the outset, Pauli brought his professor Arnold Sommerfeld in Munich a derivation of the energy components of the gravitational field valid for every coordinate system, expanded from a study by the Viennese private instructor Erwin Schrödinger. He won even greater admiration from his classmates and the other members of the institute when, as a freshman, he – following a previous talk by Sommerfeld – presented his *Remarks on the General Theory of Relativity* on 10 December 1918 for discussion at the Wednesday Physics Colloquium. From that point on he was regarded as a *"Wunderkind"*.

Sommerfeld was increasingly engrossed in the turbulent developments in the quantum theory of atomic construction emerging at the time, and was attempting to involve Pauli in this complex of issues as well. Nevertheless, as the editor of a wide-ranging volume on physics of the large-scale *Encyklopädie der mathematischen Wissenschaften*, he also planned to include a paper on the increasingly timely subject of relativity, originally to be written by Einstein. Therefore he was delighted to find in Pauli a new staff writer who could assist him in composing such an article. By summer 1919 Pauli had completed a further study *On the Theory of Gravitation and Electricity by Hermann Weyl,* which demonstrated that Weyl's theory was incapable of explaining the differences actually existing between the two kinds of electricity. This result aroused Einstein's attention, because he saw in it a further objection "not only against Weyl", "but against every other continuum theory, even against those that treat the electron as a singularity."

As is apparent in Einstein's letter to his colleague Felix Ehrenhaft in Vienna, from that point on he regarded Pauli as one of the best specialists on his ideas. When he had to cancel a lecture invitation to Vienna in December 1919 due to illness, he proposed the nineteen-year-old as his replacement: "On this occasion I do not wish to neglect to refer you to the highly gifted Mr. Pauli of Vienna, currently abiding in Munich, who could serve as a splendid replacement speaker, not to mention Messrs. Schrödinger and Thirring, of whom I am also aware that they master the chain of reasoning of the general theory of relativity completely." With this, Pauli's reputation as an expert on relativity became so established that Sommerfeld now could consign the composition of the encyclopedia article to him alone. For Pauli, who was quite familiar with the conventional material of lectures, this constituted a welcome opportunity to use the minimum three years of study until his doctorate advantageously.

He immediately undertook comprehensive studies of the literature, which brought him into contact especially with the mathematicians' methods of invariant theory and then, through his influence, took on increasing importance in modern physics. At the same time he entered into correspondence with the most important scholars of his age, like Hermann Weyl, Felix Klein, Max Born, Erwin Schrödinger and Alfred Landé, blossoming into an intensive exchange of letters. Through this correspondence, he was then able to exert a constantly increasing influence on physical research, far beyond the borders of the institute.

In September 1920, during a visit to the first postwar convention of German natural scientists in Nauheim, Pauli finally made Einstein's personal acquaintance. "Since that day you helped to identify me to Einstein," he wrote his colleague Lise Meitner on her eightieth birthday, "it was always a pleasure for me to see you."

By fall 1921, when he not only completed his studies in Munich, but also published his now-famous

article on relativity, Pauli's balanced and clear portrayal of the theory was no longer a surprise for his closer colleagues in the field. Yet Einstein's effusive review of the successful article made a stir in broader circles, through which the self-confidence of the young Pauli, already considerable as it was, grew even further. This is illustrated by a charming anecdote. Arnold Berliner, a friend of Einstein's and the editor of the journal *Die Naturwissenschaften*, is supposed to have warned Einstein that such unusual praise could easily make a scientist of such a tender age a megalomaniac. "In this case there is no such danger," replied Einstein, "because he already is."

Yet Einstein himself was soon to experience Pauli's pronounced confidence and his merciless critique of any developments in theoretic physics that appeared to him unfruitful or to lead to a dead end. When quantum mechanics had been accomplished at the end of the 1920s, thanks to Pauli's influential participation, a relativistic generalization of the theory was sought to describe the processes in atomic nuclei and in cosmic radiation. While the representatives of quantum mechanics, a field by now acknowledged by the overwhelming majority of physicists, could only imagine such a generalization on the basis of the conception of quanta, once again, Einstein set out on his own course. He imagined a unified field theory with distant parallelism, which could conform to the basic equations in order to allow a natural means of deriving the existence of elementary particles and the other discrete variables of quantum theory as well.

Yet this initially unrealizable agenda encountered Pauli's sharpest rejection. "God Almighty seems to have abandoned Einstein completely!" he exclaimed in a postcard to his friend Paul Ehrenfest; he wrote even more plainly to his companion Pascual Jordan: "The word is that Einstein spouted some terrible nonsense about new distant parallelism in the Berlin Colloquium! [...] Such rubbish

Erwin Schrödinger, Physics Nobel laureate in 1933

Wolfgang Pauli

Pauli's vehement critique, especially since he had high regard for Pauli and his scientific candor.

In a review of an article from the year 1931, Pauli again scoffed "in a variation of a well-known historical declaration" at Einstein's repeated attempts to formulate a field theory: "Einstein's new field theory is dead. Long live Einstein's new field theory!"

He rejected Einstein's proposals especially because they were not compatible with the conceptual understanding Pauli advocated: for "the value of a theory must be judged not only according to the effects it predicts, but also according to the internal characteristics of its logical structure." As a partisan of modern quantum theory, Pauli also considered as proven, "that with this a new level of physical conceptualization has been achieved, which contains those of classical physics, including the theory of relativity, as a special border case, and therefore represents a natural generalization of the latter." Therefore he also designated the "attempt to interpret the coexistence of the undulatory and corpuscular characteristics of light and matter, conditioned by the existence of the quantum of energy, while maintaining the classic, causal space-time description of the phenomenon, as just as artificial and unsatisfying as trying to explain the negative result of Michelson's experiment while maintaining absolute time and the stationary aether through some special mechanism."

is only good for impressing American journalists, not even American physicists, let alone the European physicists."

Shortly thereafter, in a letter of 19 December 1929, he deplored to Einstein the "extensive demolition of the general theory of relativity" he had caused, and declared himself willing, "to enter into any bet with you that you will have given up the whole idea of distant parallelism within the year." He, "however, adheres to this fine theory, even if you betray it." Yet Einstein was not dissuaded by

As Pauli had predicted, Einstein soon gave up his idea of distant parallelism. The "unified theory of gravitation and electricity" now advanced by Einstein and his new collaborator Walther Mayer, was granted a longer life expectancy by Pauli. After visiting Einstein in Princeton, Pauli himself even undertook an attempt to apply this formalism to the case of Dirac's wave equation for electrons. Even more than Einstein's attempts at field theory, Pauli opposed his attitude toward quantum mechanics, which he was not willing to acknowledge as the definitive solution of the problems of atomic physics. Even in Berlin Einstein had made multiple attempts to refute the interpretation of the probability of Schrödinger's wave function through thought experiments. When he then developed these thoughts into the well-known paradox with his collaborators Boris Podolsky and Nathan Rosen in 1935, reaching the conclusion that quantum mechanics in its original form must be incomplete, again it was Pauli who endeavored to confront this heresy, even mobilizing his comrades-in-arms Niels Bohr and Werner Heisenberg. "Einstein expressed his views on quantum theory publicly yet again," reads a letter to Heisenberg. "As we know, it is a catastrophe every time this happens. Because, as he concludes razor sharp, what may not be cannot be (Morgenstern). In any case I must admit that I would regard any underclassman who raised such objections to me as quite intelligent and full of potential."

In order to clarify the imminent controversy on this subject, Arnold Berliner requested that Schrödinger take a position. This is where the essay originated in which Schrödinger's famous cat made his first appearance.

Regardless of such scientific antagonisms, the personal relationship between Pauli and Einstein always remained intact. Pauli was still the only physicist from the camp of the quantum mechanists who had collaborated with Einstein on a publication. When Pauli worked at

Albert Einstein with Wolfgang Pauli in Leiden, 1926 (photo: Paul Ehrenfest)

the Princeton Institute of Advanced Study in Einstein's immediate vicinity during the war, a closer personal friendship developed from the scientific esteem they held for each other during the many walks and long conversations they shared.

Albert Einstein with students in Princeton, 1946

get that speech," Pauli later reported. "He was like a king who steps down and appoints me his successor as a kind of foster son."

Although Princeton regretted that Pauli declined the tempting offer to stay, claiming a greater affinity with European physicists, he nevertheless made a significant contribution to the maintenance of close scientific contacts between the two continents. Many of his students and assistants now went to America, most of them to Princeton, where, as a kind of patron, the aging Einstein continued to visit his old institute, now under Robert Oppenheimer's direction. Pauli himself was also among the frequent visitors and occasionally intervened to explain Einstein's dissenting point of view to other quantum physicists.

In particular, he played "Einstein's advocate" in the well-known Einstein-Born controversy about the conception of reality and determinism in quantum theory. Even Max Born, otherwise quite reserved with regard to Pauli, had to admit that Pauli was "doubtlessly a genius of the first order," who also understood Einstein's standpoint completely: "For I knew since the time when he was my assistant in Göttingen, that he was a genius, only comparable with Einstein himself, and even perhaps greater than Einstein in terms of pure science; yet he was a completely different kind of person, who did not achieve Einstein's greatness in my view."

It was Einstein in particular who advocated awarding the Nobel Prize to Pauli after the end of the war. In an improvised address during the celebration that followed the ceremony, he designated Pauli as his intellectual successor. "I will never for-

Pauli, as president of the Bern Anniversary Conference Fifty Years of the Theory of Relativity, spoke during his ceremonious inaugural address on 11 July 1955 of a turning point in the history of the theory of relativity in reporting of Einstein's death three months earlier. Also at this conference, Hermann Weyl initiated an edition of *Einstein's Collected Works*, initially to be under the control of Valentin Bargmann, Pauli and himself, which ultimately – with the usual delays – led to the institution of an Einstein Archive and a Complete Edition of his works, which quite decisively enriched our knowledge about Einstein's work and life and will continue to do so.

That Pauli was not successful in inducing Bohr to participate in this first Einstein celebration, despite the fact that he was formally invited, also casts light on the deep-seated differences in the scientific conceptions between these two researchers, who left their mark as no others on the physical worldview of the 20th century. A synthesis of quantum theory and the theory of relativity, as Einstein had envisioned back in his address on Planck's sixtieth birthday in May 1918 – and to which Pauli's persistent efforts also aspired – remains for posterity, as the intellectual legacy of these theoretical physicists.

Anton Zeilinger

Albert Einstein:
Reluctant Creator of Quantum Technology

Of the works Albert Einstein wrote in his *annus mirabilis* 1905, the two on the special theory of relativity are probably the best known. In one he introduces the theory itself; in the other, what is probably the most famous equation in physics, $E = mc^2$. Yet the first paper published by him that year *Über einen die Erzeugung und Verwandlung des Lichtes betreffenden heuristischen Gesichts-punkt* (On a Heuristic Point of View Concerning the Production and Transforming of Light) is the

nature. The way in which single measurement events are purely coincidental goes far beyond our understanding of chance in daily life or in classical physics. In the latter we at least assume that it is possible to construct a causal chain for every single event, even if this chain is not always known to us or cannot be checked empirically. This is not the case for single quantum events, such as the decay of a certain radioactive atom. Today we are convinced that no causal chain can be given for single events, not even in principle.

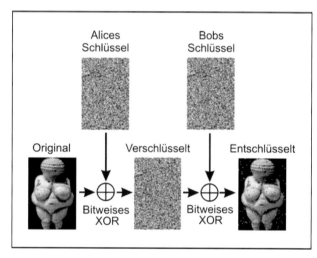

Quantum crypto-graphy. Through measurement of entangled photons, Alice and Bob generate the same key (above). The original message, the picture of Venus of Willendorf, is encrypted by mixing it bit for bit with the key. Only Bob can decode the message, since only he knows the key (graphic: Thomas Jenewein)

only one Einstein himself, in a letter of the same year, called revolutionary. In this paper Einstein proposes that light is composed of particles. This is the work for which he received the Nobel Prize in 1922. Today these light quanta are called photons. Despite this important early contribution to quantum physics, Einstein soon began to criticize the new theory, and he remained a critic for the rest of his life.

As early as 1909, at the Annual Convention of the *Gesellschaft Deutscher Naturforscher und Ärzte* (Society of German Natural Scientists and Physicians) in Salzburg, Einstein expressed his "discomfort" about the new role of chance in quantum physics. It is quite remarkable that he recognized even then, long before the formulation of quantum theory in the mid-1920s, that the concept of chance in quantum physics is of a completely new

After development of the new quantum theory by Heisenberg, Schrödinger and Dirac starting in 1925, Einstein criticized in particular the concept of reality expressed in the term complementarity formulated by Niels Bohr. Taking as an example the measurement of location and impulse (velocity) of a particle, Heisenberg's uncertainty principle tells us that it is not possible to measure both at the same time with any accuracy. While this could be understood merely as a measurement object's being interfered with by our necessarily quite coarse measurement instruments, the message is much more profound. According to Niels Bohr it makes no sense to speak of the properties of a system, if we do not actually perform a measurement that allows us to actually determine the given selected property. Thus, when we measure the location of a particle, not only is its impulse unknown, but it does not even possess a well-defined impulse. Location and impulse are complementary to each other; through the selection of the measurement instrument we can decide which of the two quantities becomes reality. This decision then precludes the possibility of the other, non-existent quantity's having any nature of reality.

After discussions with Niels Bohr, Einstein's critique culminated in the famous Einstein-Podolsky-

Rosen paper of 1935, where he analyses the physics of two particles that interact with each other. Because of this interaction, their properties are intricately connected with each other thereafter. Measuring either of the two supplies a random result, yet this measurement stipulates the corresponding property of the second system. In the same year, Erwin Schrödinger introduced the

measurement. If either one of the two properties is measured for the first particle, for instance the impulse, it happens to take on a specific value; the second, no matter how far away it is from the first, then receives a state ("is projected into a state"), which also corresponds to a well-defined impulse. Alternatively, we could also elect to measure the location of the first particle; then the sec-

Entangled photons over Vienna skyline. This demonstrates the possibility of quantum communication in cities and even to satellites (graphic: Jaqueline Gordany)

term "entanglement" to designate this phenomenon, and called it the essential characteristic of quantum physics.

In entangled systems, the quantity that is entangled does not have a well-defined property in either of the systems before measurement. To paraphrase Robert Musil, they are particles without qualities. If one is measured, it happens to take on a measurement value. But the second, located some unspecified distance away, then also takes on the state that corresponds exactly to the first. In the EPR paper these values were the location and impulse of both particles. Here entanglement means that neither of the particles possesses a well-defined location or a well-defined impulse before

ond particle would receive a state that corresponds to a well-defined location.

This is where Einstein's critique of entanglement picks up. It is rendered most clearly in his *Autobiographical Notes* of the year 1948, where he establishes that the quantum physical state we assign to the second system quite clearly depends on the measurement we perform on the first system. Yet, Einstein demands, the real, factual state of the second system must be independent of which measurement is performed on the first one. According to this line of argumentation, quantum physics thus cannot be a complete description of nature. The opposite position is represented by the Copenhagen interpretation, which was formulated primarily by Niels Bohr, but also by Werner Heisenberg and others (of course, there are fine distinctions between the different versions). This interpretation holds that, in principle, only what is actually observed can be reasonably designated as an entity. Thus an entity can be spoken of only when one actually performs an experiment that allows the physical quantity in

Werner Heisenberg and Niels Bohr

question to be determined empirically. EPR had already discussed such a position in their work, remarking that no reasonable definition of reality could tolerate it.

Especially after the work of John Bell in 1965, who showed that the philosophical position taken up by Einstein contradicts quantum physics as it can be observed in experiments, Einstein's critique led to numerous experiments, all of which clearly refute Einstein today. But just because Einstein was not right in this case, it would be wrong to underestimate his contribution. To the great surprise of everyone involved in the early experiments, it was these very experiments with individual quantum systems that led to new ideas in the fields of information processing and informa-

tion transmission, opening the door to a new technology. In these new concepts of quantum information technology, those very points criticized by Einstein – chance, entanglement and complementarity – play a central role.

The most advanced of these applications is quantum cryptography. In a procedure that builds on entangled photons, two players who want to exchange information confidentially, Alice and Bob, exchange a series of entangled pairs of photons. Through measurements of both photons they can generate a key – of any length, in principle – that consists of random numerals and is identical on both sides because of entanglement. What is essential here is that this key need not be transmitted from Alice to Bob, but rather emerges simultaneously at both locations through measurements on the entangled photons. Alice can then encrypt her message using this sequence of random numerals, and Bob, who possesses the same sequence, can decode the message easily. The encrypted image is safe from espionage because the key is purely random and is used only once. Any spy listening in is excluded by the fact that Alice and Bob switch back and forth, independently of each other, between two mutually complementary measurements. This makes any spy easy to identify.

The current state of quantum cryptography is such that systems exist today that can be used to exchange a key over distances of several kilometers. Systems for direct technical application are under development and will be ready for implementation in a few years. Recently researchers were successful in transmitting entangled photons over the Vienna skyline, between points several kilometers apart in free space. Such experiments show that it is in principle possible to overcome even the distance from the Earth's surface to terrestrial satellites with quantum cryptography. This is because the atmosphere rapidly becomes thinner at greater heights: the entire vertical distance of the atmosphere between the Earth and a satellite is equivalent to just around six kilometers of horizontal distance. Thus in a few years we

initially possessed. Through this the original photon loses its own state, which was thus transmitted to Bob's photon.

However great its importance to science fiction fans, teleportation probably will never be suitable for the transmission of objects. Rather, it will be applied to communicate between the quantum computers of the future. In such a quantum computer, information will be encoded in quantum states. Here again, entanglement and complementarity play a central role.

The development of the new field of quantum computer science is a textbook example of how problems that were originally philosophical can lead to a new technology. Albert Einstein did not criticize quantum physics

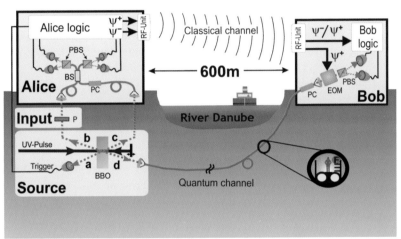

Quantum teleportation under the Danube in Vienna. Alice generates an entangled pair of photons and sends one of these to Bob. With this she can teleport him the state of a third photon (graphic: Robert Ursin)

can expect the first experiments to test quantum communications with satellites.

A particularly interesting application of quantum communication is teleportation. Here entanglement is used to transmit the state of a photon to another photon located any distance away. For this Alice and Bob generate an entangled pair of photons, one for Alice and one for Bob. Alice has another photon whose state she would like to teleport. She entangles this photon with her entangled photon. The result of this entanglement is communicated to Bob, who can thus return his photon back into the original state which Alice's teleporting photon

because it was incorrect, but because of its philosophical implications. The consequence was a series of experiments that gloriously confirmed quantum physics. The most interesting element of this is that neither Einstein nor any of the other early experimenters could have dreamed of the experiments that opened the gate to these new technologies. This also demonstrates the limits of a shortsighted research policy which is concerned only with application and implementation. Unfortunately, this is the kind of policy demanded time and again.

Michael Eckert

Einstein and Arnold Sommerfeld:
Impressions from their Correspondence

One hundred years after Einstein's *annus mirabilis* the epochal importance of Einstein's works is evident. But retrospective appreciation is one thing; the process that leads to the acknowledgement of new theories is another. We find impressive evidence of this in Arnold Sommerfeld's correspondence. Sommerfeld was among the first to recognize Einstein's genius. The correspondence these two theoreticians maintained over decades also

Arnold Sommerfeld (1868–1951), mid 1920s

offers us authentic impressions of the social and political environment of theoretical physics in the first half of the 20th century.

He is "very impressed" by Einstein and will lecture about his work in the colloquium, Sommerfeld wrote to Wilhelm Wien on 23 November 1906, after reading Einstein's work about the special theory of relativity. On 12 December 1906 he mentioned to Hendrik Antoon Lorentz, too, that he had

"studied Einstein." It is "strange to see," he continued, "how he comes to exactly the same results as you." These are the first passages in Sommerfeld's correspondence in which Einstein is mentioned. However, Sommerfeld found, as he wrote Wien in the very same letter, that Einstein's theory must "be elaborated considerably before it can be used to handle any kind of electron movements." He passed judgment on the theory of relativity in the context of the currently predominant electron theory. One year later he had become acquainted with Einstein's other works as well. The initial admiration now was joined by a trace of uneasiness. On 26 December 1907 he requested that Lorentz "express his opinion on the whole complex of the Einsteinian discussions," for it appeared to him that "something almost unhealthy lay in this non-constructible and intuitionless dogmatism. An Englishman could hardly have given this theory; perhaps what is expressed here, like in Cohn, is the abstract-conceptual manner of the Semite. I hope you will succeed in filling this ingenious skeleton of concepts with true physical life." What began with such skeptical acknowledgment for the still largely unknown Einstein, overlaid with anti-Semitic prejudice, soon turned into unconditional admiration. Sommerfeld must have made this clear, even in his first letters to Einstein while he was still working at the Patent Office in Berne, for Einstein wrote back on 14 January 1908, "no physicist has ever contacted me who is so open and at the same time so benevolent." When the two met each other personally at the 1909 Natural Scientist Convention in Salzburg one year later, a friendship emerged: "Saw a great deal of Einstein," Sommerfeld reported to his wife on a postcard from Salzburg on 22 September 1909. Einstein wrote to Munich one week later, "Now I understand that your pupils like you so much! Such a fine relationship between professor and students must be unparalleled. I intend to make you my role model."

SECOND SOLVAY CONFERENCE 1913

VERSCHAFFELT   LAUE   RUBENS   GOLDSCHMIDT   HERZEN   LINDEMANN   de BROGLIE   POPE   GRUNEISEN   HOSTELET

HASENOHRL   JEANS   BRAGG   Mme CURIE   SOMMERFELD   EINSTEIN   KNUDSEN   LANGEVIN

NERNST   RUTHERFORD   WIEN   J.J. THOMSON   WARBURG   LORENTZ   BRILLOUIN   BARLOW   KAMERLINGH ONNES   WOOD   GOUY   WEISS

Initially, their scientific debates primarily concerned electron theory, in which concepts like the dependency of mass on velocity entered physics for the first time. Sommerfeld, who had begun his own career as a mathematician, made a name for himself as a theoretical physicist in this field. In Einstein's theory of relativity he saw, as he wrote in an autobiographical outline, "my difficult and protracted studies, on which I initially placed great value, condemned to fruitlessness." In quantum theory, too, it was Einstein who opened Sommerfeld's eyes. In a letter to Lorentz on 9 January 1910,

Sommerfeld had designated himself "old-fashioned enough" to "initially offer resistance to the light quanta in the Einsteinian conception." But after a visit to Einstein in summer 1910 he became a champion of the new quantum theory.

During World War I Sommerfeld and Einstein experienced the peak of their scientific creativity: Sommerfeld with the expansion of Bohr's atomic model, Einstein with the general theory of relativity. Again, they kept each other up to date, so that their correspondence allows the reader to get a feeling for the exultation with which they report-

Photograph of the international physicist-elite at the Solvay Congress, 1913 in Brussels. Sommerfeld and Einstein beside Madame Curie

ed these achievements to each other. "But last month I had one of the most exciting, most strenuous times of my life, yet also the most successful," Einstein apologized to Sommerfeld on 28 November 1915 for not answering his letter sooner. "The magnificent thing I experienced was that it yielded not only Newton's theory as a first approximation, but also the movement of Mercury's perihelion (43" per century) as a second approximation." Sommerfeld was infected by the enthusiasm for the general theory of relativity and immediately integrated it into his lecture schedule. Conversely, Einstein also took an interest in Sommerfeld's current atomic theory: "I was very glad about your letter," he wrote to Munich on 8 February 1916, "your announcement about the theory of spectral lines is enchanting. A revelation!" He remarked on the publication of Sommerfeld's complete work on a postcard on 3 August 1916

as follows: "Your spectral studies are among my finest experiences in physics. Only through them does Bohr's idea become wholly convincing. If only I knew what little screws the Lord God uses for it!"

This written sigh already suggested what bothered Einstein about quantum theory and what he tried to make clear in his letters to Sommerfeld, over and over with different words: "If only it were possible to clarify the principles of quanta! But my hope of experiencing this is increasingly dwindling," he wrote to Sommerfeld on 1 February 1918. Then enthusiasm prevailed again: "It is quite natural that you are more interested in the spectra; it certainly has become the most hopeful area today. You have just made its fine structure accessible. It is indeed a pleasure to experience

first hand!" (Einstein to Sommerfeld, September 1918). Sommerfeld was altogether aware of the fundamental problems of quantum theory that preyed on Einstein, but was content with doing as much justice as possible to the regularities observed in experiments in his own atomic model. On 11 January 1922 he wrote to Einstein, that he had "since found wonderful, numerical laws of combinations of lines in connection with Paschen's measurements" and entrusted a student – "Heisenberg, third semester!" – with further calculations of a model for them: "Everything is going well, but at its very basis remains unclear. I can only fur-

Albert Einstein, around 1921

ther the technology of quanta; you must make their philosophy." Einstein replied, "What I admire about you especially is that you have conjured up such a large number of young talents. That is something truly unparalleled. You must have the gift of ennobling and activating the minds of your listeners."

When quantum mechanics gave quantum theory a new twist in the middle of the 1920s, it is known that Einstein kept his distance: "The Heisenberg-Dirac theories compel my admiration, but to me they do not smell of reality." (Einstein to Sommerfeld, 21 August 1926). On another occasion he formulated his stance thus: "As to 'quantum mechanics,' in terms of ponderable matter, I think that it contains just as much truth as the theory of light without quanta. It might be a correct theory of statistical laws, but an insufficient conception of the individual elementary processes." (Einstein to Sommerfeld, 9 November 1927). Einstein even had reservations about Schrödinger's wave mechanics, which Sommerfeld took into his curriculum immediately and with great enthusiasm. Although the successes of Schrödinger's theory made a "great impression" on him, as he wrote to Sommerfeld on 28 November 1926, "I do not know whether there is more to it than to the old quantum rule." His overall assessment was no less critical than for Heisenberg's version of quantum mechanics: "Have we really come closer to solving the mystery?" When Sommerfeld summarized the current state of quantum mechanics shortly thereafter in a *Wellenmechanischen Ergänzungsband* (Wave Mechanical Supplement) to his legendary textbook *Atombau und Spektralinien* (Atomic Structure and Spectral Lines), Einstein found it "quite splendid;" but again he made clear that, "the whole development does not really satisfy me, despite the great successes." (Einstein to Sommerfeld, 14 August 1930).

Considering the striking contrasts in the political stances of Sommerfeld and Einstein, especially in

Arnold Sommerfeld, 1938

the years after World War I, it is hardly a foregone conclusion that the mutual appreciation and collegial, friendly affection between the two endured. For Sommerfeld, who thought and felt nationalistically, the outcome of World War I and the conditions for peace imposed upon Germany in the Treaty of Versailles were sensed as a profound humiliation, while Einstein abhorred every form of nationalism. How close both – mutual esteem as scientists and individuals on the one hand, and dissent on issues of politics and world view on the other – occasionally came, can be seen, for example, from correspondence between Einstein and Sommerfeld in January 1922. On 11 January 1922, Sommerfeld addressed a letter to the Nobel Committee in Stockholm, in which he nominated Einstein for the Nobel Prize in Physics for this year. Einstein, he wrote, is "by far the most worthy recipient of the Nobel Prize at this time;" the "logical consistency and audacity of his thought is unparalleled in the history of physics." Aside from the theory of relativity, he continued, "Einstein's other merits are so tremendous that they alone would justify his crowning." On the very same day Sommerfeld wrote Einstein how "disgusted" he was by a "rag" – the *Schaubühne* of Berlin – that reported about an interview Einstein gave a French journalist, and which so repelled Sommerfeld that he would have loved to send a denial to a newspaper "with the observation 'from the beginning to the end, all lies.' But you will hardly give me permission." (Sommerfeld to Einstein, 11 January 1922). But Einstein explained to Sommerfeld: "It is what I said, just in French Bengal light." The journalist may have acted without his permission, but an "actual denial" was out of the question. Einstein closed the letter to Sommerfeld with a postscript, "One should respect an honest person even when he has and advocates other views than one's own."

Both Sommerfeld and Einstein acted in accordance with this principle, as their correspondence after 1922 evinces. Dissent in the question of quantum theory and in political questions did not damage their collegial friendship. Sommerfeld later distanced himself from his stance shortly after World War I. Under the impression of National Socialism, in 1934 he wrote in a drafted – but not sent – letter: "Unfortunately I can not excuse my compatriots in view of the injustice that has been done to you and many others; not even my colleagues at the Academies in Berlin and Munich. By the way, I can assure you that the abuse of the word 'national' on the part of our rulers is thoroughly curing me of the national feeling that used to be so strong in me. Now I would have nothing against Germany's perishing as a power and being absorbed in a pacified Europe." He must have perceived the last passage as too radical a break with his earlier attitude, for it was not included in the letter he sent. (Sommerfeld to Einstein, 27 August 1934).

After World War II, Sommerfeld was one of the few German scientists Einstein stayed in contact with. On 14 December 1946 he wrote to Sommerfeld: "After the Germans slaughtered my Jewish brothers in Europe, I want nothing more to do with Germans, not even with a relatively harmless Academy. This is different in the case of a few individuals who remained steadfast within the bounds of possibility. I was glad to hear that you were among these." When Sommerfeld invited Einstein to lecture in Munich, in response to a 1948 newspaper report stating that Einstein was setting out on a world tour, Einstein took this occasion to reminisce about the beginning of their friendship, "old episodes in Bern in the beer hall across from the *Zeitglockenturm*. Then later in Munich, when your wife brought a cradle for our newborn child." These memories were followed by the sentence that has since become famous, "Now I am an old fogy who doesn't make any more journeys, after having made sufficient acquaintance of people from all perspectives. The newspaper report was, of course, false as usual." (Einstein to Sommerfeld, 5 September 1948).

Milena Wazeck

"Einstein on the Murder List!": The Attacks on Einstein and the Theory of Relativity in 1922

On 5 August 1922, an article with a seemingly harmless headline appeared in the Leipzig newspaper *Neueste Nachrichten*: "Is Professor Einstein Coming to Leipzig?" The *Gesellschaft Deutscher Naturforscher und Ärzte* (Society of German Scientists and Medical Doctors – GDNÄ), one of Germany's oldest scientific associations, intended to celebrate its hundredth anniversary there with a festive gathering. The official lecture about the theory of relativity by Albert Einstein was showcased as a special event. For this reason the opponents of the theory of relativity, headed by the Nobel laureate physicist Philipp Lenard and the experimental physicist Ernst Gehrcke were well prepared: They launched a "Declaration of Protest" beforehand, in order to prevent a lecture about the theory of relativity from being held in Leipzig.

The political events of the summer of 1922 lent this "Declaration of Protest" a particularly explosive nature. The question of whether Einstein would speak about the theory of relativity in Leipzig was, in fact, anything but harmless, and did not concern the scientific defensibility or indefensibility of the theory of relativity. After Foreign Minister Walter Rathenau was murdered by nationalist extremists on 24 June 1922, the situation became ever more precarious for the committed democrat Einstein as well. Back in the spring of the same year he had come under fire by nationalist activists for his trip to the country of the erstwhile archenemy France, during which he promoted the reconciliation of the former enemies and the renewal of scientific relations. At the time the newspapers printed that Einstein had "taken up the cause of relativizing national feelings of honor." However, several months later, there were concrete threats, and Einstein resolved to withdraw from public life for a time; he even considered ending his scientific career in Berlin for good.

On 6 July 1922, in a letter to Max Planck, who was chairman of the GDNÄ at the time, Einstein canceled the lecture he promised to give at the hundredth anniversary celebration, explaining, "I am supposed to belong to the group against which assassinations from the nationalist side are planned [...] Only patience and travel can help now."

Shortly thereafter he reported to his friend Maurice Solovine: "We've had troubled times here since the abominable murder of Rathenau. I, too, receive con-

Declaration of protest against the theory of relativity, 1922

Die Leitung der „Gesellschaft Deutscher Naturforscher und Ärzte" hat es für richtig gehalten, unter den wissenschaftlichen Darbietungen der Leipziger Jahrhundertfeier Vorträge über Relativitätstheorie auf die Tagesordnung einer großen, allgemeinen Sitzung aufzunehmen. Es muß und soll dadurch wohl der Eindruck erweckt werden, als stelle die Relativitätstheorie einen Höhepunkt der modernen wissenschaftlichen Forschung dar.

Hiergegen legen die unterzeichneten Physiker, Mathematiker und Philosophen entschiedene Verwahrung ein. Sie beklagen aufs tiefste die Irreführung der öffentlichen Meinung, welcher die Relativitätstheorie als Lösung des Welträtsels angepriesen wird, und welche man über die Tatsache im Unklaren hält, daß viele und auch sehr angesehene Gelehrte der drei genannten Forschungsgebiete die Relativitätstheorie nicht nur als eine unbewiesene Hypothese ansehen, sondern sie sogar als eine im Grunde verfehlte und logisch unhaltbare Fiktion ablehnen. Die Unterzeichneten betrachten es als unvereinbar mit dem Ernst und der Würde deutscher Wissenschaft, wenn eine im höchsten Maße anfechtbare Theorie voreilig und marktschreierisch in die Laienwelt getragen wird, und wenn die Gesellschaft Deutscher Naturforscher und Ärzte benutzt wird, um solche Bestrebungen unterstützen.

Dr.-Ing. L. C. Glaser, Würzburg,
Prof. Dr. F. Lipsius, Leipzig,
Prof. Dr. M. Palagyi, Darmstadt,
Dr. L. Kühn-Frobenius, Berlin,
Geh. Rat Prof. Dr. P. Lenard, Heidelberg,
Prof. Dr. J. Riem, Berlin,
Dr. H. Fricke, Charlottenburg,
Prof. Dr. K. Strehl, Hof,
Dr. K. Geißler, Eisenach,

Prof. Dr. E. Gehrcke, Berlin,
Prof Dr. S. Mohorovicic, Agram,
Dr. K. Vogtherr, Karlsruhe,
Dr. R. Orthner, Linz,
Dr. J. Kremer, Graz,
Dr. St. Lothigius, Stockholm,
Dr. W. Nachreiner, Neustadta. d. H.,
Prof. Dr. M. Wolff, Eberswalde,
Dr. A. Krauße, Eberswalde,
Geh. Rat Prof. D. Dr. E. Hartwig, Bamberg.

stant warnings, have given up my course of lectures and am officially absent, but in truth still here. Anti-Semitism is very great."

Disgusted about the influence that right-radical circles could exert through their threats, Planck wrote to Max von Laue on 9 July, "The rabble have actually come so far that they have managed to thwart an event of historical importance for Ger-

Walther Rathenau, 1922

man science." An agreement was reached that von Laue should take over the official lecture on the theory of relativity instead of Einstein. Planck even gleaned a positive angle from this solution, as he wrote in a letter to Wilhelm Wien on the same day. This replacement may "even have the advantage of setting straight those who still believe that the principle of relativity is basically a Jewish advertisement for Einstein." Planck was convinced that separating the theory from its creator would avoid anti-Semitic attacks against the theory of relativity, and feared that an official denunciation of the attacks against Einstein would endanger the apolitical "temple of science." This strategy of artificial separation between science and politics, in an age during which the confrontation about the theory of relativity had long become politicized, ultimately culminated in Planck's consenting to Einstein's expulsion from

the Prussian Academy of Sciences after the National Socialists had seized power: Einstein himself had – according to Planck – "through his political behavior made his remaining in the Academy impossible."

The reasons behind Einstein's cancellation of the official lecture were discussed at length in public. The daily press of the Weimar Republic documented of this controversy in numerous newspaper articles during the summer of 1922. From the very start of the excessive coverage about Einstein and the theory of relativity starting in 1919 it was apparent that a newspaper's position for or against the theory of relativity corresponded strikingly with its political orientation. Accordingly, the democratic press immediately expressed its solidarity with Einstein after the death threats against him became known. The judgement of the *Volkszeitung* of Dresden on 5 August reads: "It is a disgrace for all of Germany that a world-famous scholar can be out on the assassination list by an unspiritual, reactionary mob of scoundrels and chased out of the country." On the same day, the *Berliner Tageblatt* speaks of a "moral degeneration [...] in large sectors of right-wing radicalism."

The anti-Semitic newspaper *Die Wahrheit*, on the other hand, scoffs on 23 September that it was Einstein's own fault for losing out on the opportunity to give the official lecture, "for it is not credible that there are actually crazy people out there who have such murderous intentions." Considering that over 350 political murders were committed by right-wing radicals in the period from 1919 until 1922, this can only be regarded as an attempt to play down the real danger for Einstein. The *Rheinisch-Westfälische Zeitung* of 5 August even suggests to its readers that they "interpret [Einstein's] dramatized flight as publicity [...], intended to lend new luster to an already noticeably paler star," thus implicitly accusing Einstein of having staged the threat to his life himself.

That the scientific acceptance and public enthusiasm for the theory of relativity was indebted to a

propaganda campaign by theoretical physicists, above all by Einstein, was an established fact for a number of scientists. In spite of the politically charged atmosphere, they used the occasion of the hundredth anniversary of the GDNÄ for a "Declaration of Protest" against the theory of relativity. The initiators Gehrcke and Lenard had both already made public their opposition to Einstein: Gehrcke achieved dubious notoriety as one of the speakers at the event against Einstein in the Berlin Philharmonic Hall in 1920; Lenard, who found himself in an altercation with Einstein about fundamental issues of a scientific nature at the convention of natural scientists in Bad Nauheim in fall 1920, launched a polemic shortly before the meeting in Leipzig with his "Warning to German Scientists," against the "alien spirit, ... which appears everywhere as a dark force and which leaves its mark so clearly on everything that belongs to the 'theory of relativity'."

The "Declaration of Protest" was presented as a scientific rectification, but was clearly a politi-

cally motivated statement. The signatories claimed to bemoan "most deeply the deception of public opinion, to which the theory of relativity is extolled as the solution to the riddle of the universe, and which keeps it in the dark about the fact that many ... scholars ... reject the theory of relativity as fundamentally misguided and logically untenable fiction." Moreover, the doctors and professors of physics, philosophy and mathematics hold "it to be irreconcilable with the seriousness and dignity of German science for a theory disputable to the highest degree to be conveyed to the laity prematurely and in a charlatanic manner, and for the *Gesellschaft Deutscher Naturforscher und Ärzte* to be used to support such endeavors."

Prof. Albert Einstein (left), discoverer of the theory of relativity, is here shown with Prof Ludendorff, director of the Potsdam obervatory. The famous scientist demonstrated his theory on this occasion. Prof. Einstein was virtually forced to flee from Germany owing to reactionary plotters who, he was warned, had him marked for the next public figure in republican Germany to follow Foreign Minister Walther Rathenau.

Austritt Einsteins aus der Kaiser-Wilhelm-Akademie

Eine Folge der reaktionären Hetze.

Telegramm der Neuen Berliner Zeitung.

K. Wien, 29. September.

Der „Abend" veröffentlicht nachstehende Mitteilung, die allenthalben das größte Aufsehen erregen wird: In Berliner wissenschaftlichen Kreisen wird soeben bekannt, daß Professor Albert Einstein, der gestern abend seine viel erörterte Reise nach Japan angetreten hat, nach seiner Rückkehr seine Beziehungen zum offiziellen preußischen Lehrkörper nicht wieder aufnehmen wird. Einstein hat bereits jetzt sein dienstliches Verhältnis zum Kaiser-Wilhelm-Institut teilweise gelöst und keinen Zweifel mehr darüber gelassen, daß er entschlossen ist, die Trennung zu einer vollständigen zu machen. Die Gründe, die den Gelehrten zu diesem bedeutsamen Schritt, den man wohl geradezu als einen tragischen Verlust für die deutsche Wissenschaft bezeichnen kann, bewogen haben, sind unschwer zu erraten. Einsteins Austritt aus der Kaiser-Wilhelm-Akademie ist der Ausdruck für die Empfindungen, die den Gelehrten

Der Entschluß Professor Einsteins wird nicht verfehlen, peinlichsten Eindruck hervorzurufen. Dank dem nationalistisch und „vaterländisch" sich gebärdenden Klüngel, dessen Mitläufer den Namen des Gelehrten sogar auf jene Mordliste gesetzt hatten, auf der auch Rathenau verzeichnet war, und dank den Wühlern in akademischen Kreisen, die sich zwar sanfter benehmen, allein nicht minder engherzig denken und handeln, sieht sich Einstein jetzt bemüßigt, seine öffentliche akademische Tätigkeit in Deutschland aufzugeben. Es ist dies eine Schmach vor ganz Europa, für die jeder wirklich als gut deutsch Empfindende schamrot werden muß.

Neue Berliner 12 Uhr Zeitung, 29 September 1922

September, a newspaper like the anti-Semitic *Wahrheit*, which had just played down the death threats to Einstein, now warmly presented the "protest [...] against the propaganda in favor of Professor Einstein."

The "Declaration of Protest" is heard in the democratic press, too; there, however, the dubious argumentation of Einstein opponents is unmasked in no uncertain terms as the continuation of the politically motivated attacks on the theory of relativity conducted publicly since 1920. In the *Berliner Tageblatt* of 17 September it reads: "The dogged opponents of Einstein ... thus regard it 'as irreconcilable with the seriousness and dignity of German science' when an 'unproven hypothesis' is thrown open to discussion before a forum of mathematical scientists, but they apparently deem it altogether dignified of German science to present this 'unproven hypothesis' with all of its difficult scientific issues to the attention of the laity in the Berliner Philharmonie."

The Einstein opponents' opinion that their critique of the theory of relativity was "forcibly suppressed" cannot be substantiated in view of the fact that their "Declaration of Protest" was printed in many newspapers. Yet the "Declaration of Protest" reflects that physicists like Lenard and Gehrcke, who could not take part in executing the turn from classical to relativistic physics, were marginalized within the field and therefore had no reservations about continuing their fight against the theory of relativity on a political level.

"It would appear that any resistance to the theory was supposed to be forcibly suppressed from the outset ..." confirmed a cosignatory in the newspaper *Tag* of 28 September. An anti-Semitically disposed sympathizer of the appeal goes even further. He suspects that a directed campaign is behind the intensive media coverage about Einstein and his theory, and constructs from this a Jewish conspiracy against "pure science," declaring in *Luzerner Neueste Nachrichten* of 28 October that, "the major press in Germany is almost exclusively in the hands of Einstein's fellow countrymen" and for this reason opposed "a public discussion in the exclusive interest of Einsteinianism." On 23

Andreas Kleinert

Philipp Lenard and Johannes Stark:
Two Nobel Laureates Against Einstein

Philipp Lenard (1862–1947), who in 1905 was the second German physicist to receive the Nobel Prize, was delighted when Johannes Stark (1874–1957), the winner of the Nobel Prize for Physics in 1919, was appointed president of the Imperial Institute for Physics and Technology (*Physikalisch-Technische Reichsanstalt*) in May 1933.

"A Great Day for Natural Science" is the headline of the article he wrote in honor of the event in the *Völkischer Beobachter* of 13 May 1933. Yet what pleased him even more than the advancement of his friend Stark to such an outstanding position in the German scientific landscape was that Einstein

"The alien spirit is already voluntarily leaving the universities, yes, even the country."

Three years later, in the first volume of his textbook *Deutsche Physik,* Lenard continued his attacks against Einstein, fortified with anti-Semitic insults. With his theories of relativity, the "doubtlessly pure-blooded Jew Einstein" had intended to transform and dominate the entire science of physics; in the face of reality, however, these theories were "already completely done for."

In order to understand how it could have come to this, that one of the most respected German physicists was moved to make such malicious attacks

Johannes Stark
(presumably 1930s)

Philipp Lenard,
photograph of
6 July 1937

had finally left Germany: For Lenard, this was above all a reason to celebrate. His appraisal of Stark's merits in this article nearly receded into the background in the face of its malicious invective against Einstein. Einstein, he claimed, had supplied "the most outstanding example of the pernicious influence on natural science on the part of the Jews," his "theories pieced together mathematically from good, already existing knowledge and a few arbitrary ingredients" were gradually falling to pieces. Yet hope was in sight:

against a colleague, we must go back to the beginning of the century, when the relationship between Lenard and Einstein was still characterized by personal esteem and respect for each other's achievements. Einstein's publication of 1905, which explains the photoelectric effect by means of the photon hypothesis, is full of admiration for "the pioneering work by Mr. Lenard," which showed for the first time that the energy of photoelectrons is dependent on the frequency of the incident light. In private letters Einstein did occasionally poke fun at "Lenard's whims," but he repeatedly emphasized his high professional estimation of the professor from Heidelberg: "a great master," "an original mind," "truly a genius." Even as late as 1913, Lenard considered founding a chair for theoretical physics in Heidelberg, "if a personality like Einstein were available."

The first professional differences arose in the summer of 1910, when Lenard held a lecture at the Heidelberg Academy in which he advocated keeping the concept of "aether" in physics. Einstein regarded this lecture, which was printed a short time later, as "infantile" and designated Lenard's

on the level of a scientific dispute conducted without subjective attacks directed toward the other's person.

This changed abruptly when public agitation organized by Nationalists and anti-Semites began against Einstein in August 1920. The swindler and charlatan Paul Weyland, who announced a series of anti-relativistic lectures in the Berlin Philharmonic, placed Lenard on the schedule without asking him, in response to which Einstein attacked Lenard publicly in a caustic newspaper article. Lenard, Einstein wrote, had yet to accomplish anything in theoretic physics, and his objections to

Albert Einstein in his study in Princeton, around 1938 (photo: Lotte Jacobi)

Paul Weyland

comments as "abstruse aethery" – albeit, only in a private letter to a friend. They were both still far from personal enmity. Lenard formulated his critique of the general theory of relativity for the first time in 1918, whereby he emphasized that he had no reservations regarding "the principle of relativity in its original form" (i. e., the special theory of relativity). Einstein published a reply, to which Lenard responded that the arguments brought forward by his opponent were "not convincing." Yet this confrontation, too, took place

the general theory of relativity were so superficial that Einstein did not believe it necessary to go into them. With this, the final break between the two was complete. Deeply offended, Lenard complained in a letter to Johannes Stark "about the personal element which Messrs. Einstein and Laue [...] brought into the matter and which they believe they may level against me; for my part, I behaved purely objective in my writing and [...] there is absolutely nothing I set forth which could excuse the insults these men devoted to me."

In the Nazi propaganda exhibition "The Eternal Jew", (Der ewige Jude), which opened at the Deutsches Museum in November 1937, Einstein is given the most prominent place for "Jewish Intellectualism" *(jüdischer Intellektualismus)*, next to lawyer Hans Litten (imprisoned in a concentration camp from 1933; murdered in 1938), and cultural philosopher Theodor Lessing (murdered in Czechoslovakian exile in 1933)

Einstein, he demanded, should apologize publicly for his insulting remarks. However, Einstein was not willing to do this. A notice did appear in the *Berliner Tageblatt* – shortly after the convention of German Medical Doctors and Natural Scientists, at which Lenard and Einstein faced each other in a heated discussion for the first time – in which Max Planck (1858 – 1947) and Franz Himstedt (1852 – 1933), an experimental physicist from Freiburg, declared that Einstein had authorized them to express his vivid regret about his public reproaches against the "highly esteemed colleague, Mr. Lenard." However, this apology, indirect and not even signed by Einstein himself, was not enough for Lenard, who worked himself into an irreconcilable, increasingly anti-Semitic hate against the "relativity Jew."

Johannes Stark, whom Lenard knew to share his rejection of that modern theoretical physics defamed as "Jew-spirited," was another militant opponent of Albert Einstein. But in contrast to Lenard, Stark had not concerned himself with the contents of the theory of relativity; the main target of his attacks was quantum theory. He never took issue with the special theory of relativity; indeed, in a pamphlet of 1941 with the telling title *Jüdische und deutsche Physik* (Jewish and German Physics), a work hardly to be outdone for its subjective polemics, he is fair enough to acknowledge that the photoelectric effect discovered by Lenard, according to which the velocity of electrons depends on the frequency of light, was an expected consequence of Planck's law, "as Einstein first noted in 1905." All the more violent was

Stark's bluster against the general theory of relativity, whereby his critique was restricted to very general attacks. In the first of the numerous polemics he wrote criticizing modern theoretical physics (*Die gegenwärtige Krisis in der deutschen Physik,* 1922), he reproached the theory for its "lack of physics content" and declared that its exaggeration into abstract and formal concepts, its restriction to an intellectual game with mathematic definitions and formulas finds its expression in the intentional disregard of ether. What bothered Stark even more was the "unprecedented propaganda" for the theory of relativity printed in the daily press, which had found "fruitful soil in this age of political and social revolution." Einstein "should not be spared the reproach that he did not put up any resistance to their dragging his theory out onto the fairgrounds" and that he "encouraged works by dilettantes in honor of his theory." Even Einstein's lecture trip to France annoyed him: The newspaper reports about "Einstein's trip to the French" were for him "a sad in-dication of German decay." Not until after 1933, when he had ascended to high state offices, did Stark combine his public attacks against Einstein with anti-Semitic agitation. Even the respected science journal Nature gave him the opportunity for polemics against Einstein. In an article published in 1938, *The Pragmatic and the Dogmatic Spirit in Physics,* he repeated his reproaches of 1922 to an international audience: "The relativistic theories of Einstein, which are based on an arbitrary definition of space and time co-ordinates or their differentials, constitute an equally obvious example of the dogmatic spirit." At the close of the article he finally lets the "cat out of the bag" and explains that the dogmatic spirit in physics he is battling against is of Jewish origin: "I have also directed my efforts against the damaging influence of Jews in German science, because I regard them as the chief exponents and propagandists of the dogmatic spirit."

Wolfgang Trageser

Why Einstein Did Not Go to Frankfurt

The fact that the new physics has two birthdays and consequently two birth certificates could be called an "early complementarity" in modern physics. The birth certificate for quantum theory goes back to 14 December 1900 when Max Planck delivered his famous lecture *On the theory of the energy distribution law of the normal spectrum* to the German Physical Society in Berlin. The birth of the reltivity theory occurred five years later, on June 30, 1905, when Albert Einstein published his work *On the electrodynamics of moving bodies* in the seventeenth volume of the German scientific periodical *Annalen der Physik*, the second birth certificate. Modern physics was born, and one of the contributors to its spectacular development was the new University of Frankfurt.

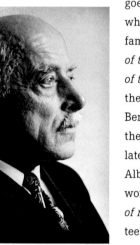

Max von Laue,
24 October 1950

Max von Laue:

A Nobel Prize Winner is the First Chair of Theoretical Physics in Frankfurt am Main

Max von Laue, a top-ranking relativity theorist with contacts to Planck and Einstein, held the first chair of theoretical physics at Frankfurt University. Von Laue was *"extraordinarius,"* or professor, at the University of Zurich when he received a letter from the Prussian Culture Minister von Trott zu Solz, offering him a position as *"ordinarius,"* or associate professor, of theoretical physics at the newly founded university in Frankfurt am Main for the winter semester of 1914–15. A few financial difficulties stood in the way, but they were quickly removed. And thus the path was clear. On 17 September 1914, Kaiser Wilhelm II appointed von Laue to the science faculty as associate professor. The University of Frankfurt had made an excellent choice. On November 11, 1915 von Laue informed the chairman of the university's board that the Swedish Academy of Sciences had awarded him the Nobel Prize in Physics for the year 1914. Two days later an article appeared in the *Frankfurter Zeitung* with the headline *Frankfurt's Nobel Prize Winner*. The article made a larger public aware that X-ray diffraction was discovered by von Laue and his colleagues Walther Friedrich and Rudolf Knipping. But the Nobel Prize awakened the interest of other universities in von Laue. On 24 July 1916, he informed the university that the Austrian Ministry of Education had offered him a chair at Vienna University. Von Laue wished to remain in Frankfurt, but only if offered a full professorship which would free him from his teaching duties. His wish was fulfilled on 1 October 1916. Dr. Richard Fleischer, a wealthy businessman in Wiesbaden and a university benefactor, offered to pay the salary of 5000 marks a year for a full professorship. The call to Vienna was turned down, but von Laue made no secret of his interest in Berlin in a letter to Adickes, the Mayor of Frankfurt. In 1903 he had written his dissertation in Berlin under Max Planck on interference phenomena in plane-parallel plates. Two years later, in 1905, von Laue began working as an assistant to Planck, also in Berlin. During this time he met Albert Einstein, visiting him in Switzerland soon after the publication of Einstein's famous treatise *On the electrodynamics of moving bodies* to discuss a few questions on the new relativity theory. In 1909, as a lecturer at the University of Munich, von Laue wrote the first textbook on the relativity theory: *Das Relativitätsprinzip* (The Relativity Principle). Von Laue wished to participate in current developments in physics – and Berlin was the best place to do so. Berlin was where the new quantum theory was being developed, and here Einstein had just presented his general relativity theory to the Prussian Academy of Sciences.

Left:
Title page of Max
von Laue's paper
*On the Principle
of Relativity*,
Brunswick, 1911

Right:
Bohr Colloquium
in Berlin 1920:
From left to right
at rear: Knipping,
Wagner, Baeyer,
Hevesy, Westphal,
Gustav Hertz;
middle row: Lenz,
Ladenburg, Hahn;
front: O. Stern,
J. Franck, N. Bohr,
L. Meitner, Geiger
and Pringsheim

Exchange: Max von Laue Goes to Berlin – Max Born Goes to Frankfurt

His desire to go to Berlin took concrete form in 1918. On 20 May 1918, von Laue informed the university that he wished to exchange positions with Max Born in Berlin. On 9 July 1918, the Prussian Ministry of Education endorsed his plan. The Board of Trustees at University of Frankfurt agreed on 16 December 1918 to have Max von Laue transferred to Berlin and Max Born succeed to his Frankfurt chair. Einstein was not uninvolved in this development. He had encouraged Born to accept the position in Frankfurt. A strong friendship connected the Born and Einstein families. Einstein did not seem to be as close to von Laue at the time, although in later years he learned to respect him not only as a physicist but as a just man of character. The friendship that developed between them in Berlin went very deep, and even survived the year 1933.

This swapping of chairs came about after "some hemming and hawing" (Born), and Max Born took up his position as associate professor of theoreti-

cal physics in Frankfurt am Main in the summer semester of 1919. Today Born is mainly known for his groundbreaking work in quantum mechanics. But he was also an enthusiastic supporter of the relativity theory and its founder Einstein. In the preface to his correspondence with Einstein he talks about his intellectual development and how his fascination with the relativity theory drew him to theoretical physics. With Max Born, the University of Frankfurt had acquired another prominent relativity theorist. Born's work in Frankfurt am Main coincided with the years of hardship after World War I when scientific institutions lacked the necessary financing. A major cause was the substantial loss in value suffered by the university trust's assets as a result of high inflation. Born's institute, in which Otto Stern and Walther Gerlach conducted their experiment on directional quantization, was no exception. Public interest in the relativity theory was very high, so in the summer semester of 1920 Born held a lecture every Tuesday from 5 to 6 pm on the relativity theory for a small fee. The difficult financial situation had little impact on their creativity, however. On the contrary, they proved that necessity is indeed the mother of invention. It is astonishing how many significant experiments were conducted in the Frankfurt institute and its small laboratory during the brief period, from 1919 to 1921, in which Born worked there as a theorist. Here is a list of some of these experiments:

Max Born

Otto Stern: Eine direkte Messung der thermischen Molekulargeschwindigkeit, Zeitschrift für Physik 2, 49–56 (1920)
Max Born/Elisabeth Bormann: Eine direkte Messung der freien Weglänge neutraler Atome, Physikalische Zeitschrift 2, 578–582 (1920)
Peter Lertes: Der Dipolrotationseffekt bei dielektrischen Flüssigkeiten, Zeitschrift für Physik 6, 56–68 (1921)
Otto Stern: Ein Weg zur experimentellen Prüfung der Richtungsquantelung im Magnetfeld, Zeitschrift für Physik 7, 249–253 (1921)
Otto Stern/Walther Gerlach: Der experimentelle Nachweis der Richtungsquantelung im Magnetfeld, Zeitschrift für Physik 9, 353–355 (1922).
Born's famous book titled *Einstein's theory of relativity*, which contained his Frankfurt lectures on the relativity theory, was first published in 1920 by Springer Verlag in Berlin. Up until the end of 1919 discussion of Einstein's relativity theory was limited to a small circle of experts.

This changed when two expeditions, organized by Sir Arthur Eddington in England, set off to northern Brazil and the Portuguese island of Principe on the African coast to observe the total solar eclipse that was to occur in the tropics on 29 May 1919. These expeditions confirmed a major prediction of the general relativity theory: the deflection of light by the sun's gravitational field. From then on physics had a new gravitation theory, and the world had a new Newton. Einstein became an immediate celebrity and a constant presence in the media.

Vilified and Praised:
Einstein in the Crossfire of Criticism

But with the "relativity hype" began the hostilities and resentment. Nationalist circles accused Einstein of promoting his own theory. Max Born was also a target of this animosity. The first edition of his book on relativity theory had included a photo and a short biography of Einstein. Some scholars objected to this, and it was pilloried by anti-Semitic parties as a typical example of Jews promoting themselves. In Berlin, Paul Weyland, who was unknown in scientific circles, founded an *Arbeitsgemeinschaft deutscher Naturforscher zur Erhaltung reiner Wissenschaft e.V.* (Association of German Natural Scientists for the Preservation of Pure Science) for opponents of the relativity theory. Even eminent physicists such as the Nobel Prize winners in physics, Philipp Lenard and Johannes Stark, sympathized with this *Arbeitsgemeinschaft*. The campaign against Einstein and the relativity theory culminated in a mass meeting in the Berlin Philharmonie on 20 August 1920. This rally horrified the educated public and Einstein's scientific colleagues. The *Berliner Tagesblatt* and other newspapers reported in several issues about the offensive launched against Albert Einstein. Einstein responded to Weyland's organization, which he called the *"Anti-*

relativistische GmbH" (Anti-Relativity Company, Ltd.), in the *Berliner Tagesblatt* on 27 August 1920. Einstein's colleagues, Max von Laue, Walter Nernst, Max Planck, Arnold Sommerfield, and others, joined forces to defend Einstein and expressed their indignation at the actions of Einstein's oppo-

NATURWISSENSCHAFTLICHE
MONOGRAPHIEN UND LEHRBÜCHER
DRITTER BAND

DIE
RELATIVITÄTSTHEORIE
EINSTEINS

VON

MAX BORN

nents. Although Einstein attempted to dismiss the unqualified attacks with humor, he was not unaffected by these incidents and wondered whether it was time to leave Berlin. The press turned these rumors into a fait accompli. This in turn moved Education Minister Haenisch to assure Einstein of his solidarity in a letter of 6 September 1920. With his response on 8 September 1920, Einstein

put an end to press reports about his departure from Berlin. In a letter to the minister he wrote that relationships with friends and colleagues bound him to Berlin, and that he would only follow a call to another university if forced by external circumstances. News that Einstein might want to leave Berlin spread rapidly, however, and universities in Germany and abroad made efforts to recruit him. The University of Frankfurt am Main was one of these contenders.

From a letter to Frankfurt mayor Voigt on 18 September 1920 – Einstein was staying with Born's family in Frankfurt during this time because he wished to attend the annual meeting of the *Gesell-schaft deutscher Naturforscher und Ärzte* (Society of Natural Scientists and Physicians) in Bad Nau-heim – we know that the University of Frankfurt had expressed its interest in Einstein. Wachsmuth's negotiations with Einstein presumably began in August 1920 after the mass meeting instigated by Paul Weyland and his organization and subsequent press reports about Einstein's possible departure from Berlin. Einstein's letter of 8 September to Minister Haenisch made the likelihood of his being appointed to a university outside of Berlin unrealistic. But in October 1920, Einstein accepted a visiting professorship in Leiden, Netherlands. For the above reasons, we can date the correspondence between Einstein and Wachsmuth to the period between 27 August to 8 September, 1920. It is amazing that the correspondence between Einstein and Born, which was very friendly and intimate, makes no mention of Einstein's correspondence with Wachsmuth nor of the possibility of his appointment to Frankfurt am Main.

Title page of Max Born's paper *Einstein's Theory of Relativity*, Berlin 1920

Dieter B. Herrmann

Einstein and Archenhold: Two Champions for the Popularization of the Natural Sciences

When Albert Einstein arrived in Berlin from Zurich on 29 March 1914, to settle permanently in the capital as a member of the Prussian Academy of the Sciences, he was certainly known among physicists, but not yet to the general public. Nevertheless, word had spread that Einstein was working on a generalization of his special theory of relativity of 1905, and that he had made great progress toward this objective. Thus it was not long before the editors of the *Vossische Zeitung* asked him to write a lengthy essay about the principle of relativity for a generally interested readership. Einstein granted this request quickly, writing an essay that appeared on 26 April 1914. Perhaps it was this article that gave Friedrich Simon Archenhold (1861–1939), the founder and director of the Observatory in Treptow, the idea of inviting Einstein to hold a lecture at the observatory, where other speakers in the past had also imparted their "first-hand knowledge" to an interested public. Einstein accepted this invitation too, and on 2 June 1915 held his first public speech in Berlin about his two theories of relativity. The talk had been advertised with the title "About the Relativity of Movement and Gravitation" and, according to a report in the *Vossische Zeitung*, enjoyed a "relatively large audience." What is surprising about these two appearances, in the newspaper and at the lectern, is the fact that Einstein – in contrast to many of his famous colleagues at the time – was willing to speak about his research to laymen at all. Equally noteworthy is that the "general theory of relativity" had not even been completed at that time, let alone published. This may have been a particular stimulus for the physicist to think through the problems once more, so thoroughly and consis-

Friedrich Simon Archenhold, photograph from 1931

tently that even non-scientists would be able to understand them.

As far as the willingness to impart knowledge to the public is concerned, Einstein's personal experience had made him extremely open. Even at the age of twelve, he claimed, he had read popular science books that had made a great impression on him. He particularly emphasized the "Popular Natural Science Books" (*Naturwissenschaftliche Volksbücher*) by Aaron Bernstein, which explained the essential methods and results of natural science in an understandable manner, limited almost exclusively to their qualitative presentation.

Einstein claims that Humboldt's large-scale "draft of a physical description of the world" entitled *Kosmos*, also inspired him in his youth. More than the specific findings that could be extracted from them, he valued the general insights such works imparted. Einstein later ascribed to popular education quite similar tasks: "The requirements of life are far too diverse to make such special training possible […] School should always have the objective of releasing young people into the world as well-rounded personalities and not as specialists […]".

Archenhold pursued very similar ideas. He saw his observatory as a building, "from which new light may always be radiated outward and new knowledge may always find its way to every individual in the nation." Archenhold owed the initial idea for his popular education efforts to his teacher Wilhelm Foerster (1832–1921), who had won Archenhold's admiration for his ability to expand the horizon of his listeners by interweaving "specialized knowledge with general aesthetic and philosophical views" in his lectures. Archenhold had accompanied Einstein's theory of relativity from its very inception in *Das Weltall*, the journal he had founded in 1900. From 1905 on, well into the period of the anti-Semitic agitation against Einstein, he repeatedly published a

The new Archenhold Observatory building in Treptower Park, 1909

A historic hall in the Archenhold Observatory

variety of articles on Einstein's research. Archenhold even held a public lecture about Einstein's theory, at which he made the unpleasant acquaintance of Paul Weyland, the man behind the scientifically completely unqualified anti-Semitic campaign of agitation against Einstein. Weyland was also the one who organized an anti-Einstein event at the Berlin Philharmonic in 1920, at which a vehement exchange of verbal blows between him and Archenhold took place. Unperturbed, Archenhold continued popularizing the theory of relativity through such efforts as the repeated showing of the instructional film *The Foundations of Einstein's Theory of Relativity* ("Die Grundlagen der Einsteinschen Relativitätstheorie") accompanied by a lecture.

After the lecture in the observatory, Einstein and Archenhold apparently became friends. They met not only at the meetings of the *Physikalische Gesellschaft*, but also in Einstein's home. On such occasions they also discussed truly pioneering projects, in which Archenhold tried to involve Einstein. After the end of World War I, for instance, the idea was to found a "Panterra" society dedicated to the realization of large-scale, international projects for science and technology as a means of preventing another war. On the occasion of Archenhold's Mars exhibition in the observatory in 1926/27, he also tried to win Einstein's participation. Yet Einstein must have recognized that in this case Archenhold was more interested in his popular name; he declined with the remark that he did not want to show up "everywhere as a symbolic bellwether with a halo."

Einstein's lecture at Archenhold's observatory was hardly a singular event in the physicist's life. Well before that, in 1911, he had addressed the *Naturforschende Gesellschaft* in Zurich, and in 1913, the 85th convention of German Medical Doctors and Natural Scientists in Vienna. Once his fame had spread to the world at large through the proof of the diffraction of light in the gravitational field of the sun in 1919, he was bombarded with invitations to lecture events and had to se-

IN DIESEM SAAL
HIELT
ALBERT
EINSTEIN
(1879-1955)
AM 2. JUNI 1915
DEN ERSTEN
ÖFFENTLICHEN
BERLINER VORTRAG
ÜBER DIE
RELATIVITÄTS-
THEORIE

Einstein memorial plaque by Hans Füssl (1979) at the entrance to the Great Hall (Einstein Hall) of the Archenhold Observatory (photo: M. Dohrmann, Berlin)

lect his appearances carefully. For instance, before leaving for the United States in 1921 he spoke at the URANIA association hall in Prague, and at a concert hall in Vienna, where he addressed an audience of more than 3000.

Especially illuminating for Einstein's attitude toward the popularization of science is his lecture held at the "Marxist Workers' School" (*Marxistische Arbeiterschule*, MASCH) in Berlin in 1931. The director of the school, Professor Dr. Johann-Lorenz Schmidt (László Radványi), had sent his wife Anna Seghers (1900–1983), later to become a famous novelist, to Einstein in Caputh to win him over for a lecture. Einstein agreed immediately. In response to his wife's interjection, "You have to decline! You were determined not to accept any more lectures," Einstein answered, "This is a completely different kind of lecture; I'm interested." The lecture was held on 26 October 1931 at the inauguration of the school year at a community school in northern Berlin and was entitled "What a Worker Must Know about the Theory of Relativity." Among the audience were the composer Hanns Eisler (1898–1962) and the writer Bertolt Brecht (1898–1956). Brecht is said to have later worked a number of suggestions from Einstein's lecture into his play *The Life of Galileo*.

Einstein was also a prominent popular science author. His works in this connection are based on a deep understanding of the difficulties, but also the necessity of a generally understandable explanation of scientific ways of thinking and results. There was indeed danger, Einstein believed, in translating from specialized language to a presentation anyone can understand, of either missing the core of the problem or feigning simple clarity through superficial description. On the other hand, for Einstein there was no doubt that scientific instruction for the public was one of the most essential subjects of education: "It is of great importance that the general public has an opportunity to receive experienced and

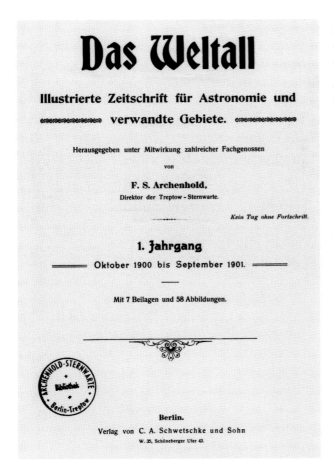

Das Weltall

Illustrierte Zeitschrift für Astronomie und verwandte Gebiete.

Herausgegeben unter Mitwirkung zahlreicher Fachgenossen

von

F. S. Archenhold,

Direktor der Treptow - Sternwarte.

Kein Tag ohne Fortschritt.

1. Jahrgang

Oktober 1900 bis September 1901.

Mit 7 Beilagen und 58 Abbildungen.

Berlin.

Verlag von C. A. Schwetschke und Sohn

W. 35, Schöneberger Ufer 43.

popular science books, two of which stand out especially. The first appeared in 1917 under the title *Relativity: the Special and the General Theory* („Über die spezielle und allgemeine Relativitätstheorie, gemeinverständlich") and was printed in more than 10 editions within a few years. To date it is doubtless one of the best introductions to Einstein's theories. Reading this work did, however, as Einstein remarked in the foreword, "presuppose quite a bit of patience and willpower on the part of the reader." The other work is *The Evolution of Physics*, a collaboration written in the U.S. along with his Polish colleague, Leopold Infeld (1898–1968). This book has since become a classic, and was intended to provide insight into "the eternal struggle of inventive human mind for a fuller understanding of the laws that govern physical phenomena."

Considering Albert Einstein's personality, his political, social and philosophical views, his energetic activity dedicated to the spread of knowledge in the interest of enlightenment comes as no surprise. It was part of that great responsibility Einstein felt that researchers owed to society, an obligation he upheld.

The ties of friendship between Archenhold and Einstein were severed brutally by the rise of German fascism. The Jewish Archenhold family was forcibly removed from its life work. Members of the family perished in the Theresienstadt concentration camp while Einstein was driven into exile.

understandable instruction about the efforts and results of scientific research ... The restriction of scientific knowledge to a small group of people weakens the philosophical spirit of a nation and leads to its intellectual impoverishment."

Besides various newspaper articles — not at all restricted to issues of his own research — we are also indebted to Einstein for several important

The magazine
Das Weltall
(The Universe),
published from
1900 to 1944,
helped popularize
astronomy

Karl Wolfgang Graff

The Automatic "Concrete People's Refrigerator" CITOGEL by Albert Einstein and Leo Szilard

In winter 1925/6 newspapers reported a tragic accident that cost the lives of a family in Berlin. A large quantity of poisonous refrigerant suddenly leaked from a refrigerating machine in the middle of the night, suffocating the entire family within minutes. The report roused Einstein's inventive spirit. Together with the physicist Leo Szilard (1898–1964) he pondered how such accidents could be prevented in the future. Believing that the refrigerant had been released by a leaking or a lasking pump, they sought a solution for a household refrigerating system without moving parts. For Einstein the solution to this everyday problem was quite novel, for at that time his household had only an icebox, which had to be filled from time to time with broken-up blocks of ice. The Einstein family did not own an electric refrigerator until they lived in their summer cottage in Caputh.

In the course of his research, in June 1927 Szilard discovered the *Reichspatent* No. 441 752, which protected the invention by Klemens Bergl and Walther Dietrich for their "Refrigerating Process using Organic Liquids without Recovery of These as Such" and was distinguished by the fact "that water-soluble organic liquids, especially methyl alcohol, ethyl alcohol and acetone, are made to vaporize, or to evaporate through simultaneous suction of air. The application of such liquids offers the advantage that the gases produced by vaporization or evaporation can be removed quite easily without any danger for the surroundings." The specification of the simple invention piqued Einstein's interest and, together with Szilard, he considered how to improve the process without violating the patent rights of the inventors.

On 14 July 1927 they applied for a *Reichspatent* for their "Refrigerating Process" and indicated in their description that "the rational refrigerating using an organic refrigerant, whose vapors were not recovered, is proposed only in Patent 441 752." Their improvement of the process of vaporizing

"methyl alcohol in particular" and removing it without recovery consisted primarily in the fact "that the vapors are dissolved in water in an absorber and the solution removed from it against the pressure of the atmosphere by a special pump." In their patent specification of March 1926 Bergl and Dietrich had described an apparatus for the production of ice or for cooling drinks,

mentioning in passing that the evaporator could also be housed in a refrigerator. Einstein and Szilard picked up on this idea and developed a "refrigerating machine" according to the process they had invented, which they submitted for a supplementary patent on 31 October 1927. The way their "refrigerating machine" worked is explained on the basis of a schematic drawing: Under the refrigerator, 11 is the storage container for the refrigerant, 12. The water jet pump affixed on the side of the refrigerator, 2/3, generates a

Photo of the Einstein-Szilard refrigerator using the absorption principle

slight vacuum, causing the refrigerant to move upward. The amount of refrigerant is set manually by means of the regulator located in the refrigerator, 13. The refrigerant vaporizes in the evaporator, 10, so that a mixture of vapor and air moves through connecting line 9 to the absorber, 1. The

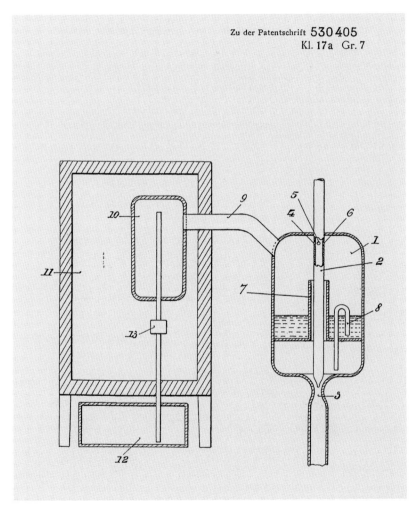

Zu der Patentschrift **530 405**
Kl. 17a   Gr. 7

water jet pump is connected to the public water supply. The water flowing into the pump is sprayed out of small openings in the feed pipe, 4 through 6, to the inner wall of the absorber, thus increasing the surface of the water for the absorption of the refrigerant vapor. The aqueous solution collects in the container 7 and is emptied when the solution reaches a certain level in the container, by means of lever 8. The water jet pump alter-

nates between drawing out the solution and the vapor-air mixture from the absorber, with the solution being discharged into the public sewer. The two inventors were able to make their first attempts to realize their ideas in late 1927, in the workshop of a Berlin factory. However, the company had to end the collaboration after just a few months due to financial difficulties. On 6 April 1927 two businessmen from Hamburg had closed a general licensing contract with Bergl and Dietrich for *Reichspatent* no. 441 752. For the development and exploitation of the invention they founded *Citogel Gesellschaft für chemische und technische Erzeugnisse mit beschränkter Haftung* in Hamburg on 20 August 1927, and were entered in the Commercial Registry as its directors. The purpose of the company was the production and marketing of the manufactures mentioned in the company name, "especially those cooling apparatus traded under the CITOGEL name [...]." The capital stock of the company totaled 20,000 Reichsmarks. The founders of the company had brought in the rights from the licensing contract with Bergl and Dietrich as a contribution in kind. In early November 1927 the *Reichspatentamt* announced the transfer of Patent no. 441 752 to *Citogel GmbH*. Toward the end of the year, Einstein and Szilard contacted *Citogel GmbH*, as starting in 1928 they would no longer have a workshop where they could perform their development work, and time was running short. On 28 December 1927 they signed a contract with the company, in the presence

Schematic drawing of the absorption refrigerator from the patent specifications D.R.P. 530 405, July 1931

of a notary public in Berlin, in which they agreed to combine their patents and other protection rights with those of the company to an aggregated complex. They further agreed that the refrigerating apparatus were to be completed for the market under Szilard's direction in a room in Berlin. Upon the closing of the contract, all parties were certain that their apparatus would be a success. In January 1928 *Citogel GmbH* announced its participation in the Spring Trade Fair in Leipzig

and raised its capital stock by 20,000 Reichsmarks on 30 January 1928.

The Technical Trade Fair took place from 4–10 March. At a double stand, *Citogel GmbH* showed its "newest automatic refrigerating apparatus without mechanical devices" and demonstrated it in action. The "people's refrigerator" was offered in two sizes of 40 and 80 liter storage capacity, at "unrivaled prices." According to Citogel's specifications, if methanol was used as a refrigerant, temperatures as low as –12 °C could be reached, and if denatured alcohol were used, just above

0 °C. "The daily methanol consumption is given by around 600 g, which would correspond to the heat of fusion of around 2 kg of ice. The corresponding refrigerator is made of concrete; between two layers of concrete is a compressed cork panel." The representatives of *Citogel GmbH* and the two inventors must have been satisfied with the outcome of the trade fair, for on March 18 the company increased its capital stock for the second time, by another 20,000 Reichsmarks. Einstein and Szilard thought about how they could improve the refrigerator further. The schematic drawing *Citogel GmbH* submitted with its application for a Swiss patent on 2 March 1928 already showed significant improvements over the portrayal in October 1927. In early May Einstein and Szilard applied for yet another *Reichspatent* for an insulation material made of several layers of paper, which made the refrigerator lighter, and probably towards the price as well.

The first difficulties for customers must have arisen in summer 1928. Upon delivery, the refrigerators were connected to water supply and sewage lines, filled with refrigerant and adjusted for operation. However, as one of Szilard's former colleagues still remembered just a few years ago, the water pressure in the supply pipe varied so widely that it was often necessary to readjust the settings. Furthermore, the small amounts of methanol needed to refill the refrigerator were much more expensive than the inventors had assumed. Despite these difficulties, *Citogel GmbH* displayed its devices again at the Leipzig Spring Trade Fair in 1929, this time at a smaller stand. After this, things got quieter and quieter for the enterprise founded with such great optimism. By the time the *Reichspatentamt* issued *Citogel GmbH* the patents Einstein and Szilard had applied for on 28 May and 16 July 1931, the company had probably already closed down operations. After giving public notice to creditors, *Citogel GmbH* was officially deleted from the Hamburg Register of Companies on 13 October 1933.

Under penalty of a fine, Einstein had contractually forbidden *Citogel GmbH* to mention his name "in connection with the refrigerating process or even to add the name Einstein to the brand name of the refrigerator." Therefore it can be assumed that the visitors to the trade fair in Leipzig did not know who had invented the Concrete People's Refrigerator CITOGEL. Several archive materials, in besides the Citogel patents and the trade fair reports of a refrigeration engineer, demonstrate

Einstein's activity as an inventor in an area far removed from his scientific activity as a theoretical physicist. His invention was original and unique, and it worked. It failed due to external conditions he could neither foresee nor affect. Even during their work on the absorption refrigerator, technology developed in a direction Einstein and Szilard did not wish to take. They recognized this and even made patentable contributions to the other field with their invention of refrigerator functioning according to the electrodynamic principle, yet even their many years of work at the research institute of the AEG concern did not yield a refrigerator for mass production.

Roger Highfield

Einstein's Women

When she first set eyes on baby Albert on the morning of 14 March 1879, Pauline Einstein became alarmed. His head was so big and so angular that she thought he was deformed. Later the slowness with which this quiet, fat child learned to speak made her fear he was also a little slow on the uptake. She need not have worried.

The first woman in Albert Einstein's life was an ample and commanding figure, and it is from her that her son seems to have inherited his fleshy

surrogate parents: Jost Winteler and his wife. And Marie, the prettiest of the three Winteler daughters, and two years older than Einstein. Their names are linked in a letter little more than two months after his arrival in Aarau. Back home in Italy for the school holidays, Einstein wrote to "my little angel" with all the yearning one would expect of a lovelorn seventeen-year-old. "Only now do I realise how indispensible my dear little sunshine has become to my happiness [...] You mean

Mother Pauline Einstein, nee Koch (1858–1920)

Sister Maja (1881–1951) in 1899

nose, as well as that unruly nest of hair. She was never indulgent and it was her domineering personality that defined much of the atmosphere of Einstein's childhood. His letters reveal her fondness for teasing him, reflected in turn in his own propensity to sarcasm. In various ways, Einstein was his mother's son.

Leaving aside the young woman who was first the target of his affections (and tantrums), his sister Maja, Einstein's first unhappy romantic attachment was to lay down many emotional themes that ran through his life. He met her on 26 October 1895, when Einstein went to the cantonal school in Aarau, west of Zurich. There he found

more to my soul than the whole world did before." He left Aarau in October to enrol at the Polytechnic in Zurich. The following month, Marie took up a post as a teacher in Olsberg, a village in the north-west of Aargau. Her letters showed the love affair with her "clever darling curlyhead" was faltering. In May 1897, Einstein wrote to Pauline Winteler to "cut short an inner struggle" and turned down the chance to visit Aarau and his "Mummy Number Two" at Whitsun. It would, he said, "be more than unworthy of me to buy a few days of bliss at the cost of new pain, of which I have already caused much too much to the dear little child."

According to Marie, Einstein would once have married her. All that prevented it, she said, was her reluctance to follow "the path of duty." Both sets of parents probably would have been very happy if a marriage had resulted, and the eventual betrothal of Maja Einstein to Marie's brother Paul can only have sharpened their regrets. Long long after the romance ended, Einstein admitted Marie still preyed on his mind. In September 1899, he referred "to the critical daughter

with whom I was so madly in love four years ago […] I know that if I saw her a few more times, I would certainly go mad. Of that I am certain, and I fear it like fire."

Mileva Marić, a Serb from the province of Vojvodina, in the north of the former Yugoslavia, was the most significant female attachment of his life. They met in Section VI A of the Swiss Federal Polytechnic School, Zurich. A friend described her as "a very good girl but too serious and quiet." Small, delicate and plain, Mileva also had a limp. Einstein, by contrast, was now a handsome teenager exuding casual charisma.

Einstein's letters to Mileva would increasingly present her as his intellectual comrade. Here at last – after his philistine parents and the unassuming Marie Winteler – was someone with the intelligence and interest to share the ideas that excited him. Einstein began to call Mileva "Little Doll" or "Dollie." She reciprocated by calling him "Johonesl," or Johnnie.

Yet Einstein was a flirt, and one of the attractions he found on a holiday in Mettmenstetten was the company of a girl three years his junior. Anna Schmid was the sister-in-law of the Hotel Paradise's owner. On his departure he left some verses in her album, which referred to kissing her and "your rascally little friend." Julia Niggli, a friend from Aarau whom he once advised on the unreliability of the male heart, was another target of his charm. In Zurich, too, Einstein was keen to seek out female company, such as Susanne Markwalder, the daughter of one of his landladies.

While Einstein's mother had at first seemed to smile on her son's relationship with Mileva, tension built. She did not seem to care that Mileva was not Jewish – neither was Marie Winteler – but she seems to have held the common German prejudice against Serbs. Pauline also distrusted Mileva as an older woman, tainted by physical handicap, who was leading her son astray.

The summer of 1900 saw Einstein and Mileva sit their final examination. Mileva failed. She returned home to Vojvodina, tired and depressed. But she was determined to retake the examination. Meanwhile, Einstein set off to join his mother and sister on holiday at the Swiss resort of Melchtal. We owe our knowledge of what happened

Einstein with his first wife Mileva, nee Marić, around 1905

next to the description he scribbled down for
Mileva with immense relish as he sat later in bed.
At that time Einstein had no job and an uncertain
future ahead of him. "So, what will become of
your Dollie now?" his mother asked him inno-
cently. "My wife," responded Einstein: "Mama
threw herself on the bed, buried her head in the
pillow and wept like a child." Pauline Einstein's

happy!" wrote Mileva. Her Johnnie was eager for
more. "How beautiful it was the last time you let
me press your dear little person against me in
that most natural way," he wrote. "Let me kiss you
passionately for it, my dear sweet spirit!"
Then Mileva discovered she was pregnant. In July
1901, she failed her exams a second time and returned to Vojvodina in low spirits but ready to tell
her parents all her "unpleasant news" –
both academic and personal. Mileva gave
birth to a daughter some time around the
end of January 1902. Even without seeing
his daughter, Einstein pronounced himself completely in love with her. But
there is no evidence that Einstein and his
daughter ever set eyes on one another.
Einstein was never to talk of her publicly.
She was probably given away. Lieserl's
birth posed a threat to Einstein's new
start as a patent examiner in Berne. He
had gained Swiss citizenship only a year
earlier, and the stigma of an illegitimate
child would have harmed his prospects

Mileva Einstein
and the two sons
Hans-Albert
(1904–1973) and
Eduard (1910–
1965), around 1914
© The Hebrew
University of
Jerusalem, Albert
Einstein Archives

greatest fear was that the couple had "already
been intimate." She continued to heap invective
on Mileva's head. "Like you she is a book – but
you ought to have a wife," she told him, adding
for good measure, "By the time you're thirty she'll
be an old witch."
This conflict with his family intensified Einstein's
passion for Mileva as he struggled to get an assis-
tantship and she worked towards resitting her
final examinations. Einstein's career worries were
eventually eased in April 1901 when an opening
presented itself at the Swiss Patent Office in
Berne. All at once – again thanks to a recommendation by friends – he was also offered a two-
month teaching post.
He summoned Mileva to a celebratory reunion at
Lake Como, near the Italian border. "How happy
I was again to have my darling for myself a little,
especially because I saw that he was equally

both in the country's civil service and in the
socially conservative environment of its capital.
Whatever shadow the Lieserl affair cast over Einstein's life, there seems little doubt that Mileva
fell under a darker one.
On his deathbed, Hermann Einstein finally gave
his permission for his son's marriage to Mileva.
The ceremony took place in Berne on 6 January
1903. There was no honeymoon, and by one account the couple returned to their new marital
home to find that Einstein had locked himself out.
Hans Albert, their first son, was born on 14 May
1904, and for the next four years the signs are
that Einstein and his wife were still good friends.
Then came a fresh opportunity in the spring of
1909. Zurich University created a new position
and Einstein was chosen to fill it. His appointment
was reported in the local papers, where it was
noticed by a certain Anna Meyer-Schmid, to whom

Einstein had composed an affectionate poem a decade earlier. The now married Anna sent Einstein a card to congratulate him, and in May 1909 received a warm reply in which he reminisced about their brief encounter. But the correspondence was intercepted by Mileva, who was outraged. Five months later he told his friend Besso

1912. This is when his father had just renewed contact with the woman who would become his second wife: Elsa.

He and Elsa were related twice over. Yet it has to be said that Elsa was not the only one of Einstein's female relatives to catch his eye, however. He had also flirted with her younger sister, Paula. In a

Einstein with his second wife Elsa, nee Einstein (1876 – 1936), around 1922

that his "spiritual equilibrium, lost on account of M(ileva)" had not returned. Even after four decades, his bitterness endured. In a letter written in July 1951, when both Anna and Mileva were dead, Einstein returned to the theme of his first wife's jealousy, typical in a woman of such "uncommon ugliness."

Further pressure was placed on the Einsteins' increasingly fragile relationship by another move in 1911 to the German University in Prague. Hans Albert told his second wife that marital discord became obvious after his eighth birthday, in May

letter, he took the opportunity to reassure Elsa that her sister had only irritated him. "She was young, a girl, and welcoming. That was enough. What remains is a pleasant fantasy."

A sham of unity still survived in September 1913, as the Einsteins took their children to visit Mileva's parents near Novi Sad. However, by then, he thought Mileva was "an unfriendly, humorless creature who herself has nothing from life and who undermines others' joy of living through her mere presence." Einstein said that he shuddered even to think of seeing Mileva and Elsa in the same room. But that was indeed the prospect, given he was to move the family to Berlin in the spring of 1914. In the summer of 1914, when the school term ended, Mileva took her sons back to Zurich. Einstein wept and had to be supported as he walked home from the railway station.

In February 1916 Einstein stunned Mileva by asking her "to change our well-tested separation to a

Elsa and Albert Einstein in Palm Springs, 1931 (photo: Irving Lippman)

divorce." She suffered a physical and mental collapse. Einstein's first reaction was that his wife was shamming in order to prevent their divorce. However, he was ready to give in for a quiet life. The following month he repeated his pledge. "I will see to it that she suffers no more disturbance on my account. As to a divorce, I have definitely renounced it."

But in May 1918, Einstein was moved to say that his wife was behaving "very nicely" and decided to play his trump card – the Nobel prize for physics.

If his wife agreed to an uncontested divorce, the money from the prize would safeguard the future of her and the children for ever. If she did not, then she would not see a penny more than what he had decided was a reasonable annual maintenance. Einstein was obliged to admit in his legal submissions that he had committed adultery. There were also references to fierce fights. The divorce decree was issued on 14 February 1919, and Einstein and Elsa were married on 2 June. The celebrations were as inauspiciously low-key as those for his first marriage.

Elsa was to suffer the same jealous furies as Mileva. In the summer of 1929, Einstein bought a plot at the village of Caputh, near Berlin, and built his own summer house there. The architect, Konrad Wachsmann, recollected how his marriage was by then under considerable strain. There were frequent rows, even talk of separation, and the reason was always the same. Women were drawn to the world-famous professor like iron filings to a magnet, he said.

Abraham Pais has referred to letters written by Einstein in the early 1920s revealing a significant romance that ended late in 1924, "when he wrote to her that he had to seek in the stars what was denied to him on earth." Pais said Einstein had experienced a profound attraction, "more emotionally felt" than his relationship with Mileva. There is reason to believe that she was one of Einstein's earliest secretaries, Betty Neumann. His friend, Dr Janos Plesch, described his friend as a man with a strong sex drive. "In the choice of his lovepartners he was not too discriminating," he wrote, "but was more drawn to the robust child of nature than the subtle society woman." By way of example, he recounted an incident where Einstein become enraptured by the sight of a girl kneading bread.

Herta Waldow, who served as his live-in maid from 1927 to 1932, also had first-hand experience of Einstein's flirtations. One of her employer's most regular escorts was Toni Mendel, a wealthy Jewish widow who cultivated Elsa with gifts of chocolate creams. There were frequent visits by Estella Katzenellenbogen, the rich and elegant owner of a florist business, who transported Einstein about Berlin in her limousine. Then there was a blonde Austrian, Margarete Lebach, who was "very attractive, lively and liked to laugh

attempt to make something lasting out of an incident," he told one friend.

With the rise of the Nazis in the early 1930s, Einstein began to consider his escape. He set up a temporary home at the Villa Savoyarde in Le Coq sur Mer on the Belgian coast, in the spring of 1933. Even at this difficult hour, however, it appears that Elsa's private humiliation continued.

Estella Katzenellenbogen, Elsa Einstein, Albert Einstein, Dr. Walter Mayer and Helene Dukas 1931 sailing to the United States

a lot, just like the professor." Herta even overheard a passionate argument about the Austrian interloper between Elsa and her daughters. Throughout his life, Einstein voiced deep misgivings about the institution of holy matrimony. He told Plesch that it must have been invented "by an unimaginative pig," and once declared to Konrad Wachsmann that it was "slavery in a cultural garment." The subject became a personal favourite, to which he returned often during the remaining years of his life. "Marriage is the unsuccessful

Micha Battsek, Einstein's godson, recalls staying there and seeing a beautiful Viennese in her early forties. This was Margarete Lebach.

The Einsteins set sail for a new life in the United States in October 1933. The first symptom of what would kill Elsa, a swelling of the eye, was noticed when she and Einstein moved into what became his last home – 112 Mercer Street, Princeton – in the autumn of 1935. The Polish physicist Leopold Infeld, who had begun to collaborate with Einstein, stated that Einstein gave his wife "the greatest

Sailing with Grete
Lebach in 1937
(photo: Lotte Jacobi)

care and sympathy." But he added that Einstein "remained serene and worked constantly." An even stronger picture of his power of detachment was painted by Peter Bergmann, another of his collaborators. Elsa's anguished cries that came out unnerved Bergmann, but Einstein was absorbed in his work. Elsa died on 20 December 1936.

Einstein gave up on the affairs and pursued serial monogamy. First, there was Margarita Konenkova, an alleged Russian spy who was 15 years his junior. "Just recently I washed my head by myself, but not with the greatest success; I am not as careful as you are," he wrote on 27 November 1945. "But everything here reminds me of you: "Almar's shawl, the dictionaries, the wonderful pipe that we thought was gone, and really all the many little things in my hermit's cell; and also the lonely nest." (Almar was apparently a pet name the two had devised for each other, made up of Albert and Margarita, and the "Nest" a room where the two would meet.)

His last girlfriend was Johanna Fantova, who he had first met in 1929, a former curator of maps in the Firestone Library at Princeton. She cut his hair. He moaned about his memory, aches and pains. They went to concerts and sailed together until the doctors took his boat away. As had been the case with his previous girlfriends, Einstein wrote her poems and letters bedecked with jokes and kisses.

Einstein touched on the prickly issue on how he got on with the most important women in his life a month before his death in 1955, in a letter of condolence that he sent to the grieving son and sister of his dearest friend, Michele Besso. "What I most admire about him as a human being," said Einstein, "is that he managed to live for many years not only in peace but in harmony with a woman – an undertaking in which I twice failed rather disgracefully."

This text is based on *The Private Lives of Albert Einstein* by Roger Highfield and Paul Carter (Faber, 1993)

Barbara Wolff

Albert Einstein and Music

"Music has no effect on my research. But both feed from the same wellspring of longing and complement each other in the satisfaction they provide."

Einstein Played the Violin

He received violin lessons before he started school, a customary practice for middle-class families at that time. It was said that his mother was as an outstanding pianist. Still, despite his mother's encouragement, it was difficult to get

Albert Einstein playing violin

him to practice. His teachers were unable to convey a love of music to the young boy. This love he would discover himself at age thirteen. Most likely it occurred around the time when his father, a bit jealous perhaps, complained that his son thought of nothing but music.

Fifty years later Einstein would write to his grandson, who also played the violin:

"I can't imagine my life without making music. In difficult times (and in happy times, too) I could cope with myself and the world because I always had this escape. It makes one free and independent."

He had said something similar to his two sons when they too reached that age: "I am pleased that you practice so diligently, my dear Tete. Music makes the best companion, one that greatly enriches and eases one's path through life." Eleven-year-old Hans Albert, however, needed encouragement from his father:

"Just don't forget your piano! One who plays well can bring much happiness to himself and to others. Play on the piano pieces that give you the greatest pleasure. [...] That's the best way to learn."

This Einstein discovered on his own: "I only started learning after I fell in love, mainly with Mozart's sonatas."

Because he considered "love in general" to be a "better teacher" than "a sense of duty," he did more than just supply his children with good advice and music books from afar. He shared with them, wherever and whenever possible after his divorce from their mother, his own pleasure in making music with others.

He shared this pleasure not only with Tete and Hans Albert. In an era in which the educated classes considered it good taste to be able to play a little music, he could always find two or three musical friends among his classmates, fellow students, and professors with whom he could regularly or spontaneously organize a small chamber concert. During his years in Zurich and Bern, those he played music with, apart from his closest circle of friends, were people with whom he had little more in common with than a love of music. In the Berlin years he had his colleagues, and when the violinist Einstein joined the physicist Einstein in his rise to fame, several great and not yet great musicians joined his circle of musical acquaintances. In addition to the many unnamed and unknown with whom Einstein played were Max Planck and Max Born, Walther Nernst and Erwin Freundlich, the young violin virtuoso Boris Schwarz, Artur Schnabel, the renowned pianist, and the cellist Francesco von Mendelssohn. Einstein played music passionately with his friend physicist Paul Ehrenfest and his children in Leiden, mathematician Jacques Hadamard in Paris, Pater Caramelli in Fiesole, violinist Fernandez Bordas in Madrid, and two young jazz musicians

gentler passages, without the crescendi and allegri," as his assistant Straus recalled.
Besides Mozart, it was Bach and several Italian and English masters of the 18th century whose compositions appealed most to Einstein in their "simplicity and their clear architectonic structure."

Einstein at a benefit concert at the Synagogue on Oranienburger Strasse, Berlin, on 29 January 1930

in Palm Springs. In Jerusalem Einstein "played with [an] officer in [High Commissioner Herbert] Samuel's apartment – far too long, because he was starved for music" and days later with the British Attorney-General, whose wife wrote: "He looked so happy while he was playing that I enjoyed watching as much as listening." Queen Elisabeth of Belgium waited for Einstein in Brussels: "Oh, come and play with us," she wrote to him years later in American exile. "Mozart desperately needs you."
The love for Mozart's music, which had a strong influence on his understanding of music, also forged the bond between the Belgian queen and Einstein. Even in his later years his improvisations on violin and piano sounded like "Mozart's

Did'nt he judge the beauty of mathematical formulas according to similar criteria?
When asked for a comment, he was stumped.
"I don't look for logic in music; I listen to music on an entirely unconscious level. [...] I never like a work if I cannot emotionally fathom its internal unity."
Yet he could not completely separate "logic" from the emotions that he associated with music. Therefore he defended his statement that Niels Bohr exhibited "the greatest musicality in the sphere of thinking" with an explanation that only appears to clarify: "If I called Bohr's thinking process musical, what I meant to say is that he approaches logical relations as intuitively as we experience music."

Then he added, almost defiantly: "After all, music is for the soul and not for the intellect." Einstein found this music for the soul also in Schubert, "because of his consummate ability to express emotions and [his] formidable talent for inventing melodies."

Wagner's compositions, which he regarded as decadent, evoked his aversion – a very emotional response. Wagner's musical personality was so indescribably repugnant to Einstein that he could listen to his music only "with great reluctance." Perhaps he felt like his secretary and confidante Helen Dukas, who "always [had] the feeling when listening to Wagner that she could hear old Adolf bellowing."

Again and again, and almost always, it is the "proportion" and "clear structure" of a composition that delighted Einstein and elicited from him, so sparing in his choice of words unless it involved encouraging or promoting an artist, the distinction "wonderful" as the highest of all praises. Obviously he did not like to talk about music. A conversation in 1930 with the Indian philosopher Rabindranath Tagore, arranged largely for the press, ran into platitudes: "The really good music, whether of the East or of the West, cannot be analyzed." That Einstein a decade later would sum up his musical preferences in a small list of printable statements for Emil Hilb was certainly only done as a favor for a friend of the family. In general, he abided by the maxim "make music, love it – and shut up!"

Even his travel diaries, in which he recorded the events of each day, seldom contain more than "a fine singer" or "a wonderful performance." We find more entries here on the music he played than on the music he listened to – aside from his comments on Japanese music, which he never understood. Tersely, he writes "Solo concert in the evening (with me)" or "Yesterday afternoon Brahms quartet with Miss Herrmann, Miss Schulz, and cellist Stegmann. Also Mozart's Divertimento."

And "Played Bach (with Mrs. Soldat and Denneke). Guests quickly left room." And, with a sigh at the end of an eventful day: "My corpse returned to Moyji, where it was dragged to a children's Christmas party and made to play the violin for the children."

What his few diaries reveal is how much of his life Einstein devoted to music. There were the frequent music sessions with others; improvisations on violin or piano – sometimes nightly and often used to find a way out of a mental impasse; the concerts he attended with a friend or stepdaughter; and then there were all the benefit concerts and similar social occasions at which the violinist with the famous name was obliged from all parties– and not least by his wife Elsa – to perform. He could not always extricate himself as quickly as he did when invited to the opening ceremony of the First International Congress for Sexual Research. "Unfortunately," he responded to the question of whether he would take on one of the violin parts in Brahm's B-major sextet, "I feel that neither my sexual nor my musical abilities are up to the performance."

It was a clever response, but it was certainly not true – at least with regard to his musical abilities. Even if widely differing statements circulate about the quality of his performances, we can assume that Einstein did not have a "stroke like a lumberjack," but, as Boris Schwarz described, "a steady rhythm, excellent intonation, and a clear and pure tone," and that he was at least a good amateur violinist. Schwarz debunked as sheer invention one of the most well-known anecdotes about Einstein the violinist: that a fellow performer allegedly once called out in exasperation "When are you finally going to learn to count to three?"

Many comments about his musical abilities exist, for even those incapable of any insight into the theories of Einstein the physicist felt capable of expressing their opinion about Einstein the violinist. "I believe he is masquerading as a scientist, and is really a musician in disguise," Shaw was overheard saying.

As early as the 1920s, Einstein's popularity spurred an avalanche of letters, many of which begin like this: "Even though we are strangers, I allow myself the liberty of writing to you. I have heard that you are very interested in the violin." Generally, these letters conclude with a request for a recommendation. Sometimes a book was

violin maker" Julius Levin gave Einstein a violin which he had restored according to a method he himself had developed.

Einstein's Violin

"Of course I must now neglect my violin. She will wonder why I never take her out of her black case," wrote Einstein the student to a friend

The trio of Albert Einstein, Bruno Esser and Francesco Mendelssohn (from right to left), 1927

enclosed. "I take the liberty of sending you my book *The Natural Method of Violin Playing*, which shows the proper way to play the violin," wrote the violin teacher Jan Woiku. Reinhard Jischke humbly asked Einstein to write a few lines about his success in making violins. The legal scholar and music writer Dr. Meissner was interested in learning from Einstein "as a top authority" whether the tone quality of a violin had anything to do with its form and curvature. Erna Jüttig-Broda sent two self-composed "Swiss peace songs." And in 1930 the "musician and

after hurting his hand in the laboratory. "I miss her very much, my old friend, to whom I confide and sing everything, often what I don't even acknowledge in my own meager thoughts."

What kind of violin did Einstein play during all those years before the public started dogging every step he took?

It was certainly not a particularly valuable or exceptionally good instrument. Einstein otherwise, averse to any form of luxury, would hardly have allowed himself a new one, not even in those days when money was "turning more and more rancid." However, that was what he did in 1920 when ordering for himself a "twin sister" of the instruments a Berlin luthier was making for his friend Ehrenfest's daughters. The new instrument turned out to be "wonderful."

Presumably it is this violin that Einstein took with him to New York in the spring of 1921: "In one hand he clutched a shiny brier pipe and the other clung to a precious violin." American reporters

tion, Einstein wrote in his letter of thanks, he could "justly and confidently" affirm. Whether Einstein enjoyed playing the Japanese violin or played it regularly is unknown. At any rate he enjoyed showing this exotic object around.

At the end of the 1920s four or five violin cases rested under the piano in the apartment in Haberlandstrasse, reported his stepson-in-law Marianoff in his Einstein biography. Even if we cannot always believe what Marianoff says, we can at least take his word seriously here, for soon afterwards Einstein's wife Elsa mentioned to Hedi Born a "roaring trade in violins. Albert has several. [...] This month he will receive another very beautiful one, from France."

Perhaps the most beautiful, possibly the most valuable instrument Einstein ever played was a Guarneri lent to him in Pasadena in the winter of 1931. While most newspapers emphasized its value – 30,000 dollars was a hefty sum – Einstein astutely put the affair in a nutshell in his diary: "It's fantastic advertising for the dealer."

What he does not mention in

Einstein with the musicians Arthur Giskin, Toscha Seidel and Bernard Ocko (from left to right), Princeton, November 1933

had no doubt that the scientist who had recently risen to prominence did not play just any old violin. That on no account should he do so was soon the opinion of numerous violin makers, restorers, dealers, and owners. In 1925, Einstein had already devoted a section in his will to his violins: violins, scientific manuscripts, and books were to go to his sons; money and securities to his widow.

In 1926 his violins, "which are generally excellent," received another companion, a gift from the Japanese instrument maker Suzuki, whose superiority compared to the other violins in his collec-

his diary is what it felt like to play this violin. Half a year later, however, he referred to the Pasadena affair in a letter to Julius Levin: "Today I fantasized for two hours on your violin and am fully intoxicated and spellbound. [...] I have the same feeling I had with the Guarneri in Pasadena, that I am not worthy of this instrument, and that a real artist should be playing it."

For several years Einstein preferred to play on Levin's restored violin. When Levin, a Jew, was driven out of Germany, Einstein, convinced of the quality of his work and eager to help, wrote a "highly favorable recommendation" for the 'violin maker' and his restoration method. This soon proved to be a mistake, when a commission of experts came to the conclusion that Levin's unprofessional technique of violin restoration irreparably damaged violins. Einstein "had to admit that the pleasure his violin gave me was based on a deficient power of judgment" and "that I [...] shouldn't stick my erudite and famous nose in something I know nothing about."

In the summer of 1934 Einstein's violin also showed signs of damage that eventually rendered it unfit to play. What instrument did he play in subsequent years for the occasional "quartet until late in the night" at his summer residence in Peconic? What violin did he play with the Hilger sisters, one of whom still raved about the encounter fifty years later in her memoirs? What violin did he accompany Gaby Casadesus with at a benefit concert in 1941? And what violin did he use when he played with assistants, students, and famous visitors in Mercer Street up through the mid-1940s?

At the end of the 1940s Einstein was still improvising, but only on the piano. Einstein's violin had finally "received a stepfather."

In 1950, in his final will, he bequeathed his violin – apparently only one was left – to his grandson.

What happened to the Suzuki violin? Who today plays the "especially beautiful one from France," which Elsa mentioned? And where is, if this was not Einstein's last violin, the violin he had had made in Berlin in 1920?

We don't know. Only the Guarneri violin has left a – false – trail as "Einstein's Italian master instrument" and crops up from time to time in press reports.

Much that is worth mentioning had to be omitted from these few pages; some things could only be briefly touched on. Einstein's relationships to the great musicians of his time, relationships in which politics often stood in the foreground and not music, remain a topic for another essay. There are also the accounts of lesser musicians who report that Einstein's assistance after the Nazis came to power was often their last chance to escape. Einstein's collection of violin music books gives one pause to think: Why do so many of them look virtually untouched?

Einstein has been analyzed and illuminated from every side by an inexhaustible number of biographies. Practically every biography has dedicated at least a few lines to Einstein the violinist. But in Einstein's archive in Jerusalem Einstein the friend of music and musicians is still waiting to be discovered.

All quotes came from documents in the Albert Einstein Archive at the Hebrew University of Jerusalem. Most of them, at least extracts, have already been published. A few quotes were taken from so far unpublished letters and diaries.

Horst Bredekamp

Albert Einstein and the Avant-garde

When anniversaries have passed, fame usually tends to fade. But Einstein's greatness will not diminish, not even after the celebration of the hundredth anniversary of the *annus mirabilis* 1905, simply because his elevation represents the reverse side of an insurmountable crisis. Drawing on the insights of Copernicus, Darwin and of

Paul Klee, *Zimmerperspective mit Einwohnern*, watercolor, 1921, 24, oil and watercolor on paper

course his own, Freud spoke of the "mortification" felt by man at having been expelled from the heart of the universe, of creation and of rational thought. But these insights were accessible to the senses and thus comprehensible, despite their deeply controversial nature. The case is different with Einstein's formulae: they extend into spheres beyond the reach of the senses. They gain their validity from thought experiments, calculations and measurements, but it is futile to try and grasp

the fourth dimension of space and time with the apparatus of the human senses. The fact that the photograph of Einstein sticking out his tongue turned him into a 20th century pop icon is partly an antiauthoritarian exoneration of the physicist whose most successful findings put a seal on the incomprehensibility of nature. Any veneration of Einstein thus embraces an element of helpless protest. His results are both great and unbearable, and this conflict will remain as long as human imagination remains tethered to the three dimensions of the world of haptic experience.

From time immemorial art has been burdened with the hope of offering a way out from the inescapability of physical laws, yet without betraying their stringency and beauty. This includes the notion that the imaginative power of art anticipated and transcended major achievements in the natural sciences. The prime example cited is one of the greatest achievements of old Europe, the central perspective. Artists well-versed in mathematics, such as Leon Battista Alberti and Piero della Francesca, were responsible for discovering the underlying rules. Again it was artists who felt confined by this type of spatial construction and skilfully transformed the existing possibilities. This critique was already beginning to emerge in the works of Sandro Botticelli, a master of the central perspective, and it continued for centuries before leading to Giovanni Battista Piranesi and cubism and the subsequent rejection of Euclidean spaces. Artists seemed to have discovered the relevant sphere of thought long before mathematicians such as Bernhard Riemann calculated n^{th} dimensions. In view of the majesty of Einstein's new image of the cosmos it was quite liberating to see the basic problem had been addressed seemingly independently – in the symbolist assumption of a fourth dimension and in the futuristic and cubist approaches of simultaneously capturing a depicted object from several perspectives. The extraordinary success of Sigfried Giedion's book *Space,*

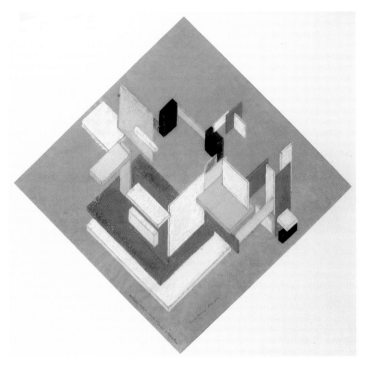

Theo van Doesburg, *Farbkonstruktion in der vierten Dimension von Raum-Zeit*, 1924, Indian ink and gouache

Time and Architecture, published in 1941, worked on this assumption. He saw a superior validity in the artistic avant-garde. Bound to cold, rational rules, the results of the natural sciences were deficient, despite their greatness. In contrast, artists who had both precise calculation as well as intuitive emotion at their command represented the all-embracing complexity of human faculties. Einstein reacted totally negatively to such interpretations and when Erich Mendelsohn, the architect of the Einstein Tower in Potsdam, sent him a copy of Giedion's book, he reacted with a sarcastic little rhyme: "It's not hard to utter something new/ When any nonsense you dare to do/ For much rarer is the season/ When the new is born of reason." At the end of the forties Paul M. Laporte presented him with his more cautiously formulated thesis on the mutual roots of cubist painting and the theory of relativity. This time Einstein's reaction was less abrasive, but still negative. But Einstein's reserve was an expression of admiration rather than underestimation. He reckoned that art is anchored in the unconscious with a similar precision

to that of science, but that science follows an image-free logic: "In science, the principle of order which creates units is achieved through logical connection while, in art, the principle of order is anchored in the unconscious". Einstein returned the verdict that Giedion had levelled as a prejudice against science, and in turn replied with a stereotype levelled against art. Despite his admiration for Einstein, Laporte refused to be deterred from the idea that, although cubist painting and the theory of relativity could not be placed in a direct relationship with each other, it could nevertheless be assumed that they both emerged from the same culture which questioned fundamental certainties including the three dimensionality of space.

In 1983 art historian Linda D. Henderson applied the precision of an Occam's Razor in her work *The Fourth Dimension and Non-Euclidean Geometry in Modern Art* in an attempt to confirm Einstein's view. She concentrated particularly on France and discovered that at least Puteaux Cubists, such as Jean Metzinger, were not stimulated by the theory of relativity but by mathematical extensions of Euclidean spaces. Representatives of this direction in art looked up to Henri Poincaré, not Einstein. His influence on the work of Marcel Duchamp has been demonstrated in numerous studies by Herbert Molderings.

One problematic aspect of Henderson's prize-winning study is that she more or less made a detour around the German-speaking and Dutch avant-garde. That meant that the group surrounding the astrophysicist Erwin Freundlich was excluded from her scope of vision. It was he who developed and translated into practice the idea of the Einstein Tower in Potsdam. Starting in 1906 the related artist, Otto Freundlich, probed the fourth dimension as time based on the special theory of relativity. He developed a two-dimensional kind of painting that differed greatly from cubism and identified space as a field in which the time dimension could unfold. Freundlich was not aiming at a visualization of the theory of relativity. He was in fact searching for an independent reflection of the questions that accompanied its arrival in the world.

The same applies to Erich Mendelsohn, who was a friend of Otto Freundlich, and had also concerned himself with the theory of relativity before Einstein's triumphant confirmation of 1919. He already designed the essential features of the Einstein Tower as a soldier during World War I. The success of this observatory can be seen as a quintessence of the whole problematic. It is regarded as one of the most significant structures in twentieth century architecture, not as a translation of Einstein's astrophysics, but because the latter inspired Mendelsohn to create forms expressing their inherent purpose.

Paul Klee experimented in painting with heightened detachment but admitted that "the notion of flowing space with the fourth imaginary dimension 'time' results in no greater clarity". However, it was precisely because of this insight that his artistic forays resulted in extraordinary examples of a highly individual imagination. Klee had contact with physics professor Felix Auerbach in Jena who had presented the theory of relativity in two works for a non-specialized audience in 1906 and 1921. Reminiscent of Edwin Abbott Abbott's satire *Flatland*, Auerbach imagined a realm of flat shadowy beings that were only able to experience two-dimensional flatness. When penetrating a sphere these beings would only be able to perceive at first a point and then a flat area extending to a semicircle. This would subsequently shrink again until it became a dot and then finally vanished. The shadowy beings would have experienced the sphere as a flat entity that moved in time. He compellingly concluded that, in analogy to this, three-dimensional human perception is equally limited in its ability to access the four-dimensional space-time zone.

Klee not only included Auerbach's ideas in his *Pedagogical Sketchbook* of 1925, he also transformed the shadowy beings into the two poetic watercolours *Zimmerperspective mit Einwohnern* ("Room perspective with inhabitants") of 1921 and *Nichtcomponiertes im Raum* ("Non-composition in space") of 1929. The flat matchstick beings are introduced into the confining two-dimensional spaces where they dissolve either into anamorphic entanglements or dynamic vanishing points. The idea of trying to make something that is incognitive understandable led to the creation of outstandingly original works of art.

It was the art historian Ulrich Müller who discovered that Auerbach acted as a mediator in transferring the theory of relativity to the Bauhaus teachers. He gained insight into the concept of "flowing space" from the discussion between Auerbach, Walter Gropius, Ludwig Mies van der Rohe and Theo van Doesburg. Van Doesburg referred to the theory of relativity in order to substantiate his utopian concept of a general revolution of time and space. His auratic construction drawings were an immediate expression of this. He depicted them in axonometric perspectives to avoid restricting the view to one aspect and allowing the layered walls to seemingly float freely in space.

This idea competed with the thoughts of Gropius and Mies van der Rohe about developing the dynamics of the moving observer and the metamorphotic presence of architecture. It was in this climate that the buildings emerged which overcame the system of supports and load-bearing elements and created interlocking wall areas to attain architectural semantics of suspension and movement that incorporated time within space. This all goes to show that it is pointless to reproach visualizations of a theory which is def ined as impossible to visualize and say they reached no satisfactory solutions. Despite the inevitability of this conclusion, it immediately bounces back

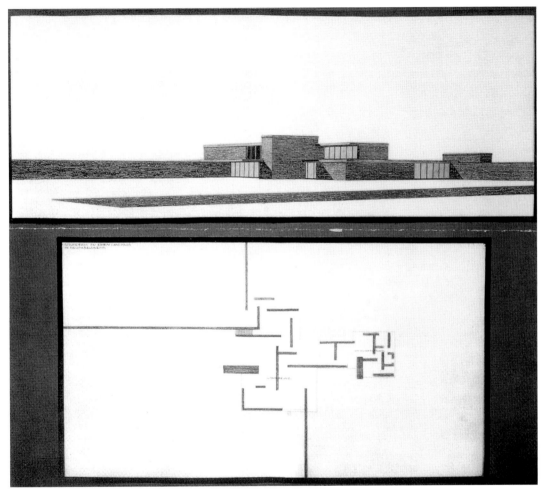

Ludwig Mies van der Rohe, *Country house in brick, elevation and ground plan*, 1923/24, photograph, Städtische Kunsthalle Mannheim

because people instinctively demand that every theory should be rendered in visual terms. The best attempts to come to terms with the fourth dimension as spacetime are well aware that it is not the theory that can be grasped, but the impossibility of its concrete representation. This is not sophistry. It is more a reflection on the morification mentioned at the beginning. After being freed from the idea that art enjoys supremacy over science, as Giedion maintained in his rousing essay, and freed of the expectation that art could provide a successful translation of the theory of relativity, the independent qualities of art become visible: qualities that resulted from reflections on a worldview that can be understood but still cannot be grasped as it remains intangible.

Dieter Hoffmann

Einstein's Berlin

"Berlin is the city to which I am most attached by personal and scientific relationships." Einstein wrote this to the Prussian Minister of Culture and Science Conrad Haenisch in the spring of 1920. This assertion was hardly just flowery language, and is confirmed not least by the fact that Einstein did not turn his back on the city until it fell under the flag of the National Socialist dictatorship – although there certainly had been enough occasions in the years before to leave Berlin and emigrate from Germany. So what were these human and scientific relationships that bound Einstein to Berlin so closely for nearly two decades?

The Family

When Einstein moved to Berlin in the spring of 1914, it was not noble science alone that brought him to Berlin – here, as in so many other cases, a woman added a strong force of her own to the academic enticements. It was his cousin Elsa Löwenthal, with whom he had become reacquainted during his first visit to Berlin in the April of 1912, and whom he had come to revere as more than just a relative. Elsa was not the only family member who lived in Berlin at the time, either. Upon his first visit he had found accommodation with his uncle, Jakob Koch, his mother's brother, at Wilmersdorfer Strasse 93 in Charlottenburg. Uncle Jakob had lived in the imperial capital since 1901 and earned his living as a merchant. Another relationship the Einstein family had to Berlin was that Albert's sister Maja had studied Roman languages at Berlin University in 1906/07. Since women were not yet admitted to university in Prussia at the time, these studies were actually illegal, and only possible due to her tolerant teacher Adolf Tobler. In addition to the Koch family, another part of the paternal branch of the family had moved back to Berlin at the turn of the century – hence Rudolf Einstein, a cousin of Einstein's father, is also listed as a merchant in the Berlin address register since

1909. In 1896, his daughter Elsa married the Berliner Max Löwenthal, with whom she had her daughters Ilse and Margot. After their divorce, Elsa then moved to Haberlandstrasse 5, in the bourgeois Bayerisches Quarter of Berlin-Schöneberg, where her parents lived and which also became Einstein's address from September 1917 – two years before his relationship to his cousin became "legalized" through marriage. Before that Einstein had occupied a third-floor apartment at

Albert Einstein in his study, around 1920

Wittelsbacherstrasse 13 in Berlin-Wilmersdorf – "completely secluded and yet not lonely thanks to a female cousin's care," as he wrote to his friend Zangger. In his first apartment, at Ehrenbergstrasse 33 in Berlin-Dahlem, he had been able to enjoy the solicitude of Mileva and the children for a few more weeks more. After his marriage failed definitively and Mileva returned to Switzerland with their sons, in December 1914 he gave up the apartment, which had now become too large for him alone, and moved to Wilmersdorf.

Albert Einstein
in the library on
Haberlandstrasse,
around 1929

Library commemorates this highlight of Berlin science history. Otherwise, however, Einstein had a rather ironic, distanced relationship to contemporary academic life, reporting to his friends about the academy sessions that "most of the members restrict themselves to displaying a peacock-like grandiosity; otherwise they're quite human [...]" Part of the "appointment package" with which Einstein had been lured to Berlin in 1914 was a professorship at the Friedrich Wilhelm University, today's Humboldt University – complete with all academic rights, yet without the obligation to lecture regularly. Einstein exercised his right to lecture more or less on a regular basis only at the beginning of his residence

Albert Einstein in the library on Haberlandstrasse, around 1929

Workplaces

Another factor contributing to this move was certainly the fact that the plans to construct a new building for a Kaiser Wilhelm Institute for Physics were suspended after the outbreak of war, so that it no longer made sense to live in Dahlem, the domicile of the Kaiser Wilhelm Institute. Thus Einstein gave up not only his apartment, but also his office in Haber's Kaiser Wilhelm Institute for Physical Chemistry and Electrochemistry (Faradayweg 4–6); from that point on his residential address also served as his office. His actual employer was the Prussian Academy of Sciences, but as this was a scholarly society without its own laboratories and institutes, its members met only for regular lectures in the Academy's official rooms on the street wing of the State Library on Unter den Linden. The main purpose of these meetings was for the Academy members to present their research – including Einstein, who reported on the 4th of November 1915 about the conclusion of his work on the general theory of relativity. Today a plaque in the passageway of the State

in Berlin, yet he also used the university's lecture halls to report about his work in public lectures. Sometimes so many people came to these lectures that only the *auditorium maximum* was large enough to seat everyone – a plaque in the vestibule of what is today the university cinema (at the

Prussian Academy of Sciences on Unter den Linden, photographed after 1945

Dorotheenstrasse entrance) serves as a memorial to Einstein's work in Berlin. Einstein did not only lecture about his research to students and academic circles, however; as early as June 1915 he lectured in the Archenhold Observatory, a public observatory founded in Berlin-Treptow (Alt-Trep-tow 1) in 1896, on "The Relativity of Motion and Gravitation" – here, too a memorial plaque was installed on the occasion of the Einstein centenary in 1979. Later, in the 1920s, Einstein also addressed the Marxist Worker's School MASCH – speaking in a community school in the north of Berlin for the opening ceremony of the academic year about "What the Worker Must Know About the Theory of Relativity."

Certainly one of the places in Berlin (and its surroundings) where Einstein left his mark is the Einstein Tower on the Telegraphenberg in Potsdam. After a British expedition to observe a solar eclipse in spring 1919 had confirmed the deflection of light by gravity, one of the consequences of the general theory of relativity, Einstein suddenly became world famous, there were efforts in Germany, too, to support relevant research. With state sponsorship and funds from what was known as the "Einstein Foundation," the tower was built by the architect Erich Mendelssohn and inaugurated in the winter of 1924/25. The Einstein Tower and its creator soon became hallmarks of modern, expressionist architecture. The tower did not fulfill its scientific expectations, however, as it did not succeed in providing evidence for the spectral red shift in the field of gravity, yet even today the tower is an internationally renowned site of experimentation in solar physics.

But Einstein was not only a gifted thinker and theoretical physicist; he also cultivated a great love of experimentation and was a passionate inventor. He pursued both activities in Berlin, too – immediately after his arrival he took advantage of the excellent experimental research conditions of the Imperial Institute for Physics and Technology

(PTR) in Berlin-Charlottenburg (Abbestrasse 2–12) as a guest employee, and in the winter of 1914/15, together with the Dutch physicist Wander Johannes de Haas, succeeded in proving the existence of Amperé's molecular currents (the Einstein-de Haas effect). A highlight of his work as an inventor was the a refrigerator he and the Hungarian physicist Leo Szilard developed for industrial production at the research institute of AEG in Berlin-Reinickendorf (Holländerstrasse 31–34) in the mid-1920s. The Einstein-Szilard refrigerator never made it to serial production, however; only a few prototypes were built.

In other respects, too, Einstein was by no means a removed scholar detached from the world. In addition to his activities at the Academy and his duties as its director, he also took on other leading science duties in Berlin – for example, he was a member of the board of trustees of PTR and the

Astrophysical Institute in Potsdam. During World War I, from 1916 to 1918, he even took over the office of Chairman of the German Physics Society. Since the Society's offices were located in the Physics Institute of the university on Reichstagsufer at the time (the institute was totally destroyed in World War II and is now the location of the capital studio of German's first state television network, ARD-Hauptstadtstudio), Einstein's work is also linked with this world-renowned institute – in great part because he participated regularly in the famous Wednesday Colloquia of the Society and held several lectures there. Furthermore, in the June of 1929 he was awarded the Society's highest honor, the newly established Max Planck medal. This award was not presented at the Physics Institute, however, but at Harnack-Haus in Dahlem, where the Golden Anniversary of Planck's Doctorate was celebrated, the occasion for establishing the medal.

Political Activities

Not only Einstein's achievement as a physicist was exceptional in comparison with that of his academic colleagues, but also his political behavior, and it, too, left its traces in Berlin. Even during World War I he did not join in the general war craze, but spoke out for a rapid end to the war and for international understanding as a signatory of the proclamation "To the Europeans"; he was also involved in the pacifistic *Bund Neues Vaterland* ("New Fatherland Alliance"), later to become the *Liga für Menschenrechte* ("League for Human Rights"), at whose events he occasionally appeared as a speaker. The meetings of the Bund, and later the Liga, were held at various venues in central Berlin. In 1923 he was also one of the founding members of the pro-Soviet "Society of Friends of New Russia," whose activities he supported up to his emigration – appearing at the Soviet Research Weeks in Berlin in 1927 and helping

to organize a lecture event the previous year, on the 2nd of November 1926, at which the Soviet mineralogist Alexander J. Fersman spoke at the *Preussisches Herrenhaus*, now the seat of Germany's federal council, the Bundesrat (Leipziger Strasse 3–4).

The first bearers of the Max Planck Medal, Max Planck and Albert Einstein, Berlin on 28 June 1929

Einstein's commitment to democracy and pacifism and his world fame, which set in at the beginning of the 1920s, also made him the object of campaigns of chauvinistic and anti-Semitic agitation. Besides this, the latter not only channeled the opposition to Einstein's fundamentally pacifistic and democratic views, but also became the reservoir of those physicists who regarded modern physics, and particularly Einstein's theory of relativity, as a degeneration of common sense and "typically Jewish deception." Among their spokesmen were the physicists Philipp Lenard and Johannes Stark, who, however, abandoned any basis for a serious scientific debate with their campaign. In the summer of 1920 these resentments provided fertile soil for an anti-Einstein event arranged with great public propaganda in the Berlin Philharmonic (the building was located at Bernburger Strasse 21–22a and was destroyed during World War II). Einstein himself participated in this meeting, taking a stand in a sarcastic newspaper commentary "About the Anti-Relativistic G.m.b.H." three days later, on the 27th of August. The Philharmonie was also the location of Einstein's final public lecture in Berlin – on the 16th of October 1932 he spoke there about the theory of relativity at the invitation of the Union of Jewish Student Associations and donated his lecture fee to a fund for needy students.

Under the impression of the considerable Jewish migration from eastern Europe, in Berlin Einstein again became more aware of his Jewish identity and devoted himself to Judaism. After pledging his efforts to the Zionist movement starting in 1920, even traveling to America with Chaim Weizmann in 1921 to solicit donations for the founding of a Hebrew University in Palestine, in 1924 he became a member of the Jewish Community in Berlin. Part of his pro-Jewish commitment that was marked more by solidarity than faith was a benefit concert he held for the Welfare Centre of the Jewish Community in the Great Synagogue, today's Centrum Judaicum in Berlin-Mitte (Oranienburger Strasse 28–30) on the 29th of January 1930.

Circle of Friends

Einstein was hardly a socialite in the classic sense, yet he was certainly integrated into Berlin's society life. His acquaintances included Walther Rathenau, Harry Graf Keßler, Gerhard Hauptmann, Graf Arco and other representatives of the social and political establishment of the Weimar Republic and he seems to have enjoyed being invited to the tables of high society. One of his closer friends was the medical doctor Janosch Plesch. Plesch was Einstein's personal physician,

but above and beyond medical consultations they also cultivated intensive and frequent association. Plesch's splendid residence in Gatow, known today as Villa Lemm (Rothenbücher Weg) even became an occasional retreat for Einstein, where he sought refuge from his family, but also from the many other turbulences bearing down upon him. This was the case in 1922 when the anti-Semitic and nationalist agitation against him reached such a climax that he was forced to feel personally threatened; he also spent his fiftieth birthday there hiding from the profusion of uninvited well-wishers and journalists. He is said to have usually resided in one of the garden pavilions on the lake, where he also played music, much to the enjoyment of passing boaters. He was even closer to his friend Moritz Katzenstein, who worked as a surgeon at Berlin's premier hospital, the Charité. Einstein also cultivated friendly relations with his physicist colleagues, especially with Max von Laue and his family. Laue, who lived in Zehlendorf (Albertinenstrasse 17), may have been his closest confidant, very close to him not only scientifically but also personally – he was also the only German scientist with whom Einstein corresponded regularly after his emigration. A fatherly friend was Max Planck, at the time the most important and influential physicist in Berlin. Planck once called Einstein his most important discovery, for it was he who recognized the young Einstein's physical genius at a very early date and it was primarily due to his initiatives and influence that Einstein was brought to Berlin in 1913, where he was granted excellent working conditions. Not only did Planck and Einstein often seek scientific counsel from each other, they were also united by their love for giving private concerts at home, to which Planck regularly invited guests to his home in Berlin's Grunewald (Wangenheimstrasse 21). Another place where Einstein indulged in his love for giving concerts was the home of the architect Erich Mendelsohn, who had built himself a home in Pichelsdorf (Am Rupenhorn 6) that was just as avant-garde as the Einstein Tower. Mendelssohn's wife was a trained concert cellist, which is said to have limited her pleasure in Einstein's company as a musician. Although not much is known as yet, Einstein also seems to have cultivated contacts to the mathematician and chess world champion Emanuel Lasker. Lasker lived "just around the corner" from the Einsteins (Aschaffenburger Strasse 6a), and Einstein wrote a warm message on the occasion of his sixtieth birthday as well as a very sympathetic foreword to the first biography of Lasker (1952). The two were also bound by their love for the surroundings of Mark Brandenburg, where both possessed summer homes, in Thyrow and Caputh, respectively.

In 1929, on the occasion of his fiftieth birthday, Einstein, who had already used a small plot in a garden colony in Berlin-Spandau back in the early 1920s and whose predilection for rural tranquility and seclusion were common knowledge, was supposed to be presented with a house in Neu-Kladow as a gift from the Berlin city government. However, there was such a great conflict about this gift in the city council and parliament that the whole issue became an embarrassment so that Einstein ultimately rejected it. Instead he bought himself a plot of land with a view of the Havel in Caputh near Potsdam (Am Waldrand 15–17) and had the young architect Konrad Wachsmann build him a summer home, which he moved into in late summer in 1929. He enjoyed the tranquility and isolation there for the next three summers, and also made it a meeting place for friends and prominent visitors such as the Indian writer Rabindranath Tagore or Heinrich Mann. "The sailing boat, the distant view, the solitary autumn, the relative tranquility" made Caputh "paradise" for Einstein. When he left Caputh in the autumn of 1932 to return to Haberlandstrasse before starting out on his annual research residence in America, he took leave forever: from Caputh, Berlin and from Germany as well.

Wolf-Dieter Mechler

Einstein's Residences in Berlin

Albert Einstein comes to Berlin for the first time in mid-April 1912. He has been a physics professor in Prague for one year, but has already been offered a chair at the ETH Zurich. In Berlin Einstein conducts not only conversations with his colleagues

A visit with consequences, for Elsa and Albert Einstein begin a closer relationship.

After his election to the Prussian Academy of Sciences in July 1913 and his confirmation by Wilhelm II in November, which he accepts in December 1913, Einstein must give up his residence in Zurich and find accommodation in Berlin for himself, his wife Mileva and their two sons Eduard and Hans Albert. With the vigorous support of their friends Fritz and Clara Haber, not Albert, but Mileva sets off on the housing search. In late 1913 the marriage is in a sort of chronic crisis; the move to Berlin will decide whether there is a chance of continuing the relationship.

The Habers lived in Dahlem, which, until the creation of greater Berlin in 1920, is an independent estate far outside the gates of the city, in the immediate vicinity of the Kaiser Wilhelm Society institutes being created at the time.

Above: Einstein's first apartment in Berlin was located at Ehrenberg-strasse 33; photograph from autumn 2004 (photo: Monika Gauer)

Right: Drawing of the façade of the building at Ehrenberg-strasse 33, 1910

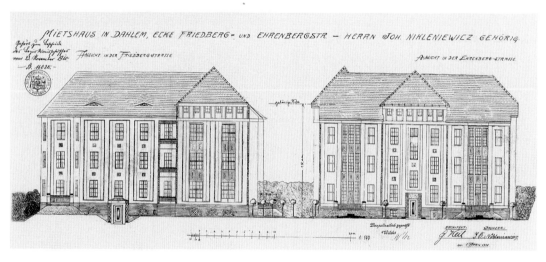

Planck, Nernst, Haber and Warburg, but also meets his uncle Rudolf Einstein and his daughter Elsa, formerly Löwenthal, his cousin. Both live at Haberlandstrasse 5, a large apartment building in the Bavarian Quarter in Schöneberg. This address will be linked with Albert Einstein all over the world for fifteen years, but initially he is just a visitor in the third and fourth floors there.

Thus it is no surprise that the Einsteins take up residence there as well. As of 1 April 1914, the start of Einstein's service at the Academy, Mileva rents an apartment at Ehrenbergstrasse 33, but we do not know on which story it was located. Einstein never saw this apartment before his arrival in Berlin, or, more precisely, in Dahlem. On 29 March 1914 he arrives in Berlin and stays

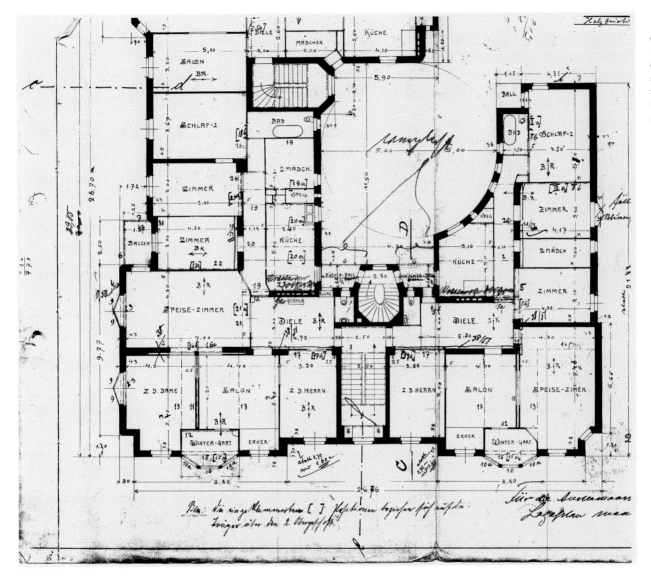

Ground plans of the apartments on the first and second floors of the building at Ehrenbergstrasse 33, 1910

for a few days with his uncle Jakob Koch at Wilmersdorfer Strasse 93 first, until he can move in to the new apartment. It is still being renovated, prompting Albert to the praise, "My new landlord is very respectable. He is arranging everything to look very nice." The landlord was the housepainter Johann R. Nikleniewicz of Lichterfelde, for whom the architect Gustav Keil had designed and constructed the building in 1910. The large, attractively renovated corner building is Albert Einstein's only residence in Berlin that is still preserved today.

Although Mileva rented the apartment, it is mid-April before she and their sons follow from Zurich. But family life never really gets off the ground in the new home; the incompatibility of the partners becomes exceedingly clear and the marriage breaks up for good. Mileva leaves Berlin with the children by late July 1914 and returns to Switzerland. Protracted, sometimes fierce confrontations follow, finally finding a conclusion of sorts with the divorce in February 1919. Einstein does not last long alone on Ehrenbergstrasse. In the meantime he has established a new relationship to his cousin Elsa, who still lives in the Bavarian Quarter with her two daughters Margot and Ilse. In order to be nearer to Elsa without compromising her position, in October 1914 Einstein moves from Dahlem to Wilmersdorf, to

Drawing of the façade of the building at Wittelsbacherstrasse 13, 1913

the building at Wittelsbacher Strasse 13. Einstein rents a seven-room apartment on the third floor of the simple apartment building built by the architect and contractor Franz Abbé in 1913/14. The building, which later belonged to Jewish owners and thus had to be sold at a low, government-dictated price in 1940, was destroyed in World War II. Einstein, who, by the way, has a telephone line both here and in Dahlem, remained resident at Wittelsbacher Strasse for almost three years.

In early 1917 he falls ill twice in quick succession, first with jaundice and then with stomach and intestinal ulcers. He finds the rest and care so urgently needed at his cousin Elsa's apartment. In December 1917 Einstein reports to his Swiss confidant Zangger, "Since the summer I have gained around four pounds thanks to Elsa's care. She cooks everything for me herself, as this has proven necessary. This is possible because I live in the apartment next door to hers (for the interim)." The

apartment used "only for the interim" is located on the fourth floor of Haberlandstrasse 5. Again, Einstein presumably had little to attend to in changing his living situation. In order to demonstrate the deepened connection to Elsa spatially while still keeping up appearances, the plan emerges for Einstein to rent the recently vacated apartment right next to Elsa's. His furniture must have been moved in on 1 September 1917, while Einstein was traveling. To his friend Michele Besso, he writes on 3 September: "I don't know yet whether I will be going anywhere with Elsa, that's up to her. Thus my address is Haberlandstrasse 5. The move appears to have been accomplished already." Although the second apartment was rented only on paper, the couple does not move in together officially until after Elsa and Albert marry on 2 June 1919. Today it is no longer possible to say exactly when Einstein gave up his own apartment, presumably as of 31 May 1919, for at the courthouse he listed a hotel as his residence. The building Haberlandstrasse 5, a large structure on the corner shared with Aschaffenburger Strasse 17, was erected in 1907/1908 by the architect and builder Otto Eisfelder. It had two staircases and two entrances: the representative one with an elevator on Haberlandstrasse and the service staircase with access from the Aschaffenburger Strasse side.

Now Albert Einstein lives with his wife Elsa, her two daughters Margot and Ilse and the maid Herta in the eight-room apartment Elsa has resided in for years. This apartment had windows to Aschaffenburger Strasse, while Einstein's identical interim lodgings from 1917 until 1919 were located on the Haberlandstrasse side. As was his famous study, which was constructed for him in the building's attic in February/March 1922. The "study" consisted of three rooms: 12 square meters for book storage, an 8.5 square-meter closet, and the actual study 17.5 square meters in size. In 1927, when building inspectors found out about the construction performed without a permit and de-manded that Einstein vacate his study on structural and hygienic grounds, Einstein stated in a "formal request for exemption" to the "Police Commissioner of Berlin" why a separate working area was necessary for him. "I had to overhaul the room and had a special entrance to it built from the stairway, so that I am able to devote myself to my studies outside my apartment, in which I am disturbed far too often. For this I have expended a considerable sum, which is of great consequence for a moderately paid civil servant – university professor." One of these expenses was the window, which used to slant along with the surface of the

roof, and had been enlarged and converted to a dormer window. The study and the book storage space had previously been partitions without dividing walls; the closet had been part of a larger

Einstein's first entry in the Berlin registry of 1915 has him listed at both apartments

Ground plans of the apartments on the third floor of Wittelsbacher Strasse 13, 1913

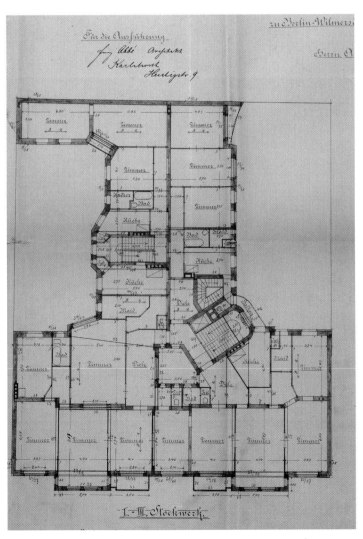

room, which presumably functioned as servants' quarters. This room was divided for Einstein and, like the other two rooms, enclosed with dividing walls. In this newly created refuge Einstein received scientist friends; here he dictated and discussed.

To the officials' reproach of insufficient hygienic facilities, as neither running water nor a toilet were present, he responds with Einsteinian irony and by staking his reputation: "The room is to be used by me personally, and not by others. A possible lack of hygienic could only have consequences for me; however it is certainly known to me that most people in Berlin have to work in much more adverse rooms. [...] As a noted scholar and teacher at the university I have a moral right to special consideration of this exemption regarding my study." Berlin police headquarters could not resist this argumentation, and thus in late July 1927, two months after the petition, Einstein is permitted to continue using the illegally converted study in the

Drawing of the façade of the building at Haberlandstrasse 5, 1907. The window to Einstein's study, before its expansion, is marked.

Ground plan of Einstein's study, photographed in 1927

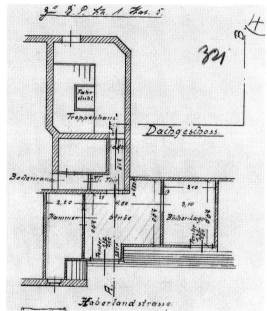

attic of Haberlandstrasse 5, "in consideration for the public interest in his research activity." for 3 Reichsmarks worth of administrative fees.

The study's furnishings were rather austere, with a desk, a bookcase, an armchair and two chairs, in contrast to the conjugal apartment, where a bourgeois, ornate Wilhelminian style predominated. No echoes of the Modern age can be discovered on any contemporary photographs.

The apartment building with the famous address Haberlandstrasse 5 was located in the "Bavarian Quarter," a bourgeois section of Schöneberg which was home to much of the Weimar Republic prominence. It also had a far greater share of Jewish residents than the rest of Berlin, earning it the occasional nickname "Jewish Switzerland." Among Einsteins' neighbors were the Rabbi Leo Baeck, the social psychologist Erich Fromm, the theater critic Alfred Kerr, the champion of women's education Alice Salomon and the photographer Gisèle Freund, to name just a few. Many Jewish residents of the Bavarian Quarter succeeded in fleeing Germany after 1933; over six thousand were deported to the extermination camps and murdered.

Albert and Elsa Einstein emigrated to the U.S. with her daughters Margot and Ilse. They moved out of the apartment on Haberlandstrasse; the furniture followed them into exile. All cash and stocks in Albert and Elsa Einstein's possession were confiscated by the Gestapo. Because of its Jewish namesake, Haberlandstrasse was renamed

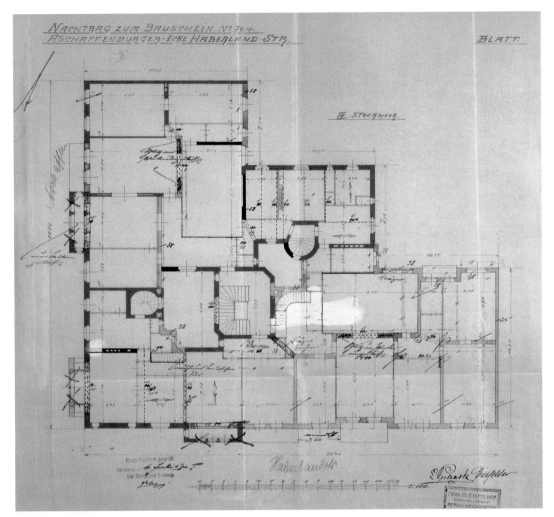

Ground plans of
the apartments
on the fourth floor
of the building at
Haberlandstrasse 5.
The couple's apart-
ment is on the
Aschaffenburger
Strasse side, 1907

Nördlinger Strasse in 1938. In 1996 its old name
was restored. Several buildings were destroyed
in World War II, building no. 5 among them.
Albert und Elsa Einstein set foot on Berlin soil for
the last time on 6 December 1932, when they start
their journey by train to Antwerp, where they set
sail to the U.S. on the ocean liner SS Belgenland.
On the ship during the return trip to Europe,
Einstein declares his resignation from the Academy of Sciences and thus from a Germany under
Nazi rule on 28 March 1933. Up to his death he
never returned to Germany again.

Erika Britzke

Einstein in Caputh

"Depth of thought does not thrive when surrounded by hustle and bustle. That's why life in the big city is no good for researchers and students."
(Albert Einstein 1924)
Albert Einstein had travelled the whole world and had held guest lectures at almost all of the European universities when, in 1929, he withdrew to the wooded seclusion of Caputh, not far from

of the rural community of Caputh. An additional piece of woodland had to be purchased to help the wooden house, which was quite generously designed by Berlin standards, to blend in with the local environment.
The architect, Konrad Wachsmann, proved himself a perfectionist in the planning, prefabrication and completion of the Einstein house. He struc-

The Einstein house in Caputh, Am Waldrand 15–17, around 1930 (photo: Konrad Wachsmann)

Berlin. A long-cherished wish for a plot of land near the River Havel finally came true. A summerhouse was built which met his needs for simplicity and functionality. But contrary to reports in the press and radio, it was not a present from the city of Berlin. Nor was it built, as originally planned, within Berlin, but on the outer perimeter

tured the project and set up the necessary supplier contracts in just three months. In a trial run he first had the house assembled in an assembly hall belonging to the company of Christoph & Unmack in Niesky. The parts were then transported to Caputh and assembled on the finished foundations. Both the outer form and the interior of the

Interior view,
vestibule and
staircase to
the upper floor,
downstairs, in
the background
Einstein study
(photo: Konrad
Wachsmann)

The Summerhouse and its Guests

"I find it simply splendid being in the new wood-
en house. Despite the fact that it's made me broke.
The sailing boat, the expansive views, the lonely
autumn walks, the relative peace, it is a paradise."
(Albert Einstein 1929)
Einstein was able to enjoy living and working in
peace in the charming Brandenburg landscape
between the late summer of 1929 and December
1932. He found relaxation in his walks through
the neighboring pine woods, sailing on the lakes

Margot and Ilse,
Einstein's
stepdaughters,
on the terrace

wooden house are designed in simple, clear lines.
Functional criteria and Einstein's personal wishes
were decisive in the overall design.
Konrad Wachsmann wrote in his memoirs of 1978,
how he reflected Einstein's wishes when designing
the house. "Mrs. Einstein found a suitable plot of
land fairly quickly. I went to look at it immediately. It was difficult to reach, but it lay in a peaceful
area, on a hill with a gentle slope, surrounded by
pine trees, and it provided generous views.
It was not far to the landing stage. Mrs. Einstein
then gave me a description of the house that
Albert Einstein would like. [...] The building material had to be brown stained wood, the windows
were to be white with a narrow metal ledge and
white wooden shutters. Like typical French windows that are relatively narrow but extend from
the floor to the ceiling. [...] Apart from this, there
had to be sufficient open and covered, wind-pro-
tected verandas, so that as much time as possible
could be spent outdoors in the fresh air."

Einstein and a colleague making the *Tümmler* ("dolphin") ready to sail

View of the back with stairs to the garden and terrace

Einstein with his son Hans Albert and grandson Bernhard on the terrace steps in Caputh
© The Hebrew University of Jerusalem, Albert Einstein Archives

of the River Havel, talking with friends or playing his violin. Einstein enjoyed life in the countryside together with his wife Elsa, his two step-daughters Margot and Ilse, son-in-law Rudolf Kayser, his assistant, the mathematician Dr. Walther Mayer, private secretary Helen Dukas, and their house-keeper Herta Schiefelbein.

There was neither telephone nor radio in the summerhouse. And the Einsteins did not own a car either. When traveling to Berlin Einstein took the mail bus from Caputh to Potsdam and then changed to the train. Occasionally friends sent a chauffeured car to pick him up. The immediate neighbors helped out if there were any urgent telephone messages.

In his summerhouse Einstein met with scientists, writers, artists, family members and many other people. The guest book has only partly survived, so that not all of the visitors are noted there. But many are also referred to in literature and documents or captured in photographs. For instance the physicist Arnold Sommerfeld wrote that he visited Einstein for the last time in summer 1930. Photographs show Einstein with the Bengali poet Rabindranath Tagore, with physicists, relatives and other guests in Caputh. It is more than likely that the renowned physicists Max Planck, Walther Nernst, Max von Laue, Erwin Schrödinger, Leo Szilard, Max Born, Paul Langevin, Felix Ehrenhaft

Einstein and Erwin Schrödinger in Caputh

Einstein's favorite 50th birthday present: The Tümmler

and the well-known chemists Fritz Haber and Chaim Weizmann were guests in the summerhouse. Famous guests from the contemporary cultural sphere included Heinrich Mann, Käthe Kollwitz, Erich Kleiber, Max Liebermann, Arnold Zweig, Alfred Kerr and Gerhart Hauptmann. In 1932 Abraham Flexner came from the USA to visit Einstein.

Other visitors to the summerhouse included representatives from the peace movement, who asked him for his support. In the summer of 1931 the pacifist theologian J. B. T. Hugenholtz from Holland held conversations with Einstein with the aim of founding an international peace center in Den Haag. The young writer Anna Seghers visited Einstein to persuade him to give a talk to workers about the theory of relativity. The writers Heinrich Mann and Arnold Zweig asked him for signatures for mutual public statements. In the summer of 1932, Einstein signed an appeal together with Heinrich Mann and Anna Seghers warning of the "terrible danger of increasing fascist trends" and calling on the major workers' parties to join forces in the *Reichstag* elections. This was probably Einstein's last political act in Germany.

Einstein sent many letters from Caputh to a great variety of countries. He proved himself a European and a world citizen. Mail went to George Bernard Shaw, Romain Rolland, Henri Barbusse,

Sigmund Freud, Maxim Gorki, Niels Bohr, Maurice Solovine, Queen Elisabeth of Belgium, Count Harry Kessler, Robert A. Millikan, Philipp Frank, Marie Curie, Max Born, to his sons in Switzerland and many other people. In his letter to Sigmund Freud on 30 July 1932 Einstein asked: "Is there a way of freeing human beings from the disaster of war?" And in greetings to Maxim Gorky marking his sixty-fifth birthday on 29 September 1932 he

from Potsdam-Babelsberg, described a sociable get-together at the Einstein summerhouse. The gathering included Einstein's youngest son Eduard and the couple from next door, Elsbeth and Adolf Stern. The Sterns and Einsteins knew each other from Berlin. It was from them that the Einsteins bought the first piece of land in Caputh, and the two couples owned a mutual right of way to the road, Waldstrasse. The Einstein family's address

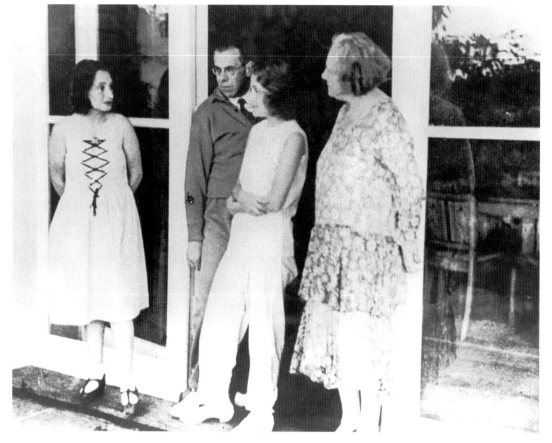

Elsa Einstein with her daughters Ilse and Margot and son-in-law Rudolf Kayser in front of the terrace door

wrote: "May your works have an ennobling influence on humanity, no matter what shapes political organization may assume. The decisive factor in destiny will always remain what the individual feels, wants and does."

The summerhouse also acted as a magnet to the large circle of relatives. Einstein's younger sister Maja visited her brother on several occasions. In his diary in October 1930 their uncle, Rudolf Moos

was Waldstrasse 7 (now Am Waldrand 15–17), and access was otherwise only possible via a narrow forest track.

The mainstay in the household was the housekeeper, Herta. She narrates that the eldest son Hans Albert arrived in Caputh on a motorbike in the early summer of 1932. Snuggled up in blankets in the sidecar was two-year-old grandson, Bernhard, who now lives near Berne.

Einstein at the window of his study (photo: Edda Reinhardt)

Children from the Jewish children's home also tell of their visits to their famous neighbour. Herta was also allowed to invite her mother and her cousin to Caputh, and the two of them were warmly welcomed by the Einsteins.

The lifestyle in Caputh, including the food, was simple. Einstein loved it when the mushrooms he had gathered in the autumn woods were cooked for their meal.

Relationships with the neighboring families were friendly. Mrs. Einstein bought fruit and vegetables, such as runner beans, tomatoes, strawberries or asparagus, over the garden fence.

In November 1932 the Einsteins bought the neighboring plot with the little garden house, so that his sons from Switzerland or other guests could stay overnight. But this dream never came true.

The Einstein House as an Institute

"Mr. Einstein is not only one of many outstanding physicists, Mr. Einstein is the physicist whose works published at our Academy have deepened physical insights in our century, and the significance of this can only be compared with the achievements of Johannes Kepler and Isaac Newton." (Max Planck 1933)

As Einstein worked scientifically at home, his summerhouse was also his institute between 1929

View of Lake Templin from the Einstein house

and 1932. The Kaiser Wilhelm Institute of Physics, of which Einstein was the director, had two other employees, the Viennese mathematician Dr. Walther Mayer and the private secretary Helen Dukas. When foreign guests were visiting, Mrs. Elsa Einstein would be the interpreter.

Einstein's study-cum-bedroom acted as the director's office. Helen Dukas, who had worked for Einstein since 1928, usually came to Caputh twice a week. A small table with a typewriter and chair were then placed in the study for her next to the bookcase. From Caputh Einstein corresponded with numerous colleagues addressing scientific questions, for instance the French mathematician

Elie Cartan and the Danish physicist Niels Bohr. The many special imprints from the meeting reports of the Prussian Academy of Sciences document shows just how much work was done in Caputh. Here are a few examples:

1. On the unified field theory (1929)
2. Two strict statistical solutions of the field equations of the unified field theory (1930)
3. Unified theory of gravitation and electricity (1931)
4. Semi-vectors and spinors (1932)

Einstein's scientific and private life often intermingled in his writings. For example, one page is covered with formulae plus the comment:

"I'm so looking forward to Caputh and the independent life there."

Conversations with trusted colleagues, such as Max Planck and Max von Laue, concentrated on new scientific topics and decisions about the work of the Prussian Academy of Sciences. Numerous letters and documents bear witness to this.

The first entry in the Einstein summerhouse guest book dated 4 March 1930 and the last entry dated 29 November 1932 both bear the signature of Max von Laue. Von Laue was the first German scientist to visit the young Einstein in 1907 while he was still working at the Swiss Patent Office in Bern. In 1910 he wrote the first book about the theory of relativity. On his final visit to Caputh he asked Einstein for a contribution for the 70th birthday of Arnold Berliner, the publisher of the scientific journal *Die Naturwissenschaften*.

In one photo Erwin Schrödinger can be seen leisurely talking with Einstein next to the bookcase in the study. In another one Schrödinger and Einstein are shown on Einstein's sailing boat. On the reverse side of a small photo for Antonina Vallentin's book *Le Drame d'Albert Einstein* Elsa Einstein noted in 1931: "Albert as a little rascal." The combined refrigerator patent by Leo Szilard and Einstein was also formulated, together with Szilard, in the summerhouse.

Members of the board of trustees of the Einstein Tower on Potsdam's *Telegraphenberg* (Telegraph Hill) also met with Einstein in the summerhouse. The last woman to visit the Einsteins in Caputh was master violin maker Margarete Kielow from Potsdam, who came on 6 December 1932. Next to the cases that were already packed for the journey to Pasadena, Einstein spent an hour playing and assessing a violin that had been crafted

by master violin maker Erich Kielow. This assessment is most likely the last document to be handwritten by Einstein in Germany.

The ship's passage to Pasadena, from which Einstein was never to return to Germany, began on 10 December 1932 in Antwerp.

The Einsteins never saw their summer paradise again.

"Einstein, the little rascal" – the dedication on the back of this picture – with guests on the garden terrace, around 1930

Britta Scheideler

Democrat with an Elitist Self-Image:
Albert Einstein Between 1914 and 1933

Almost everyone who was anyone in the world of science and art in the German Reich, including Max Planck, Conrad Röntgen and Fritz Haber, was among the ninety-three signatories of the proclamation "To the World of Culture!" in October 1914. In this proclamation they legitimated the freshly launched First World War and German militarism with the need to protect German culture, provoking great dismay internationally. Shortly thereafter, the "Appeal to the Europeans" was composed as a German counter-manifesto, binding scientists and and Romain Rolland. In the words of the Einstein biographer Albrecht Fölsing, Albert Einstein was "unique among natural scientists […] as a politically motivated intellectual." On such occasions, Einstein's enormous popularity from 1919 on lent very special weight to both his voice and his membership – as Einstein mocked himself – "as bigwig of renown" – in numerous pacifistic associations supporting the republic or pacifistic goals, such as the *Liga für Menschenrechte* ("League for Human Rights"). Besides emotional pacifism, the

Revolution in Berlin. Speaker on the ramp of the Kronprinzenpalais, one hour after the declaration of the Republic on 9 November 1918

artists to a "common world culture" and a commitment for peace and international understanding. However, this proclamation was signed by not even a handful of individuals, among them a physicist still unknown to the broader public, Albert Einstein. Among his colleagues in the natural sciences, Einstein also stood largely alone in his support for the revolution of 1918 and as an active defender of the democratic Weimar Republic. In countless public appeals Einstein took a stand for pacifism and democracy, often together with writers and artists oriented toward the political left, like Alfred Döblin, Heinrich Mann, Käthe Kollwitz, Erich Mühsam, Carl von Ossietzky, Henri Barbusse philosophical core of this extraordinary involvement can be summarized in the conviction Einstein so often expressed, the belief in the "human personality as the highest human value" and in the dignity, in the equivalence in principle and in the right of every individual to self-determination. He saw these threatened long before the National Socialists took power in 1933, by militarism, nationalism and every kind of dictatorship: "My political goal is democracy. Everyone should be respected as a person and none idolized." At the same time, however, Einstein distinguished between the "common rabble" and the "vulgar masses" on the one hand and the "intellectually

and morally superior part of the nations" on the other, complaining that the "intellectual and moral elite [exerts] no direct influence on the history of the world today." Another striking contrast to the image of Einstein as the defender of democracy and as a politically motivated intellectual is presented by the retrospective assessment of the physicist Werner Heisenberg in 1955, that Einstein's commitment should not actually be designated as political, but rather as moral and even quite naive: "Through his adolescence he was linked with the 19th-century belief in progress, and his essays reflect an image of the world that [...] could always become better if people were ready to relinquish their previous prejudices and rely on their reason. Despite negative experiences Einstein was not willing to abandon this ideal. In the political sphere this attitude was manifested in an almost naive belief in the possibility to resolve political problems through good will alone."

Considering Einstein's preeminent patterns of explaining social developments (and aberrations) and his proposals for their resolution, it seems that Heisenberg was not all wrong. In his paper *My Opinion About the War*, published in 1916,

Einstein recommended treating greed for power and avarice as contemptuous vices, just like hate and contentiousness: "Every well-meaning person should work hard on himself and in his personal circle to improve in these respects. Then the weighty plagues that afflict us today will disappear as well." By "weighty plagues" Einstein meant nothing less than war, which he wanted to prevent in the future by improving the morals of each individual.

This fixation on the moral individual, neglecting all social and political circumstances, is also reflected in Einstein's appropriately titled essay of 1930 *As I See the World*. In this work he declares his allegiance to a late-idealistic ideal of personality, determined by "kindness, beauty and truth." As this ideal, as Einstein elaborates in 1932, includes "man's selfless, responsible devotion to the service of the community" and general welfare, the union of such individuals can only lead to a harmonious society. This would eliminate conflicts of interest, and thus automatically also the need for any politically institutionalized balance of interests. In this ideal conception of society, there was thus not really any room for politics, understood as the rational behavior of different actors to establish their respective values and interests. As long as he lived, Einstein remained true to his view expressed in 1933, "that the fate of the community is primarily determined by the level of its moral standards."

Of the 1918 revolution Einstein hoped that, instead of the power elites and interest cliques of the Kaiser's empire, now the "healthy will of the people" would decide Germany's fate. However, when the economic, social and political crisis led to fights over distribution, revolts, and renewed a militarism and nationalism starting in 1919, amounting to a failure of his ideal of a harmonious society, he no longer spoke of the "healthy will of the people," but rather of the "common rabble, that [was led] by mass suggestion" and

Albert Einstein (center) and the professors Langevin (France; right) and Smith (England) at an anti-war demonstration in Berlin in 1923

"hollow passions" and can be made into the "un-resisting tool" of interest groups. Einstein contrasted them with those who are "benevolent and prudent" and "intellectually and morally superior," and capable of acting in the interest of the general welfare. This fixation on the moral individual and a harmonious society had far-reaching conse-

Albert Einstein surrounded by journalists, New York, 1930

quences for Einstein's political thought and activity. For it was irreconcilable with the recognition of competing societal groups in a modern, pluralistic democracy. He thus did not recognize conflicts of interests or values as legitimate and directed into orderly political paths, but rather as morally depraved.

The fixation on the moral individual also corresponded with Einstein's proposals for the solution of social problems. They are aimed at the "ennoblement of man" through improved school education and the study of science and art. As Einstein

wrote to the Soviet writer Maxim Gorki in 1932, the education of the individual enjoyed higher priority than changing the social and political structures: "May your work continue to ennoble men, whatever form their political organization may take. Destiny will allways be decided by what the individual feels, wills and does." Since the mid-1920s, Einstein expected the solution to urgent problems, such as military disarmament, to emerge from the enlightenment and mobilization of the masses through the "intellectual leaders of all countries," among which he doubtlessly counted himself. Einstein shared this claim to leadership as an instance of moral orientation with most of the intellectuals of the Weimar Republic. Just as he did, they understood "education" to be above all the "formation of the personality," such that the concept of the "spiritual leader" covered outstanding intellectual and moral capabilities. Just as widespread in left-intellectual circles of the Weimar Republic was the high assessment of the individual creative personality, which was believed to be threatened by industrial mass society. This corresponded to a deprecation of the "masses," which Einstein expressed in 1930 in his conviction that only the "creative, independently thinking and judging personality" creates values for the community, while the "herd as such remains dull in thinking and dull in feeling." This elitist pattern of thinking comes to bear when he writes to the psychologist Sigmund Freud in the early 1930s, "Political leaders and governments have their position thanks in part to violence, in part to their election by the masses.

They cannot be regarded as the representation of the intellectually and morally superior part of the nations. However, the intellectual elite does not exert any direct influence on the history of nations today." Nevertheless, this intellectual elite was not supposed to exert any political power, but only "a significant and whole same moral influence on the solution of political problems by taking up the proper positions in the press." Characteristic of the ambivalence of elitist and democratic traits in Einstein's political thought is that, at around the same time, he emphatically called for the individual's right to self-determination and political emancipation. Thus in 1932 he protested against the restriction of freedom of the press through emergency decrees, as he claimed they degenerated the state into an "association of underlings". Einstein's own political involvement, which always avoided links to any political party, was marked by his doubt in the reconcilability of power and morals, by his fear of compromises, of corruption and the – not to be overlooked – consequences of political action, which kept him from taking on any direct political responsibility.

Einstein's conviction that it was the task of intellectuals to lead and educate, and to exert a healing moral influence on the resolution of political issues, remained unbroken even after 1933. He explained the National Socialist's coming to power in part through the masses' manipulability and determination by instincts, and on the other hand with the failure of the intellectual elite in their role as leaders: "Our representatives of science fail in their duty to defend moral and intel

lectual values [...]. This is the only reason why individuals of base and inferior nature have been able to seize power." However, Einstein found no explanation for the behavior of the "intellectual leaders," of whom, for instance, nearly all members of the Prussian Academy of Sciences swung onto the Nazi course in 1933 by approving Ein-

ALBERT EINSTEIN
MEIN WELTBILD

1934

QUERIDO VERLAG AMSTERDAM

Title page from Albert Einstein: *Mein Weltbild*, Querido Verlag, Amsterdam, 1934

stein's withdrawal from the Academy. Helplessly, he wrote from exile on 5 April 1933 to the Dutch astronomer Willem de Sitter: "The only curious thing is the utter failure of the so-called intellectual aristocracy."

Kenji Sugimoto

Einstein and Japan

At the age of 43, Albert Einstein was invited to Japan by Saheniko Yamamoto, the editor of Kaizosha Publishing House. He traveled around the country with his second wife Elsa from 17 November through 29 December 1922, visiting numerous imperial Universities and holding lectures about the principle of relativity. It was on the way to Japan that Albert Einstein learned he had been awarded the Nobel Prize for Physics.

Albert Einstein's Travel Route in Japan:

10 October (Sunday):
Departure on the Kitanomaru from Marseilles (France)
10 November (Friday):
News that he had been granted the Nobel Prize (still at sea)
17 November (Friday):
Arrival in Kobe with the Kitanomoru; Kyoto (Hotel Miyako)

18 November (Saturday):
Sightseeing in Kyoto, Einstein was in Sekigahara, at Hamana Lake, and saw Mt. Fuji from the train. Arrival in Tokyo: An enthusiastic reception greeted Einstein at Tokyo Station.
19 November (Sunday):
Keio University: First public Lecture on the special and general theories of relativity, interpreter: Professor Jun Ishiwara (Ishiwara had studied under Albert Einstein in Europe).
20 November (Monday):
Einstein was received warmly at the Imperial Academy. He held a welcome address. Einstein attended a Japanese theater.
Sightseeing in Asakusa with Morikatu Inagaki (interpreter)
21 November (Tuesday):
Einstein attended a Chrysanthemum festival. Einstein was received by the Empress at the Imperial Palace.

Elsa and Albert Einstein in Japan, 1922 (photo: Meiji Seihanjo)

22 November (Wednesday):
Visit to Kaizosha Publishing House
Temple Zoujou
Lunch with Sanehiko Yamamoto: Sukijaki
Meiji Shrine
23 November (Thursday):
Einstein listened to a concert of Japanese music
at the Tokyo Music School
24 November (Friday):
Stroll through the Ginza shopping area
Lunch: Tempura
Second public lecture in Kanda Seinenkaikan,
on space and time in the theory of relativity
25 November (Saturday):
University of Tokyo, first special lecture
Tour of a Japanese theater
Einstein enjoyed Geisha performance and Kyoto
cuisine at Hiranoya.
26 November (Sunday):
Noh Theater visit. Einstein was delighted.
Visit to a Japanese publishing house (Maruzen)
27 November (Monday):
University of Tokyo, second special lecture
Dinner with Yoshichika Tokugawa
28 November (Tuesday):
Reception at Hitotubashi University
University of Tokyo, third special lecture
29 November (Wednesday):
Waseda University
Ochanomizu Women's University
University of Tokyo, fourth special lecture
30 November (Thursday):
Visit to a concert of music of the Japanese court
University of Tokyo, fifth special lectures
debate on relativity theory: Uzumi Doi writes
Einstein two letters.
1 December (Friday):
University of Tokyo, sixth special lecture
Commemorative photograph (Einstein with
Japanese scientists)
Einstein received gift of an autograph album.
Einstein played violin at Hotel Kaiser

Einstein at
Nagoya train
station

2 December (Saturday):
Reception at the Technical University of Tokyo
Sendai
Meeting with Professor Molisch
3 December (Sunday):
Sendai: third public lecture, interpreter: Professor
Keichi Aichi
Sightseeing in Matuschima
Visit to the Zuiganji Temple
Poet Bansui Tuchi gave Einstein a poem
Einstein enjoyed the moonlight
Visit to Tohoku University
Einstein signed his name on the wall with a
calligraphy brush
4 December (Monday):
Trip to Nikko with the Japanese Cartoonist Ippei
Okamoto
Einstein stayed at Nikko Kanaya Hotel.

5 December (Tuesday):
Trip to Lake Tyuzenji
When Einstein fell in Nikko, he declared "That's gravitational activity."
6 December (Wednesday):
Sightseeing in Nikko
Return to Tokyo
7 December (Wednesday):
Journey from Tokyo to Nagoya
In the train, Einstein wrote his *Chat about My Impressions of Japan.*

10 December (Saturday): Fifth public lecture
Sightseeing tour of Kyoto
11 December (Sunday):
Journey from Kyoto to Osaka Osaka:
Sixth public lecture
12 December (Monday):
Visit to Nijou castle
Einstein laughs about the "Tiger" on a sliding door
13 December (Tuesday):
Seventh public lecture in Kobe
Return to Kyoto

Commemorative photo of Einstein with Japanese scientists in the garden of the University of Tokyo, 1 December 1922

8 December (Thursday):
Nagoya Castle
Einstein entertained by the Golden Dolphin
Fourth public lecture
9 December (Friday): Visit to Atsuta Shrine
Journey from Nagoya to Kyoto
Arrival in Kyoto (Hotel Miyako)

14 December (Wednesday):
Einstein held speech at a reception for students.
Toshima Araki held an address in good German
15 December (Thursday):
Sightseeing in Kyoto
16 December (Friday):
Trip to Lake Biwa.

17 December (Saturday):
Einstein played piano at Hotel Nara.
18 December (Sunday):
Visit to Todaiji Temple with its giant Buddha
statue, the Kasuga Shrine, Nara National Museum,
and Nara Park

23 December (Friday):
Journey from Miyajima to Moji (North Kyushu),
Einstein stayed at Mizui Club.
24 December (Saturday):
Hakata: Eighth and final public lecture
Interpreter: Jun Ishiwara

Albert and Elsa Einstein on the stage before a public lecture

19 December (Monday):
Sightseeing at Horyuji Temple
Journey from Osaka to Miyajima (Hiroshima)
20 December (Tuesday):
Arrival in Miyajima, Ikutushima Shrine, Mt. Misen
21 December (Wednesday):
Einstein took a walk on the beach in Miyajima
22 December (Thursday):
Einstein got food poisoning

Einstein stayed at a Japanese hotel (Sakaiya) and
slept in a traditional Japanese bed
25 December (Sunday):
Kyushu University
Einstein met Professor Hayari Miyake again (Miyake
had treated Einstein when he took ill on the ship)
Reunion with Professor Ayo Kuwaki, who as the
first Japanese physicist in Bern while Einstein was
living there
Christmas celebration with children in Moji

26 December (Monday):

Einstein played piano at the Mizui Club

Einstein climbed Mt. Otani

27 December (Tuesday):

Sightseeing by boat in Kanmon

28 December (Wednesday):

A reception at the Chamber of Commerce
and Industry.

29 December (Thursday):

Trip from Mizui Club to the wharf

Einstein wore a red headband and took part

Einstein's Impressions in Japan

Einstein on Japan

"It is a peculiar Japanese tradition not to express
one owns feeling and emotions, but to stay
calm and composed under all circumstances.
Herein may lay the deep sense of the Japanese
smile, which seems so mysterious to Europeans.
For a foreigner like me, it is not easy to look
into the Japanese soul.

Every detail has a sense and a meaning. Even
people, with their
picturesque smile,
bows, and how they
sit down – all things
one can admire but
not imitate. You
will attempt in vain,
stranger! – But
Japanese food won't
agree with you, it's
better to just watch."

Einstein on Nature

"Nature and man
seem to be unified,
constituting a unity
of style that does not
exist anywhere else.
Everything from this
country is delicate
and cheerful, not ab-
stractly metaphysical
but always closely

bound up with what has been given by nature."

Einstein on Japanese Music

It seems to me that Japanese Music is like a
painting with an unexpected emotional impact.
I have the impression that it is a stylized representation of human emotional expressions, or
like sounds of nature that evoke a reaction in
the human soul, like the birdsong or the ocean's
roar. Japanese music made the greatest impres-

Albert and Elsa
Einstein at Kyushu
University
in Fukuoka,
24 December 1922

in preparation of rice cakes

Einstein went on board the Hatanamaru

3:00 p.m.: departure from Moji to Palestine,
Spain and then Germany

30 December (Friday):

Einstein wrote a letter to poet Bansui Tuchi

sion on me when it served to accompany a play or a mute act (dance), especially in Noh Theater.

Einstein and the Atomic Bomb

Einstein loved Japan and the Japanese people. But 23 years later, the atomic bomb would define his further relationship with Japan. When Einstein heard from his secretary Helen Dukas the tragic

news, *Atom Bomb dropped on Japan*, Einstein could say no more than "Alas!" Later he admitted: "Had I known that the Germans were not working on an atomic weapon, I wouldn't have done anything for the bomb."

When Einstein met the first Japanese Nobel laureate in Princeton, Dr. Hideki Yukawa, he apologized with tears in his eyes. Hideki Yukawa became a signatory to the Russell-Einstein Manifesto against atomic armament and the threat of atomic war to mankind.

In the year 2005 we commemorate:
the 60th anniversary of the atomic bombing of Japan;
the 50th anniversary of the signing of the Russell-Einstein manifesto;
the 50th Anniversary of Albert Einstein's death.

We Japanese address the following message to the whole world:
Never again Hiroshima, never again Nagasaki!

Left:
Albert Einstein at a lecture at the University of Tokyo

Above:
Elsa and Albert Einstein in a Japanese house

Alfredo Tiomno Tolmasquim

Einstein's Journey to South America

Einstein traveled to South America in 1925, while he was at the height of his fame and was making trips both within Europe and further afield. After one of the proofs of the theory of relativity, the light deflection near the Sun, was announced in

Itinerary of Albert Einstein's South America Tour (Argentina, Uruguay, Brazil), entries from 22 and 27 March 1925 © The Hebrew University of Jerusalem, Albert Einstein Archives

1919, Einstein's name, which was already familiar in European scientific circles and in Germany at large, spread the world over, reaching people from countless nations, those of South America included. In this region, newspapers published long articles on the revolutionary theory, and only a few months later Latin American scientists started to publish books and articles on the theory of relativity.

Physics in Argentina was considered among the most advanced of the southern hemisphere, and the country's universities had invested to attract different foreign scientists as a means of consolidating the local scientific community. It was one of these scientists, Jorge Duclot, who took the lead in proposing to the University of Buenos Aires's Board of Governors that they invite Einstein to give a series of lectures, but it didn't go ahead. As it turned out, it was eventually the Jewish cultural institution, the Hebrew Association, which contacted Einstein's wife, Elsa, to find out how likely he was to accept. She replied that as a rule he only accepted invitations extended by scientific institutions to prevent his name being used for other causes against his will.

The Hebrew Association then contacted the University of Buenos Aires, suggesting that they made the invitation, and provided the institution with part of the funding required. Parallel to this, they made the same suggestion to the University of Montevideo, and told Isaiah Raffalovich, the rabbi from the Rio de Janeiro's Jewish community, to proceed likewise with the Brazilian university. Despite Einstein's caution, he was fully aware that the invitations he received had been coordinated by the local Jewish communities, and in the case of Rio de Janeiro, the invitation was even signed by the rabbi himself on behalf of the heads of the engineering and medicine faculties.

The journey was long: 20 days' voyage to Buenos Aires and a total of three months alone away from home. Elsa did not accompany her husband this time as she had done previously on other trips outside Europe.

Einstein had little hope of meeting important peers within the local scientific communities; his motivations were different. Firstly, due to his worldwide recognition, he felt somewhat responsible for mending the rift between scientists caused by the First World War and divulging the theory of relativity. He sometimes sardonically referred to himself as a "messenger of relativity." The second issue was his involvement in Jewish communities. At the time, he was actively involved in the construction of the Hebrew University. His visit to Palestine three years before had reinforced his support for Zionism and at the time he was particularly concerned about involving other scientists in the same cause. And finally, but no less importantly, he was curious to find out about other cultures. His previous trips to Japan, Palestine and the USA had opened his eyes to different realities from those he knew so well in the old world of Europe. What would a journey to these tropical nations in the southern hemisphere be like?

Einstein set off from the port of Hamburg on 5 March in cloudy weather. The voyage was an opportunity for him to retreat from the problems he was facing in Europe. He had the time and peace of mind to read, play the violin in an impromptu quartet put together on the ship and have long conversations with friends made on board. What he actually feared was the harassment which inevitably awaited him on his arrival.

His first contact with South America was positive. The ship docked in Rio de Janeiro for one day before sailing on to Argentina. Einstein was received

there by a welcome committee of journalists, scientists and representatives of the Jewish community. Despite the hullabaloo, especially in the press, he did not feel put out. Everything was new. He wrote about this time in Rio in his diary: "Botanical garden as well as flora in general surpass the dreams of a 1000 nights. Everything lives and grows, so to speak, before your eyes. The mix of people in the streets is exquisite. Portuguese – Indian – Negro, and all their transitions. Driven plant-like, subdued by heat. Wonderful experience. An indescribable abundance of impressions in a few hours."

But this enchantment was to be short lived. The crossing of the La Plata river from Montevideo in Uruguay to Buenos Aires in Argentina put paid to his hopes for three months of peace and quiet. He was awoken in the middle of the night by journalists seeking an interview with him. In the morning, his berth was invaded by a gaggle of journalists and photographers, and he eventually managed to transfer the interview venue outside, explaining that it was impossible to breathe with so many people inside his berth.

Einstein spent one month in Argentina. He gave a series of lectures on the theory of relativity, visited some Argentine institutions and universities and participated in a session at the National Academy of Exact Sciences. His daily movements were covered in the press, which even published transcripts of his talks. He also had close contact with the local Jewish community: he visited a number of institutions, gave a talk at the Hebrew Association and took part in a ceremony at a large theatre in the capital city celebrating the creation of the University of Jerusalem, which was founded while he was in

Einstein during a lecture in the University School of Buenos Aires, Argentina (*Colegio Nacional de Buenos Aires de Universidad de Buenos Aires*)

Argentina. The same could not, however, be said of the country's German community. On the eve of Einstein's arrival, a newspaper published his article entitled *Pan-Europe*, where he contrasted Europe's cultural unity with a small-minded patriotism which sought to limit intellectuals to the political boundaries of their States. To many, still marked by the recently ended war, this smacked

The Jewish Society takes its leave of Einstein at Sunchales station in Rosario, Santa Fe province, Argentina

of treachery. Even the Argentinian-German Cultural Institute, which had provided financial help to ensure the success of the trip, withdrew from the scheduled events. The community's contact with Einstein was limited to a reception offered by the German consul, at his residence.

After Argentina, Einstein went on to Montevideo, where he stayed for a week. He was impressed by this small country, which reminded him of Switzerland. Apart from the compactness of the capital city, which pleased him greatly, Uruguay had undergone a national modernization and reform process, with the introduction of the secret ballot, the establishment of a secular state providing religious freedom for all, and the setting up of a comprehensive welfare structure. Unlike in Argentina, Einstein had good contact with the

German community living in Uruguay. He was received at a pleasant afternoon tea at the German Club and took part in a reception put on by the German Consul in Uruguay.

Though his stay in Uruguay was pleasant enough, Einstein could no longer stand "swinging on the trapeze." He would have given anything to avoid his third port of call, Rio de Janeiro, but it was not in his power to do so. He was lonely, surrounded by unfamiliar faces, even if they were friendly and kind. He was also afflicted by the heat of early fall in the southern hemisphere. The notes in his diary of the journey on the way to Rio de Janeiro illustrate well his mood after forty days wandering around the continent. "It is becoming rather hot. In addition, rather heavy and badly cooked food. One sleeps badly [...] All the scientific ideas which came to me in Argentina are proving to be useless. Weather is bad to moderate."

When he reached Rio de Janeiro on 4 May there were people looking forward to meeting him. As always, there was a welcoming party to greet him, and his sole request was that there be plenty of free time for him to rest and enjoy the city. Even so, he was not relieved from giving two talks on relativity and participating in a number of formal engagements, such as meeting the country's president and going to a reception organized by the German Consul. In Brazil, he was also received with open arms by the German community, which offered a dinner in his honor at the German Club.

By the end of his journey, Einstein was "more dead than alive." In his view, there had been a lot of fuss but no real interest, with the saving grace of a few weeks' rest during the transatlantic voyage. He found the people far more concerned with superficial issues, like clothing, than with serious, important problems. Despite its madness and its politicians, Europe was still better and more appealing. One highlight was the enthusiasm with which he had been received by the Jewish community. He knew he was seen as a symbol of union and hoped his visit would help strengthen

man affairs, and everywhere he went he was considered a "wise German" and hardly any mention was made of his Swiss nationality. His visit helped encourage more interest in German culture than that of any other scientist could have. His simple,

Albert Einstein with engineering students in Montevideo

solidarity among Jews and support the project to construct the Hebrew University. But he swore to himself that he would never make another journey which did not have a strictly scientific purpose.

As for the German diplomats, Einstein's trip to South America was extremely important for Ger-

friendly and somewhat transcendental manner, as the Consul in Argentina described it, helped set up empathy with the local people. Einstein returned to Berlin on June 1. His next journey outside Europe only took place five years later, in 1930, when he spent a few weeks at the California Institute of Technology, in the U.S.A.

Circe Mary Silva da Silva

The Theory of Relativity in Brazil: Reception, Opposition and Public Interest

How can a normal expedition to observe a solar eclipse bring with it such important consequences for science that people still talk about it one hundred years later? In 1919 a scientific expedition returned with a result unexpected by science. Astronomers from Greenwich and Rio de Janeiro had organized this expedition with the goal of observing and photographing the solar eclipse expected on 29 May. The atmospheric conditions at the location of the observatory were especially good on this day, and a few months later the noteworthy report circulated – proclaimed by the Royal Society in London – that the results of the expedition were in keeping with Albert Einstein's theory of relativity. His name appeared in the largest headlines of daily newspapers in cities all over the world, and his name became famous even in Brazil.

In the same year the scientist Manoel Amoroso Costa wrote his first article about Einstein's theory in the daily *O Jornal*. However, this article would hardly have encountered the interest of a broad public.

The dissemination the theory of relativity experienced in the following period cannot be traced back to the mere fact of its applicability to certain phenomena, however, but primarily to the fact that it led to a new understanding of physics and brought new, fundamental concepts in its wake. The theory of relativity became the embodiment of a new image of science and Einstein's ideas a symbol of modernism.

University education does not enjoy a long tradition in Brazil. The first university was founded in the 1920s by uniting several faculties in Rio de Janeiro. The possibilities for research were especially limited. Only a minor portion of research was financed by the public sector, and the government did not formulate science policy. In such an intellectual milieu there were very few possibilities for research in the exact sciences. Scientific studies were concentrated in the field of medicine and the biological disciplines. A policy for the systematic promotion of the sciences was not formulated until the beginning of the 1950s.

Einstein's theory of relativity initially attracted little notice in this situation. It was first propagated in Brazil through lectures by visiting French mathematicians and physicists. Of particular importance in this connection were visits by Emile Borel (1922) and by Jacques Hadamard (1924), who had personal contact with Einstein and had already acquired deep sympathy for his theory. Moreover, Einstein's own visit in Rio de Janeiro in 1925 drew a great deal of attention to the theory of relativity and contributed to Einstein's reputation as an extraordinary scientist.

On 6 May 1925 Einstein held a public lecture at the Rio de Janeiro Engineers' Club entitled "Remarks on the Contemporary State of the Theory of Light," in which he reported about the difficulty of proving the quantum nature of light. However, he was not convinced that his audience had understood him, for he wrote in his diary:

"At 4 ½ first lecture in Eng. Club in crowded hall with street noise through open windows. Understanding already impossible due to the acoustics. Little scientific sense. Here I'm a kind of white elephant for the others, for me they are monkeys.

In the following year a visit by Marie Curie (1867–1934) reinforced interest in modern physics. At this time she was already famous and had already been awarded the Nobel Prize twice. In Brazil she was appointed a corresponding member of the Brazilian Academy of Sciences – a title the French Academy had yet to grant her. In 1928 the physicist Paul Langevin (1872–1946) also visited Brazil.

Manoel Amoroso Costa, mathematician, engineer and professor at the Polytechnic Academy of Rio de Janeiro, co-founder of the Brazilian Academy of Sciences. Held lectures and wrote the first book about the general theory of relativity in 1922

He was a friend of Einstein's and one of the most important people who contributed to the propagation of the theory of relativity in France.
The theory of relativity met with interest in Brazil, especially among instructors at the Polytechnic Academy and the Marine Academy in Rio de Janeiro und in São Paulo, who began to write scattered articles and books on this theory. The first was Manoel Amoroso Costa. In 1922 he held a series of lectures at the Polytechnic Academy and published an introduction to the theory of relativity as a book in the same year. This was the first book on the subject published in Brazil. It dealt with not only the general idea of the general theory of relativity, but also went into the mathematical details of the theory in its discussion of the tensor calculation and of cosmology.
French scientists played a special role among the scientists whose work gained influence in Brazil. This role had its roots in the importance of the French school system and the French system of universal education for the development of education in Brazil. This influence also explains the occasional resistance to the theory of relativity that developed in Brazil. A movement of anti-relativists and Einstein opponents existed in France as in Germany. In Brazil the situation was similar. For instance, Licinio Athanasio Cardoso, a professor at the Polytechnic Academy of Rio de Janeiro, was a known opponent of Einstein's theory.

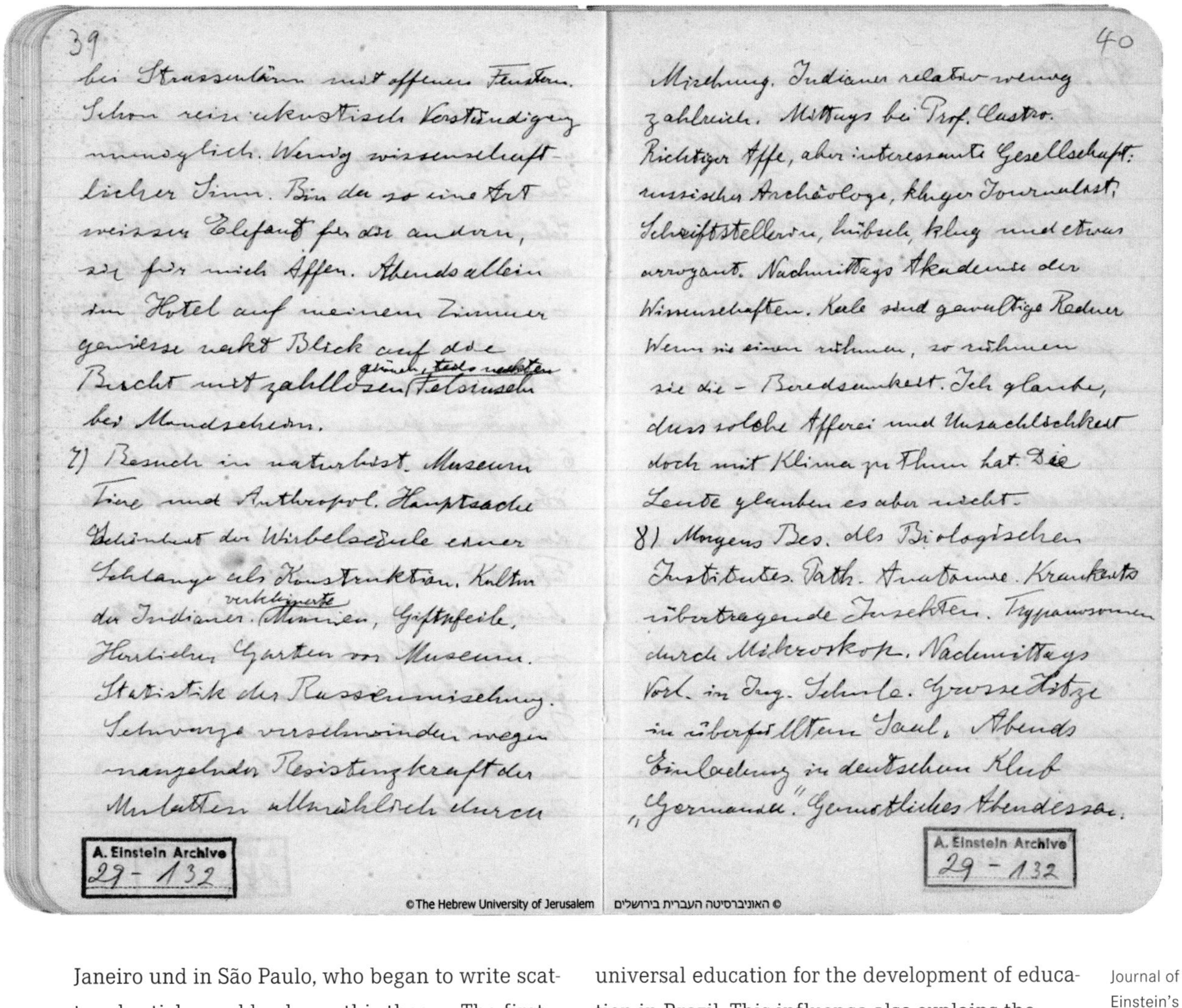
39
bei Strassenlärm mit offenen Fenstern.
Schon rein akustisch Verständigung
unmöglich. Wenig wissenschaftlicher Sinn. Bin da so eine Art
weisser Elefant für die andern,
sie für mich Affen. Abends allein
im Hotel auf meinem Zimmer
genüsslich nackt Blick auf die
Bucht mit zahllosen Felsinseln
bei Mondschein.
7) Besuch im naturhist. Museum
Tiere und Anthropol. Hauptsache
Schönheit der Wirbelsäule einer
Schlange als Konstruktion. Kultur
der Indianer. verkleinerte Mumien, Giftpfeile,
Herrlicher Garten vor Museum.
Statistik der Rassenmischung.
Schwarze verschwinden wegen
mangelnder Resistenzkraft der
Mulatten allmählich durch

A. Einstein Archive
29 - 132

40
Mischung. Indianer relativ wenig
zahlreich. Mittags bei Prof. Castro.
Richtiger Affe, aber interessante Gesellschaft:
russischer Archäologe, kluger Journalist,
Schriftstellerin, hübsch, klug und etwas
arrogant. Nachmittags Akademie der
Wissenschaften. Kerle sind gewaltige Redner.
Wenn sie einen rühmen, so rühmen
sie die – Beredsamkeit. Ich glaube,
dass solche Afferei und Unsachlichkeit
doch mit Klima zu thun hat. Die
Leute glauben es aber nicht.
8) Morgens Bes. des Biologischen
Institutes. Path. Anatomie. Krankheits
übertragende Insekten. Trypanosomen
durch Mikroskop. Nachmittags
Vort. in Ing. Schule. Grosse Hitze
in überfülltem Saal. Abends
Einladung in deutschen Klub
„Germania". Gemütliches Abendessen.

A. Einstein Archive
29 - 132

Journal of Albert Einstein's South American trip. Shown here is the note about his lecture to the Engineering Club of Rio de Janeiro on 6 May 1925. © The Hebrew University of Jerusalem, Albert Einstein Archives

During Einstein's visit to Rio de Janeiro, he wrote an article entitled *Imaginary Relativity* in the daily *O Jornal*.

In this article Cardoso asked: "Why can I not recognize the new truth of this theory? Why can I not recognize the superiority of this theory?" He was convinced that the ideas of the relativists were lacking a sound theoretical foundation. His opposition made ridicule about the theory of relativity popular. Caricaturists used his arguments to make

Einstein's theory fascinated people because it opened up discussion of ideological questions and topics that had always occupied man, such as the question of the meaning of the "world-whole," of the certainty and uncertainty of recognition and of the origin of our concepts of time and space. The kinds of questions the theory of relativity raises were never the exclusive domain of scientists, but are rather among those questions that every human with reason poses at some time.

Impressão do prof. Einstein, re-gistrada pelo desenhista do Instituto de Manguinhos, para O JORNAL

Hence the ideas of Einstein's theory suggest themselves to everyone, but at the same time they are intimidating, not least because of the complicated mathematical formulae used to express them. Einstein's theory occupied not only researchers in the exact sciences. Also particu-

Einstein caricatures from Brazilian daily newspapers

fun of Einstein, by falsely attributing the colloquial meaning to the concept of physical relativity. Like Licinio Cardoso, Theophilo Nolasco Almeida, professor at the Marine Academy of Rio de Janeiro, also opposed the theory of relativity. He repeatedly made clear that he was not enthusiastic about the new theory. In 1930 he published a short book with the title: *Einstein versus Michelson*, in which he writes that Einstein was not a seaman, and if he had ever been one, only as a passenger on a steamer; neither was he an airman, for if he had been one, he would have been more careful with aether and air, and would know that the direction in aether can be none other than the diagonal in space. Almeida knew that he had not swayed the foundations of Einstein's structure with his interpretation, but he was convinced that Einstein and scholars like him "obscure, strike dumb and delude" human reason.

larly interested in the subject was the jurist, Francisco Cavalcante Pontes de Miranda (1892–1979). He discussed not only the physical concepts of space, time and matter, for example, in a lecture entitled *Conception of Space* held in 1924 at the *V. Congresso Internazionale di Filosofia* in Italy, but also applied the physical concepts in the sociology of law. In particular, he used the physical concept of space in order to introduce the concept of social space and to derive implications: "A social space exists only where there is matter, where a social energy is present."

In the 1920s the favorable opinion of the theory of relativity ultimately prevailed over the opposition, but the number of published articles and books remained limited. The theory of relativity was not introduced into the curriculum of the universities until the 1930s. There was one exception, however: in the *Colegio Pedro II*, a classical secondary school in Rio de Janeiro, the theory of relativity was taught starting in 1929.

As in all countries, in Brazil the way the theory of relativity was taken up in the scientific circle was different from its reception by the man on the street. Newspapers attempted to impart an idea of the new theory, particularly through art and humor. Einstein became a legend in the 1920s after receiving the Nobel Prize. Caricaturists used all kinds of stylistic means to make his conceptions of science more understandable to people. How can the work of a genius be presented to the entire public? Einstein was generally characterized by a combination of humanity and remarkable peculiarity. Just as in other countries, in Brazil, too, many caricatures of Einstein were published in periodicals. A number of them attempted to portray Einstein as a simple man; others as a synthesis of mathematical formulas and a human figure. Einstein and the theory of relativity were even used in advertising, as the following ad in the daily *O Estado de São Paulo* shows: Einstein in Berlin! "Gentlemen: All over the world I have found that my "theory of relativity" is confirmed, except in São Paulo, where the House of the Lottery, Antonio Prado Platz 5, sells great fortune every day!! This is a unique and absolute case and therefore worthy of mention."

Advertisement from the newspaper *O Estado de Sao Paulo*

C. V. Vishveshwara

Einstein and India

Einstein never came to India although he visited the neighboring country of Ceylon or Sri Lanka as it is called today. Nevertheless, he had sufficiently extensive interaction with many Indians through personal meetings and correspondence. The records of these meetings, as well as the letters themselves, reveal the extraordinary and multifaceted personality of Einstein.

as the only possible weapon to combat McCarthyism. In 1931, when Gandhi was attending the Round Table Conference on Indian Constitutional Reform in London, Einstein wrote him a letter, which read:

"You have shown by all you have done that we can achieve the ideal even without resorting to violence. We can conquer those votaries of vio-

Mohandas Karamchand Gandhi in July 1946 with Jawaharlal Nehru (left), Vallabhbhai Patel (right)

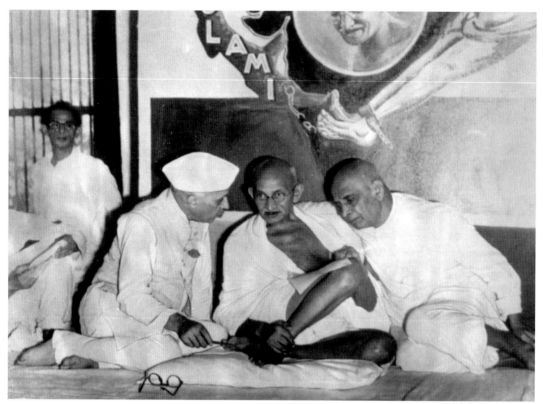

First and foremost there was Gandhi. For Einstein, Mohandas Karamchand Gandhi, the Mahatma or the Noble Soul, as he was called by millions of Indians, occupied a position of highest esteem. We are told that on the walls of Einstein's study at his Princeton home hung the portraits of the four people whom he probably admired most: Newton, Faraday, Maxwell, and Gandhi. Einstein strongly advocated, but not uncritically, Gandhi's ideas of non-violence and non-cooperation as the most effective means of achieving world peace. He went so far as to recommend non-cooperation

lence by non-violent method. Your example will inspire and keep humanity to put an end to a conflict based on violence with international help and cooperation guaranteeing peace in the world. With this expression of my devotion and admiration I hope to be able to meet you face to face."
And Gandhi replied from London:
"I was delighted to have your beautiful letter. It is a great consolation for me that the work I am doing finds favor in your sight. I do indeed wish that we could meet face to face and that too in India at my Ashram."

Albert Einstein and the first Indian Prime Minister Jawaharlal Nehru at a meeting in the USA, October 1949

as exhibited by the exchange of messages that followed.

The second Indian of high standing whom Einstein met was Jawaharlal Nehru, the first Prime Minister of independent India. Nehru had great admiration for Einstein. Several letters were exchanged between the two men both before and after Nehru became the Prime Minister. Discussions carried out in these letters touched upon a number of topics including the political situation in China and Russia, world peace, and the situation of the Jews the world over. In 1947, Einstein wrote a seven-page impassioned letter in which he compared the plight of the Jews to that of the untouchables in India. He wrote:

"May I tell you of the deep emotion with which I read recently that the Indian Constituent Assembly had abolished untouchability ..."

It is a pity that the two great men were not destined to meet *face to face*, as both of them put it. On the other hand, Einstein did meet two other eminent Indians. First of them was Rabindranath Tagore, whom Einstein jokingly referred to as Rabbi Tagore, the poet-philosopher who won the Nobel Prize for literature. The two met thrice and two of those meetings have been recorded. These transcripts are quite interesting to read. The first one ranges over a number of subjects such as the place of man in the universe, beauty, truth, reality, and the Divine. The second dialog starts off on the nature of causality, the two men talking away with hardly any mutual understanding. At one point, Tagore mentions the flexibility of the Indian musical system and Einstein seems to pounce on this topic, which the two could discuss meaningfully. Although both seemed to be unhappy to some extent with the lack of mutual comprehension, there was definitely warm rapport between the two

Albert Einstein with Rabindranath Tagore, Berlin 1930 © The Hebrew University of Jerusalem, Albert Einstein Archives

Albert, Elsa and
Margot Einstein
with Rabindranath
Tagore and his wife,
Professor and Mrs.
Mahalanobis and
Dr. Loewenthal in
Berlin, 1930

In his five-page reply, Nehru remarked, "It is a privilege and an honor to be addressed by you and I was happy to receive your letter, though the subject is a sad one."

Nehru visited Einstein in Princeton in 1949. Professor Amiya Chakravarthy, who arranged the meeting between the two, later wrote to Einstein: "Mr. Nehru repeatedly told us that the great expectation of his life has been fulfilled. For years he had cherished the hope of meeting you, and it seems that at last he had the privilege at a critical stage of his own life when he feels that he stands at the crossroads and must now take the right direction for India in a world split by power blocs. Evidently, his conversation with you that day has strengthened his conviction that India must stand outside the two big blocs ...

That afternoon may well affect the course of thought and action in India ..."

One is left wondering about the deliberations of that afternoon which might have shaped the future of India. The answer may lie in the papers left behind by Nehru himself. On the other hand, for all we know, it may be a fragment of history lost forever.

Several Indian scientists wrote to Einstein. Needless to add, the most celebrated correspondence in this category was the one between Satyendra Nath Bose and Einstein that culminated in Bose-Einstein statistics. An excellent account of this important saga has been given by John Stachel (John Stachel, *Einstein from 'B' to 'Z'*, Birkhäuser, 2002). The gradual change in Bose's own personal attitude towards Einstein appears to be reflected in the manner in which Bose addresses the latter in his three letters: "Respected Sir, Revered Master, and Dear Master!" Bose's regard for Einstein is further highlighted by a photograph that still exists, which shows Bose in his study with a framed picture of Einstein on the wall, adorned with a garland. A common Indian practice of expressing one's highest degree of reverence. Several other scientists corresponded with Einstein, but none of them on scientific subjects. Their concerns seem to have centered essentially on nuclear disarmament and peace. A letter from Meghnad Saha, the discoverer of the Saha equation dealing with the thermal ionisation in stars, discusses calendar reform, a rather unusual subject indeed.

As is well known, Einstein used to receive innumerable letters seeking advice, his opinion on politics, philosophy, wrong theories, as well as requests for his autograph and best wishes. He patiently answered many of these and never failed to respond to student queries. This was true in the case of letters coming from India as well. There are examples displaying charming simplicity and innocence on the part of those who wrote them. For instance, a practitioner of native medicine sent Einstein a copy of his tract *Elements and Atoms in Greek and Indian Thought* requesting Einstein's opinion. It is astonishing to note that Einstein took time to study it and send his serious,

critical comments. In 1936, Einstein received a letter from a small town in India with best wishes for his 58th birthday and offering as a birthday gift, coffee powder or tea leaves as per Einstein's preference. We do not know which of the two gifts Einstein chose!

To sum up, Einstein's brush with India – the records of his meetings as well as his interactions and his correspondence with a number of Indians belonging to different strata and professions – reveals his ideas, his opinions, and his own personality within a framework which is different from that which one commonly encounters in the Einstein lore. A deeper and detailed study in this regard may very well prove to be most rewarding. We began this article with Einstein and Gandhi. Let us conclude it on the same note. On the occasion of Gandhi's seventieth birthday in 1939, Einstein paid him the moving tribute: "Generations to come, it may well be, will scarce believe that such a one as this ever in flesh and blood walked upon this earth."

Undoubtedly that remark applies equally well to Einstein himself.

Ze'ev Rosenkranz

Albert Einstein and the German Zionist Movement

In the aftermath of the First World War, Albert Einstein's attitude towards Jewish nationalism and the Zionist movement went from one of disdain to intense involvement. At the same time, Einstein became a highly significant figure for the Zionist movement and one of its major emblematic icons. How did this radical transformation come about?

sition at the German University of Prague in 1911, he initially declared that he was "*konfessionslos*" (non-confessional). Only when the Austro-Hungarian authorities made it clear to Einstein that he had to declare his allegiance to a recognized faith, did he resignedly declare himself as belonging to the "Mosaic" faith.

On the road in the United States to collect donations for the Hebrew University in Jerusalem. Group photo on board the "Rotterdam." From left to right: Ben Zion Mossinson, Albert Einstein, Chaim Weizmann, Mendel Ussishkin, March 1921

Albert Einstein was born into an assimilated German-Jewish family which was not religiously observant. Nevertheless, he received Jewish religious instruction at home and his family seems to have inculcated in him a strong sense of Jewish identity. During these years, Einstein's social milieu – his extended family and friends – were predominantly Jewish.

However, in his early adulthood Einstein revealed strong assimilatory tendencies. He married a non-Jew, the Serbian physicist Mileva Marić (1875–1948) in 1903, and when he was appointed to a po-

Although Einstein met with members of Zionist circles in Prague during his stay there in 1911–1912, he does not seem to have been influenced by them ideologically. Indeed, four years later, he disparagingly described one of their leaders, Hugo Bergmann (1883–1975), as belonging to a "*zionistisch verseuchte[r] kleine[r] Kreis* […]" of "*weltferner Menschen*" (a small Zionist-infested circle of unrealistic people). Notwithstanding, one Einstein biographer, Philipp Frank (1884–1966) has argued that during his period in Prague, Einstein's interest in Jewish affairs was starting to be rekindled.

Einstein's first documented public action related to the Zionist movement occurred in December 1918, when he agreed to participate in a provisional committee towards creating a Jewish Congress in Germany, co-initiated by the *Zionistische Vereinigung für Deutschland* (ZVfD).

Various factors led to this radical shift in Einstein's views. Firstly, similar to other German-Jewish intellectuals of his generation, Einstein was deeply impacted by the plight of Eastern European Jewish

to this sharp turn in his ideological preferences: he was getting a divorce from his first wife Mileva during this period; he found access to his children difficult as they were in Switzerland living with Mileva; and his mother was terminally ill throughout 1919 and would eventually die in early 1920. Einstein may have been feeling quite vulnerable at this period in his life and may well have been looking for a new cause to support. Thus, he may have been "ripe" for "conversion" to Zionism.

Albert Einstein and his wife Elsa with the mayor of Tel Aviv, Meir Dizengoff and the Zionists Bezalel Jaffe and Ben Zion Mossinson at a reception in honor of Einstein being made honorary citizen, in front of the town hall in Tel Aviv, 8 February 1923

refugees in the aftermath of World War I and the ensuing new wave of anti-Semitism in Germany. He was especially concerned about the plight of Eastern European Jewish students who were subjected to various restrictions. In addition, as a consequence of having been one of very few German academics to oppose the War, he felt a considerable degree of alienation from German society. There were also turbulent personal events in his life around this time which may have factored in-

This vulnerability may have been picked up by Kurt Blumenfeld (1884 – 1963), who was head of the propaganda department of the ZVfD and who, in his writings, claimed to have had a major influence on Einstein in his move towards Zionism. Blumenfeld was the major proponent of the "second generation" of German Zionists. This generation rejected both the assimilationist approach of German-Jewish liberals and the philanthropic approach of the first generation of German Zionists, who saw Zionism primarily as an option for their less fortunate co-religionists from the East rather than for themselves.

In his memoirs, Blumenfeld describes the mobilization of Einstein for the Zionist movement as follows: in light of the opportunity for new developments in Palestine in the aftermath of World War I, he and fellow Zionist leader Felix Rosenblüth (1887–1978) drew up list of German Jewish intellectuals they wanted to interest in Zionism. In February 1919, Blumenfeld invited Einstein to one of his lectures. After the lecture he noticed a "transformation" in Einstein when he remarked: "*Ich bin gegen Nationalismus, aber für die zionistische Sache* (I am against nationalism, but for the Zionist cause)." Blumenfeld subsequently met with Einstein several times and (according to him), Einstein's interest in Zionism increased gradually over time.

However, the first contemporary documentation for this major change in Einstein's views can be found as early as March 1919 in a letter to Einstein's colleague and friend Paul Ehrenfest (1880–1933): In the aftermath of war and revolution in Germany and Berlin, Einstein stated that "the issue from which I derive most joy is the realization of the Jewish state."

In the meantime, plans within the Zionist Organization (Z.O.) in London for the establishment of a Jewish university in Jerusalem were progressing. In October 1919, Einstein wrote to physicist Paul Epstein (1883–1966) that "the Zionist cause is very near to my heart" and that he has had recent meetings with Zionist leaders. He was especially interested in the project to establish the Hebrew University. That same month, Einstein was invited by the Z.O. to attend a conference of Jewish scholars on plans for the University. Einstein accepted and even recommended other European Jewish scholars.

The next month, Einstein gained international renown after sensationalist headlines in the An-
glo-American press proclaimed the verification of his general theory of relativity. This did not pass unnoticed by the Zionists in London and he was described by Hugo Bergmann as "the hero of the day." Indeed, it was (the previously maligned) Bergmann who sent Einstein copies of *The Times* articles on the verification of the theory. So, intriguingly, the Zionists had "discovered" Einstein prior to his world-wide fame and they were even the ones who informed him of his newly-acquired celebrity. On 14 December 1919, Einstein also became a household name in Germany when his image appeared on the front cover of the *Berliner Illustrirte Zeitung.*

There were various factors for Einstein supporting the Hebrew University project: first and foremost, he saw it as a refuge for Eastern European Jewish students who were unable to study in the East or in Germany; secondly, he was aware that his newly-acquired fame could be utilized for the cause; thirdly, he was deeply interested in the establishment of the Hebrew University because it was an academic project. Einstein saw intellectual aspiration as an important Jewish ideal and the academic nature of the project added to its appeal. It is clear that in the early stages of his support for Zionist projects, it was the Hebrew University which was of primary concern to Einstein within the spectrum of Zionist causes.

The next big step in Einstein's support for Zionist causes was his joint trip – together with the head of the Zionist Organization, Chaim Weizmann (1874–1952) – to the U.S. in 1921. The main declared purpose of the trip was to raise funds for the planned medical faculty at the Hebrew University. There were also other factors for the visit: e.g., Einstein's desire to meet with American colleagues and to disseminate his theories in the U.S. Einstein received a tumultuous welcome from the (mostly Eastern European) Jewish masses in the U.S., especially in New York.

In early 1923, Einstein and his wife visited Palestine on their return trip from the Far East. During their twelve-day visit, they toured all over the country and Einstein was given the royal treatment and hailed as a Zionist hero. In Jerusalem,

In 1925, Einstein was chosen to be a member of the Hebrew University's first Board of Governors and chairman of its first Academic Council. In the late twenties and early thirties, major differences of opinion emerged between Einstein and the

Albert Einstein speaking to the Jewish Student Conference in Germany, around February 1924

Einstein gave (what is considered as) the inaugural scientific lecture of the Hebrew University at its future site on Mt. Scopus.

Einstein saw Palestine first and foremost as a refuge for Eastern European Jews. He did not expect the majority of Western Jewry to settle there, including himself (although he was offered an academic position there at the Hebrew University several times). He envisioned Palestine becoming a spiritual center for world Jewry, especially as a means to strengthen their cultural identity and social cohesion. In his references to a national entity in Palestine prior to the Holocaust, he usually referred to a Jewish homeland, but not necessarily to a state.

University's chancellor, J. L. Magnes (1877–1948) regarding the way in which the University was being administered. Einstein was opposed to what he saw as the excessive influence of American philanthropists on the handling of academic affairs – he advocated a German model for the University whereby the academics would have the decisive say. Major reforms were instituted at the University in 1935 to satisfy Einstein's demands for change. One expression of his long-lasting ties with the University was the bequest of his personal papers and literary estate to the Hebrew University in his last Will and Testament of 1950.

From the late 1920s onwards, Einstein advocated peaceful co-existence between the Arab and

Kurt Blumenfeld holds an address at the 25th Convention of Delegates to the Zionistic Association for Germany, Berlin 2–4 February 1936

Jewish inhabitants in Palestine – indeed he did not believe that the Zionist movement had any moral justification or practical possibility of establishing a homeland there without such cooperation with the Arabs.

How can we define the role Einstein fulfilled within the German Zionist movement?

The Zionist movement's mobilization of Albert Einstein for their cause was a huge public relations coup – especially for the German Zionist Federation. From 1919 to 1933, he became their most emblematic figure and they utilized his world-wide fame in claiming that a genius supported the Zionist cause. Intriguingly, Einstein was inducted or "converted" to Zionism by the German Zionists in 1919, but contacts between him and world Zionist leaders were established soon afterwards. The hero's welcomes he was given during his trips to the U.S. and Palestine made Einstein an iconic fig-

ure for the entire world Zionist movement. Within the German Zionist movement, Einstein's mobilization on their behalf was a major accomplishment both internally within the German-Jewish community and externally within the German political scene. Landing such a renowned genius gave the Zionists much needed prestige both vis-à-vis their assimilatory co-religionists and the general German public.

Einstein was not a typical German Zionist. Age-wise, he was roughly a contemporary of the second generation of German Zionists. Like them, he did come from an assimilatory family and received an academic education. However, in strong contrast to them, he did not join a Zionist student association during his student years. In general, Einstein was much less of a "joiner" than the average Zionist supporter or leader (this was a pattern of behavior that Einstein would repeat vis-à-vis other political causes he supported such as pacifism and socialism). He did support various Zionist goals and specific Zionist projects (most notably the Hebrew University), but he did not identify himself completely as being a Zionist and he does not seem to have been viewed as fully belonging to the Zionist movement by its leaders either. Though he fulfilled an iconic and symbolic role, he did not totally belong to the movement. There is a certain symmetry between Einstein not fully identifying himself as a Zionist and the Zionist movement not seeing him as completely one of theirs. Indeed, Einstein often referred to himself as a *"jüdischer Heiliger."* By the same token, he could also have described himself as a *"zionistischer Heiliger"* – a saintly, untouchable Zionist icon.

After his emigration to the U.S. in 1933, Einstein's symbolic role for the Zionist movement declined for a number of reasons: his decision to move to the U.S. rather than to fulfill the Zionist "Lebensplan" of moving to Palestine, the gradual reduction in his public appearances, and various internal developments within the Zionist movement, to mention just a few. As a consequence, Einstein's iconic role for Zionism has largely been forgotten in the two major centers of contemporary Jewry, Israel and the U.S.

Christian Dirks

The Scapegoats' Attorney: Albert Einstein and his Commitment to the Cause of the Eastern Jews

Albert Einstein's espousal of the cause of the Eastern European Jewish immigrants at the turn of the year 1919/1920 marked the beginning of his intensive involvement with questions of Jewish identity. Although he had shown little interest in religious issues since his youth, the 'Jewish Question' took on increasing importance for him in the face of growing anti-Semitism during and

Jewish cellar shop in the Berlin *Scheunenviertel*, around 1920 (photo: Ernst Thormann)

immediately after the end of the First World War. One climax of Einstein's new Jewish awareness was his public espousal of the interests of the Eastern Jewish minority in Berlin.

The Eastern Jews had been the target of anti-Semitic agitation in Germany since the 19th century. In 1879 the national-socialist historian Heinrich von Treitschke wrote a polemic against the Eastern Jews, triggering off the so-called Berlin anti-Semitism dispute: "Year after year hordes of ambitious youths selling trousers flood across our eastern border from the inexhaustible Polish cradle. One day their children and their children's children will dominate Germany's stock market and newspapers [...]."

The term 'Eastern (European) Jew' – as opposed to 'Western Jew' – was first used at the beginning of the 20th century to denote Jews who had been fleeing pogroms in Czarist Russia since the 1880s and had settled in Germany. However, this term for Jews who had immigrated from Eastern Europe and were also called 'Polskies' or 'scroungers,' only became generally accepted in the course of the First World War. The combination of two negative concepts, 'Eastern' and 'Jew,' characterized the prejudice-laden use of the term 'Eastern Jew.' The Eastern Jews were regarded as a threat and a public nuisance by large sections of society. An 'Eastern Jew issue' developed out of this in the wake of the First World War when, for one thing, many German soldiers first came into contact with Eastern Jews in occupied Poland and, for another, tens of thousands of Polish Jews were recruited to work in German wartime industry.

In anti-Semitic publications the Eastern Jews represented a contrast to the emancipated and assimilated German Jews. The language of anti-Semitism used biological terminology as early as the 1920s, calling the Eastern Jews 'parasites,' 'cancerous tumors' or 'vermin.' They were denounced as racketeers, usurers and war profiteers and held responsible for the increasing housing shortages in the cities, for rising unemployment and the economic depression – the classic scapegoat role.

Thus anti-Semitism, anti-Slavism, anti-Bolshevism and a widespread rejection of the new, democratic social order entered into a fatal alliance.

Advertising for
the settlement of
Palestine, 1919

By 1922, an estimated 150,000 Eastern European Jewish workers had emigrated to Germany, some voluntarily, some under duress, some illegally. Many of them were recruited during the First World War to do forced labor in German munitions factories. As a rule they belonged to the lower-middle and lower classes. Except for a small number these Eastern Jews had left the country again by 1923 because, for many of them, Germany was only a stopover on their way to the United States.

In contrast to the assimilated German Jews, most of the Eastern Jews in German cities lived isolated from the majority of society. In Berlin it was in the so-called 'Scheunenviertel' near Alexanderplatz where many Eastern European Jews chose to settle, and this district took on the character of a Polish 'shtetl.' The Eastern Jews stood out as Jews simply because of their appearance: caftans, skull caps, side locks and the Yiddish language made them look like foreigners, and this was intensified

Textil vendor
in front of the
Münzglocke,
Münzstrasse in
Berlin-Mitte, 1930
(photo: Ernst
Thormann)

by their failure to adapt to German culture and the way they held on to traditional family ties. They were constantly threatened with deportation and victims of arbitrary state action. In March 1920, two hundred and eighty-two Eastern Jews were arrested after a raid in Berlin's *'Scheunenviertel'* and interned in a camp at Wünsdorf near Zossen. They were subjected to what the Zionist *Jüdische Rundschau* newspaper called 'anti-Semitic loutish behavior.' They were beaten with

as a danger to the level of assimilation already achieved and looked down on this minority with scorn and contempt. Others recognized in the Eastern Jews a kind of archetype that was reviving religious traditions, and considered them an enrichment of Jewish life. In any case, the difference between assimilated, bourgeois German Jews and the orthodox Eastern Jewish proletariat was perceived by society as a contrast between 'necktie-Jews' and 'caftan-Jews.'

Grenadier Straße in 1933 (photo: P. Buch)

rifle butts and not given enough to eat. All but three of the internees had to be released after a few days for lack of evidence. Other internment camps in Stargard / Pomerania and in Cottbus/ Sielow were harbingers of a development that was to end twenty years later in Auschwitz.

The Eastern Jews also triggered different reactions within the Jewish community. Some feared them

The Workers' Welfare Office of Jewish Organizations in Germany (AFA), an umbrella organization of Jewish pressure groups founded in 1918 to support the Eastern Jews living in Germany, looked after this group of people and gave them social, material, political and journalistic support. It was the AFA that asked Einstein to make a statement on behalf of the Eastern Jews. At the climax of anti-Eastern Jewish agitation at the turn of the year 1919/1920, Einstein offered his protection to this discriminated minority. Einstein's very first public statement on a Jewish topic was a reply to an article in the *Berliner Tageblatt*. Entitled *Immi-*

SCHRIFTEN DES ARBEITERFÜRSORGEAMTES DER
JÜDISCHEN ORGANISATIONEN DEUTSCHLANDS

HEFT II

OSTJUDEN IN DEUTSCH-LAND

PHILO-VERLAG / BERLIN 1921

gration from the East. The Russians in Berlin, the article claimed that some 70,000 people, predominantly Russians, were flooding into German cities 'in hordes.' The article was a wild mixture of half-truths and anti-Semitic prejudices and blamed the Eastern Jews for current economic, social and political problems. In his reply (which was also published in the *Berliner Tageblatt*), Einstein composed a defense of the Eastern European immigrants while at the same time denouncing the growing number of anti-Semitic attacks by right-wing extremists in Germany.

The 'demagogic agitation' against the Eastern Jews was, Einstein claimed, simply an attempt to distract from the true causes of the disastrous economic situation. He appealed for a more objective discussion and pleaded for sympathy for the East-

ern Jewish immigrants who had been forced to flee "because of the dreadful situation in Poland" and who were only seeking temporary sanctuary "until they are given an opportunity to move on. Hopefully, many of them will find a real home as free sons of the Jewish people in the newly emerging Jewish Palestine." Einstein warned that only the "poor and unfortunate" were affected by state measures, people who "have found their way to Germany at the cost of tremendous human sacrifice and are seeking work here." Only this harmless, innocuous group, "a small defenseless fraction of the population" would "fill the concentration camps" – a topic under public discussion at the time – and "degenerate there physically and mentally," he feared. "Public conscience has become so blunted to the appeals for humanity that people no longer directly perceive the appalling injustice that is about to be perpetrated." Einstein finished his article with a reference to the fatal effects that the anti-Semitic discussion would have abroad. The expulsion or internment of the Eastern Jews would be seen as further evidence of supposed German 'barbarism.' Einstein was convinced that in Eastern European Judaism there lay sleeping "a wealth of the finest human talents and productive energy that has nothing to fear from a comparison with the higher civilization of the Western Jews." And elsewhere he said: "These people still have a healthy national feeling which has not been destroyed by the atomization and dispersion of the individual."

Shortly before this written statement on the Eastern Jews debate, Einstein, who had been a relatively unknown scientist up to then, had achieved world fame after an English expedition appeared to confirm his general theory of relativity. He used this newly-won popularity to further a cause that was close to his heart. Einstein's declaration of solidarity with the Eastern Jews was also a criticism of the lack of support for the Eastern European immigrants by most German Jews. In sub-

East European Jews in Germany, an information brochure by the AFA, 1921

sequent years he repeatedly took a stand in the debate on the Eastern Jewish immigrants and was also critical of certain political currents within German Judaism, such as those represented by the *Centralverein deutscher Staatsbürger jüdischen*

Selling books in the *Scheunenviertel*, 1930 (photo: Ernst Thormann)

Glaubens. It is characteristic of Einstein that here, too, he sided with the underdog, with the helpless and the underprivileged.

This solidarity with the Eastern Jews was one of the elements of his Jewish identity, which did not begin to manifest itself until after the First World War. As a result of this commitment, he, together with some Jewish and some non-Jewish colleagues at the *Friedrich-Wilhelms-Universität in Berlin*, organized special lectures for Eastern Jewish

students, who were not allowed a regular course of study at the university.

A retrospective comment clearly shows how important his identification with the Eastern Jews was for his further development and his conception of himself as a Jew: "There [in Berlin] I saw the misery of many young Jews. [...] This particularly applied to the Eastern Jews, who are subject to constant harassment. I do not believe that there are very many of them in Germany. Yet their presence has become an issue that increasingly preoccupies the German public. In meetings, conferences and newspapers there are calls for their immediate removal or internment. A shortage of living accommodation and the economic depression are used as arguments to justify these harsh demands. The facts are carefully exaggerated to turn public opinion against the Eastern Jewish immigrants. The Eastern Jews are being made the scapegoat for certain ailments in present-day economic life in Germany, which, in reality, are the aftermaths of the War. The attitude toward these unfortunate refugees who have fled from the hell that Eastern Europe is today has become an effective political weapon used successfully by demagogues. When the government was considering measures to be taken against the Eastern Jews, I stood up for them and pointed out in the *Berliner Tageblatt* the inhumanity and irrationality of these measures. [...] These and similar experiences awakened Jewish national feelings in me."

His commitment to the cause of the Eastern Jews is a key to understanding Einstein's relationship to Zionism. His interest in the development of Palestine was always closely linked to the aim of creating a seat of learning for Eastern European Jews. The founding of a Jewish university in Jerusalem thus met with Einstein's emphatic support. He always had his own sad experience in the back of his mind, because in Berlin he had "[...] seen countless examples of how perfidiously and unkindly fine young Jews are treated here and how they are prevented from getting an education." For Einstein, the founding of a Jewish state was a contribution to solving the issue of the Eastern Jews. For him the Jewish state epitomized a "refuge for the oppressed," a "unifying ideal" and a "means of recovery for Jews throughout the world." However, according to Einstein, Palestine was not so much "a refuge for Eastern Jews" as "the embodiment of the re-awakened national sense of community of all Jews."

Einstein continued to closely follow the fate of the Eastern European and Russian Jews in the following decades. Numerous organizations devoted to helping the Eastern Jews could always count on his support. He remained deeply bound up with the situation of Eastern Jewry all his life.

Hanoch Gutfreund

Albert Einstein and the Hebrew University

On 1 April 1921 Albert Einstein arrived in the U.S. for the first time. He came with a group of Zionist leaders under the stewardship of the Head of the Zionist Organization, Israel's first president-to-be, Chaim Weizmann (1874–1952). Together, they were to take a six-week coast-to-coast tour to promote the cause of the Jewish national home and to raise funds for the establishment of a Jewish university in Jerusalem. The following day, *The New*

Albert Einstein with Chaim Weizmann, chemist, president of the World Zionist Organization (1920–1930 and 1935–1946), from 1929 chairman of the Jewish Agency, first President of the State of Israel (1948–1952) New York, April 1921 © The Hebrew University of Jerusalem, Albert Einstein Archives

York Times reported that thousands of people had waited four hours to welcome Professor Einstein to America.

On the group's arrival, Chaim Weizmann told reporters: "Professor Einstein has done us the honor of accompanying us to America in the interest of the Hebrew University of Jerusalem. Zionists have long cherished the hope of creating in Jerusalem a centre of learning in which the Hebrew genius shall find full self-expression and which shall play its part as interpreter between the Eastern and

Western worlds." To this, Albert Einstein added: "I know of no public event which has given me such delight as the proposal to establish a Hebrew University in Jerusalem." Indeed, on another occasion, Einstein referred to the establishment of the Hebrew University as "[...] the greatest thing in Palestine since the destruction of the Second Temple." This visit to the U.S., his first trip outside Europe, made a lasting impression on Albert Einstein. He was received everywhere as a hero. Thousands of American Jews greeted him with cheers, with American flags, with blue and white flags of the Zionist movement, and escorted him in a motorcade from the Harbor to the City Hall. That was the case in every city he visited. The *New York Times* reported on his visit to Cleveland on May 25: "He was greeted by a near riot of fan frenzy, a military band and a motorcade of two hundred cars. He was saved from possible serious injury only by strenuous efforts by a squad of Jewish war veterans who fought the people off in their mad efforts to see him."

Years later Einstein wrote: "It was in America that I first discovered the Jewish people. I have seen any number of Jews, but the Jewish people I had never met either in Berlin or elsewhere in Germany. This Jewish people, which I found in America, came from Russia, Poland and Eastern Europe generally [...]. I found these people extraordinarily ready for self-sacrifice and practically creative. They have, for instance, managed in a short time to secure the future of the projected University in Jerusalem, at any rate so far as the Medical Faculty is concerned."

Two years earlier, in 1919, Einstein accepted Weizmann's invitation to join the Zionist movement and, shortly after, *The Times* of London described him as an ardent Zionist who had promised to participate in the effort to establish a Hebrew University in Jerusalem. In fact, Einstein's partnership in

the Zionist program was most evident in his association with this initiative. He became Weizmann's partner in elevating this initiative to a project of highest priority on the Zionist agenda. He perceived the proposed University as the stage on which the inventiveness of the Jewish people and the Jewish quest for learning would come again into prominent play.

Einstein's first major effort to promote the university project was the 1921 trip to the United States. To appreciate how important this was to him, one

should mention that in order to make this trip Einstein gave up an invitation to participate at the Solvay Conference which was about to take place in April 1921. The Solvay Conferences constituted the most prestigious stage on which the great ideas and advances in physics were presented and debated. Einstein was the only German physicist invited to the first such conference after World War I. Yet, when Kurt Blumenfeld (1884–1963), a leading figure in the German Zionist Organization, invited him on behalf of Chaim Weizman to join the Zionist mission to the U. S., he immediately agreed. In a letter to his friend, the Dutch physicist Hendrik A. Lorentz (1853–1928), Einstein thanked him for arranging the invitation to participate at the Solvay Conference, and apologized for rejecting it because the Zionists were planning to establish a university in Jerusalem and they believed that his trip to the U. S. might influence rich Jews to contribute more generously. He felt that it was his duty to help: "As this venture lies close to my heart, and as I, as a Jew, feel a duty to contribute, as far as I am able, to its success, I accepted." Lorentz responded by expressing his understanding for Einstein's decision and wishing him success. Einstein's German colleague and

Jubilant reception for Albert Einstein in New York, 1921

Einstein's first visit to Palestine on his return from Japan in February 1923. Albert and Elsa were guests of the British High Commissioner Herbert Samuel in Jerusalem

Opening ceremony
of the Hebrew
University in
Jerusalem, 1925

friend, Fritz Haber (1868–1934), was not so understanding. He criticised Einstein's decision to go
to the U.S., invoking patriotic arguments, in the
name of his debt to German science and his loyalty to his German colleagues. He even used the
argument that Einstein's decision might generate

antisemitic sentiments against German Jews.
Einstein was undeterred. He explained that his
decision stems from his obligation "[...] to step in
for my persecuted and morally depressed fellow
tribesmen, as far as this lies within my power".
He categorically rejected any allegations of disloy-

alty: "no sensible can accuse me of disloyalty towards my German friends. "

Between 1921–25, Einstein's frequent lecture tours took him elsewhere in the world and Weizmann arranged for him to plead the University's cause in Jewish communities as distant as Singapore. In 1922, on a stopover to Japan, Einstein raised funds for the Hebrew University in Singapore. Addressing the Jewish community there, he

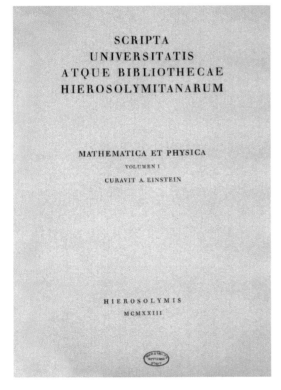

said: "One may ask – why do we need a Jewish University? Science is international but its success is based on institutions owned by nations. Up to now as individuals we have helped as much as possible in the interest of culture and it would be only fair to ourselves if we now, as a people, add to culture through the medium of our own institutions." In that address he argued, as he did on many occasions, that a Jewish University was a necessity at a time when many European universities imposed quotas on Jewish students or completely excluded them from certain fields of study.

On the way back from that trip to Japan, Einstein made his only visit to Palestine. There, in a British police academy hall on Mount Scopus, he delivered the University's first-ever scientific lecture. The event was chaired by the Zionist leader Menachem Ussishkin (1863–1941) who concluded his introductory remarks by saying: "Professor Einstein, please rise to the podium that has been waiting for you two thousand years." Einstein began his lecture in Hebrew and then apologized for being unable to continue in the language of his own people, and resumed in French. Those present heard in Einstein's voice the birth song of the long anticipated Jewish University.

In 1921, four years before its official opening, Einstein had already clearly voiced his opinion on the desired structure and mission of the University. In particular, he supported Weizmann's position in the ongoing controversy over the nature of the university to be. Both of them advocated the notion that, before providing teaching programs and granting academic degrees, the university should establish itself as a first class research institute.

Einstein inspired the entire Jewish world and many in the academic world with his vision of the Hebrew University. In the mission statement that he published for the 1925 opening of the University, he wrote: "A university is a place where the universality of the human spirit manifests itself," and he expressed the wish that "... our University will develop speedily into a great spiritual center which will evoke the respect of cultural mankind world over." As a gesture to the university on the occasion of its official opening he gave his name as editor of the first scientific publication in physics and mathematics within the *Scripta Universitatis Atque Bibliothecae Hierosolyminarum* and he donated to the university the manuscript of his monumental paper on the general theory of relativity published in 1916.

Albert Einstein edited the first scientific publication of the Hebrew University Jerusalem in the field of physics and mathematics © The Hebrew University of Jerusalem, Albert Einstein Archives

At the University's opening in 1925, Einstein joined its Board of Governors and became founding chairman of its Academic Committee. During the formative years of the University, Einstein had fierce disagreements on issues of academic policies with the University's Chancellor Judah L. Magnes. This was not only a dispute between two strong personalities, but also a confrontation between the European and American perception of university governance. These disagreements led to Einstein's resignation from all his functions at the University in 1928 and to a review of his concerns and objections by the governing bodies of the University, which ultimately ended in Einstein's favor. By that time, he had already settled in Princeton, but his commitment to the Hebrew University remained and he resumed his official involvement in its affairs in 1935. University officials and representatives continued to consult with him on such issues as presidential powers and academic appointments, while Hebrew University professors on fundraising tours carried his personal letters of introduction. Indeed, his correspondence shows the careful attention to detail that could only arise from much deliberation and concern. Einstein helped in the University's major expansion drive that followed World War II, serving as chairman of the AFHU's national council in 1947 and as its honorary president in 1951. He was conferred an honorary doctorate by the University in 1949.

In 1950, Einstein gave profound expression to his lifelong commitment to the Hebrew University: he bequeathed his own true wealth – in this case intellectual: his personal papers and literary estate – to the University. Together with his library, which the University received in 1987, they today make up the Albert Einstein Archives which constitute a cultural asset of supreme importance to mankind. Its holdings are unique – they consist of numerous manuscripts, prolific correspondence, and a large variety of additional material about Einstein. The material in the archives sheds light on the multifaceted aspects of Einstein's scientific work, his political activities and his private life.

Richard H. Beyler

The Physics Community in the National Socialist Era

The German physics community faced a series of intense pressures during the National Socialist regime. Racial and political purges had profound quantitative and qualitative effects. Moreover, demands for ideological conformity sparked controversy about the content of physics, above all the theories most closely associated with Einstein. It would be an incomplete picture, however, to suggest that physics was simply a victim of Nazism. Some members of the German physics community sought to defend the concept of the freedom of science and maintained a measure of professional autonomy. There were, conversely, also instances of cooptation and cooperation with the regime. Scientific institutions tended to sacrifice the careers of individuals towards the end of institutional self-preservation. By the late 1930s, explicit ideological attacks on theoretical physics were diminishing; this was in part because leading scientists had convinced key actors within the Nazi regime of the importance of modern physics for the modern state.

Purges

The effect of racial and political purges mandated by the National Socialists was particularly significant in physics, but the purges were not directed against physics *per se*, or even against science more broadly. Exclusionary measures affected broad areas of German culture and society and began to appear soon after the seizure of power. The most important enactment for science was the "Law for the Restoration of the Career Civil Service" of 7 April 1933. This law required the dismissal from government service of "non-Aryans" – defined as persons with at least one Jewish grandparent – and of persons whose political loyalty could not be guaranteed. There were exceptions for officials who had held their positions from 1914 or earlier, who had served on the front during the World War, or whose fathers or sons had

been killed during the war. Since German universities were state institutions, the law applied to university faculty; it applied also to the staff of governmentally funded research laboratories such as the *Physikalisch-Technische Reichsanstalt* (PTR) and most institutes of the *Kaiser-Wilhelm-Gesellschaft*. From 1933 on, political and racial scrutiny also entered into the selection process for academic scientists; for example, a political evaluation from the National Socialist Teachers' League became a standard part of the dossier for any candidate for faculty appointments and promotions.

Max Planck, as President of the Kaiser Wilhelm Society, inaugurates the Kaiser Wilhelm Institute for Metals Research, 21 June 1935

The quantitative toll is hard to determine exactly, in part because many scientists officially resigned rather than being formally dismissed; estimates have ranged from around 15 to 25 percent. Even more difficult to quantify is the way the Civil Service Law stifled the careers of potential young scientists before they even had a chance to begin. The qualitative effects arguably went even further than the loss in sheer numbers, as exhibited by the fact that among the dismissals and resig-

Physics textbook
by Philipp Lenard,
Deutsche Physik,
Munich 1937

More typical was the response of Max Planck (1858–1947), the doyen of German physics and, by virtue of his roles as president of the KWG and permanent secretary of the Prussian Academy, a preeminent figure in German scientific organization. Planck was certainly no supporter of Nazism, had been a close colleague and friend of Einstein, and more than once sought to alleviate or at least delay application of the Civil Service Law in specific cases. In general, however, Planck urged a cautious but cooperative approach towards the government, believing that institutional continuity was more important than the fate of any individual, that loyalty to the state was more important than political disagreement, and that the fervor of the Nazis would eventually cool.

Ideological Struggles

Alongside the dislocations caused by the dismissal policies, there were also attempts in the Nazi era to transform the content of physics along ideological lines. A loose coalition of physicists, whose most prominent members were the Nobel-prize winning experimentalists Philipp Lenard (1862–1947) and Johannes Stark (1874–1957), saw in the Nazis' rise to power a chance to exercise a lingering resentment against several recent theoretical developments: quantum theory, particularly in its indeterministic interpretation, and above all the relativity theory of Albert Einstein. The controversy also reflected intra-professional rivalries in which some experimentalists resented what they saw as an overweening prestige granted to theoretical physics. At a certain level, too, the resentments were the result of a kind of nostalgia for the "classical" physics of the pre-relativity and pre-quantum era.

Lenard's and Stark's active opposition to relativity theory dated back to at least the early 1920s; in the intervening years, their agenda in physics merged with a growing identification with the

nations were eleven then or future Nobel prize winners in physics.

Einstein's resignation from the Prussian Academy of Sciences was clearly meant as a public protest against Nazi policies, and was clearly perceived as such. The Göttingen experimentalist James Franck (1882–1964) and Fritz Haber (1868–1934), director of the Kaiser Wilhelm Institute (KWI) for Physical Chemistry, also resigned their posts in publicly demonstrative fashion. Most dismissed scientists, however, left their posts more quietly. Likewise, among those who were subject to the purges, there was a range of responses, but overt protests were relatively infrequent. The departure of Erwin Schrödinger (1887–1961) from Berlin in 1933 – and, yet again, from Graz after the Austrian *Anschluss* – is one of the few examples of someone not subject to dismissal policy surrendering his post out of disapproval for the Nazis' policies.

Nazi cause. These advocates of *Deutsche Physik* or "Aryan physicists" held that a true understanding of nature rested on an intuitive (*anschauliche*) appreciation of observed phenomena; conversely, they saw many of the recent trends in theoretical physics, and relativity theory in particular, were too abstract and too mathematically formalistic,

and therefore unintuitive (*unanschaulich*). These were allegedly characteristically Jewish traits, and their prevalence in the field was due not only to the mere presence of Jews in the field, but also their supposedly undue influence. Einstein was particularly targeted in this regard. Although sometimes called a "movement," the Aryan physicists never formed a unified organization, and not infrequently had serious disagreements

Werner Heisenberg,
in the mid-1950s
(photo: Eric Schaal)

among themselves. Lenard, after 1933, functioned largely as a symbolic figurehead. Stark took on a more active role, and built up something of a personal empire in the German science administration: he became president of the PTR and head of the Reich Research Association (RFG); however, he was unsuccessful in a bid to become chair of the German Physical Society (DPG).

Beyond rhetorical attacks, the Aryan physicists used Nazi Party connections to influence faculty appointments. One protracted struggle arose over the succession to the chair of Arnold Sommerfeld (1868–1951), who had trained a generation of German atomic theorists at the University of Munich. As Sommerfeld neared retirement, it was clear that he favored as successor his one-time student Werner Heisenberg (1901–1976). Heisenberg, however, was made the subject of highly unfavorable articles in Nazi publications, in which his professional connections with Einstein received prominent scrutiny. Sommerfeld's post eventually

went in 1939 to a rather undistinguished applied mathematician.

But in many respects, by this time the power of "Aryan physics" was already waning. Stark had been forced out of his RFG post in 1936. No less an authority than Heinrich Himmler (1900–1945) stopped the journalistic attacks on Heisenberg, and although he failed to gain the longed-for position in Munich, he received an equally (if not more) prestigious appointment in Berlin in 1942. A discussion between leading adherents of conventional modern and "Aryan" physics affirmed that theory was a necessary part of the scientific enterprise, and declared that relativity theory could again be part of the curriculum – as long as Einstein's name was not mentioned. By the late 1930s, key power centers in the National Socialist state, including SS and military authorities, had become concerned about the state of German scientific research, particularly as it related to defense concerns. Mainstream physicists were increasingly successful in presenting the case that they had more to offer here than their ideological rivals.

Continuity and Cooptation

Despite these tumults, in many respects German physics was able to maintain some continuity of professional identity, and in some respects even prosper. One evidence of this was the fact that in 1937 the KWI for Physics, hitherto essentially an institute only on paper was, finally, actually constructed. The counter-campaign against "Aryan physics" relied largely on arguments about the practical utility, and above all the military potential, of physical research – as argued, for example, in a report from the DPG to the Reich Education Ministry in 1941. Rocketry and aerodynamics were two physics-related fields which received generous state support in the context of rearmament and the war. Nuclear research was an ex-

Arnold Sommerfeld, around 1930

ception which proved the general rule of scientific cooptation into the military effort: scientists and military officials both were interested in the future possibilities of nuclear energy, but due to material constraints and perceived technical difficulties other projects received higher priority during the war.

Physicists' ideological responses to Nazism also ranged over a wide spectrum – even apart from the Aryan physics controversy. Experimentalist Max von Laue (1879–1960), for example, hardly concealed his contempt for the Nazis. Planck and Heisenberg, to take two other instances, were hardly sympathetic to Nazism, but they saw national loyalty as critical and, as noted above, were willing to make certain accommodations with the regime to preserve their conception of science as an institution. Yet another kind of reaction was seen with the brilliant young theorist Pascual Jordan (1902–1980). His 1936 text *Anschauliche Quantentheorie* can be read as a refutation of the "Aryan physics" thesis; in other texts, he promoted the application of science to warfare and tried to draw connections between the findings of modern physics, the principle of authoritarian dictatorship, and the supposed downfall of parliamentary democracy. In short, the experience of the physics community as a whole during the Third Reich cannot easily be described as either repression or collaboration: it was a complicated intermixture of both.

Michael Schüring

Albert Einstein and His Fellow Expellees from the Kaiser Wilhelm Society

When the Max Planck Society for the Promotion of Sciences was founded as the successor to the Kaiser Wilhelm Society in two western occupation zones in 1948, German science was in the midst of a deep crisis of legitimacy. Far too many scientists had worked willingly for the goals of National Socialism, be it in merging race ideology into the biological sciences, in weapons research or in providing cultural-scientific justifications for the National Socialist claim of European dominion. A number of Einstein's younger physicist colleagues who remained in Germany after 1933 were the subjects of discussion among the expelled scientists. With respect to the behavior of Werner Heisenberg (1901–1976) and Carl Friedrich von Weizsäcker (born 1912), major distrust prevailed in the circles of foreign scholars. The Max Planck Society thus counted on members whose integrity was generally not questioned abroad, among them its patron and namesake, but above all Einstein's friend and colleague of many years, Max von Laue (1879– 1960), as well as the famous radiochemist Otto Hahn (1879–1968). As the first president of the successor organization, Hahn invited a small group of prominent former colleagues to become non-resident scientific members of the Max Planck Society. Hahn must have suspected that he would have little success with Einstein. In his letter he expressly mentioned Einstein's Jewish colleagues from the Kaiser Wilhelm Society who had accepted the invitation, such as the geneticist Richard Goldschmidt (1878–1958, Kaiser Wilhelm Institute for Biology), the physiologist Otto Meyerhof (1884–1951, Kaiser Wilhelm Institute for Medical

Otto Meyerhof
(1884–1951),
expelled in 1938,
Director at the
Kaiser Wilhelm
Institute for
Medicine Research,
Heidelberg
(photo:
Lotte Jacobi)

Research) and the physicist Rudolf Ladenburg (1882–1952, Kaiser Wilhelm Institute for Physical Chemistry). Further Hahn informed Einstein about the composition of the Max Planck Society Senate, and he emphasized that with the persons mentioned "any kind of revival of National Socialist tendencies in our new society is out of the question." However, in Einstein's view, that was not the point. In his response to Hahn he argued much more fundamentally:

"I perceive it as painful that I have to send a refusal to you of all people, that is, one of the few who stayed honorable and did their best during these evil years. But it is the only way. The crimes of the Germans are truly the most horrible that the history of the so-called civilized nations have to show. The attitude of the German intellectuals – viewed as a class – was not better than that of the rabble. I cannot even detect regret or the honest will to make that little good that could be made good after the gigantic murdering. Under these circumstances I feel an irresistible aversion against being involved in any kind of thing that embodies a piece of German public life, for the simple need to remain clean. You will certainly understand and know that this has nothing to do with the relations between us two, which have always been a pleasure for me."

Einstein remained the only scientist who rejected the Max Planck Society, which, of course, had hardly written to all of the around 100 persecuted members and staff of its own accord. But what induced Einstein's colleagues to carefully re-approach their former employer? In his answer, Otto Meyerhof let Hahn know what he though about the re-founding:

"Since this institution is ordained to function as the heir of the Kaiser Wilhelm Society, I would like to state that I regard my fourteen-year position as scientific member and director of the Kaiser Wilhelm Society in Dahlem and in Heidelberg as the most successful period of my scientific existence, that I remember these years and the working community with the colleagues there most fondly and that I am pleased to resume these relations again."

In fact, belonging to the old sphere of action meant much more to Einstein's colleagues, as the Kaiser Wilhelm Society had become an important and innovative research institution soon after its founding in 1911. The postwar network of the Max Planck Society, mediated through scientific

their own suffering and losses through dismissal and expulsion.

Few of the expelled scientists had enjoyed such good working conditions after expulsion as Albert Einstein. Many received quite meager pensions due to their late arrival, had to eke out a living with scholarships and consulting contracts, and battled both to keep their life work from falling into oblivion, and to acquire compensation from the general administration of the Max Planck

Left: Richard Goldschmidt (1878–1958), expelled in 1935, Director at the Kaiser Wilhelm Institute for Biology, Berlin

exchange, certainly harbored the potential to integrate exiles as well. However, it would be misguided to recognize only an indulgent will to reconcile in the exile's standard formulations of collegial esteem. Other motives were often concealed behind the façade of friendly recognition, which the German colleagues either did not understand or all-too-willingly interpreted in their own favor. Behind the re-establishment of contacts was the desire for moral reparations, and above all the hope for a clear sign acknowledging

Society. The highly esteemed biochemist Carl Neuberg (1877–1956, Kaiser Wilhelm Institute for Biochemistry), who, in contrast to Einstein, did not leave Germany until the last possible minute in early 1939, was one of the most prominent figures struggling desperately for acknowledgement. One of the most significant differences to Einstein was Neuberg's attitude toward Germany. He had been

Albert Einstein und Rudolf Ladenburg (1882–1952, until 1932 at the Kaiser Wilhelm Institute for Physical Chemistry) in Princeton

highly decorated in World War I, was a fervent patriot and later in the United States suffered criticism for his pro-German attitude. He read German newspapers, kept in touch with his successor Adolf Butenandt and even accepted a guest

cally. After all, having his greatness recognized early is no difficulty for someone who once studied astronomy; besides, Planck prepared us all for Einstein, as the Messiah prepares us for world redemption."

Lise Meitner (1878–1968), expelled in 1938, Kaiser Wilhelm Institute for Chemistry, Berlin

research residency in Germany in the early 1950s. Twenty-five years of belonging to the predecessor organization, Neuberg wrote to Hahn, "remains a strong bond." At times this bond was disproportionately stronger than the bond to his fellow colleagues in exile. Although Einstein and Neuberg were of the same generation and had belonged to Berlin's scientific elite within the very same scientific institution for over two decades, they never met in exile. Presumably, they would have had little to say to each other. In 1955 Neuberg passed judgement on Einstein, telling a colleague that "His (Einstein's) political attitude, or rather his somewhat foppish flirting with very left-oriented mentalities, is only possible because he, as once Pavlov in St. Petersburg, has been granted unconditional freedom to do what he likes politi-

Aside from the political background, Einstein's indifference to the Max Planck Society shows how little his world resembled the world of Neuberg, Meyerhof and Goldschmidt. Einstein's Kaiser Wilhelm Institute existed only on paper; he never had a large scientific operation with laboratories and staff to direct. For the directors who had ruled their institutes like minor princes, a world collapsed after 1933. Einstein, by contrast, shifted from one world to the next, apparently without any major melancholy, glad for the few things he could take with him.

Through the clear break he also saved himself the confrontations that befell other émigrés, despite all efforts at conciliation with their German colleagues. Above all, Lise Meitner (1878–1968), the important physicist at Otto Hahn's Kaiser

Carl Neuberg
(1877–1956),
expelled in 1939,
Director at the
Kaiser Wilhelm
Institute for Bio-
chemistry, Berlin

has never left me. But what is left of this sphere, and how does it look in the minds of the younger generation? Added to this is the circumstance that I have neither the gift nor the inclination to not want to know things because they are too oppressive. I followed all of the terrible events the Hitler system entailed quite precisely and attempted to understand their reasons and consequences, and that means that even today I presumably have a different attitude to some problems than the majority of my German friends and colleagues. Would we be able to understand each other?"

This question did not concern Einstein. Lise Meitner, whom he had always esteemed highly, nevertheless undertook a task

Wilhelm Institute for Chemistry, and Otto Meyerhof for many years made efforts towards dialogue with the Germans, in order to arrive at a mutual understanding of the recent past. With so many suppression mechanisms and defensive strategies at work, this was a thankless task. When Meitner's former colleague at the Kaiser Wilhelm Institute, Fritz Strassmann, asked her whether she was ready to return to Germany, she refused, "although the longing for my old sphere of action

that became a painful experience for many exiled scientists. She had taken it upon herself to help her German colleagues, and in doing so was confronted again and again with the circumstance that "the Germans still do not grasp what happened." Above all, many German scientists at the time did not understand that their refusal to lay open their past degraded their exiled colleagues to supplicants and troublemakers. Einstein was spared this experience.

Barbara Picht

Succor and Political Action:
How Einstein Related to Emigration

Albert Einstein arrived in New York on 17 October 1933. His biographer Albrecht Fölsing reports that the Mayor of New York was already waiting with a parade to welcome the famous arrival at the 23rd Street pier: It was campaign time, and the Jewish vote was at stake. However, a launch had approached the steamer *Westernland* long before docking, taking on board Einstein, his wife Elsa, his secretary Helen Dukas and his assistant Walther Mayer. The mayor waited in vain while

public as much as possible to facilitate their undisturbed scientific work. Shortly after his arrival, Einstein described Princeton as "a wonderful spot, and yet at the same time an extraordinarily amusing ceremonial Gotham of tiny, stilted demigods. But by offending respectability it is possible to secure a pleasant tranquility for myself, so that is what I am doing."

However, on occasion Flexner went too far in his attempts to protect Einstein from undesired publicity. Motivated by jealous pride in the prominent new member of the institute, and to restrict Einstein's creative power and fame for the benefit of the institute, he opened mail addressed to Einstein and even rejected invitations in his name. One example of such an occasion took place in early November 1933, when President Roosevelt asked Einstein and his wife to visit the White House. All the same, Einstein found out about the invitation. He hastened to ensure the First Lady that he was very interested in meeting the president, and vigorously –

Albert Einstein at the laying of the foundation stone for the first building of the Institute for Advanced Study in Princeton on 22 May 1939 The others pictured, from left to right: Alanson Bigelow Houghton, Lavinia Bamberger, Anne Crawford Flexner, Abraham Flexner, J. R. Hardin, Herbert Maas, Harald Willis Dodds

Einstein and his companions were rushed through the formalities of immigration and driven to Princeton. This "secret arrival" had been organized by Abraham Flexner, the director of the Institute for Advanced Study in Princeton, who in the August of 1932 had succeeded in obligating Einstein for half-year residencies from fall to spring at his still young, later so famous institute. Here outstanding scholars were offered positions free of teaching duties, providing them with the opportunity to dedicate themselves completely to research. According to the conception of Flexner and his donors, the Jewish American Bamberger family, the members of the institute were to be shielded from the

and successfully – protested against such outrageous patronization on the part of the institute's curator. On 24 January 1934 the Einsteins dined with the Roosevelts.

The New York Rabbi Stephen Wise, a friend of Einstein's who had initiated the meeting, had hoped that it would increase public attention to the fate of the Jews who had fled Germany. While this hope was not fulfilled, Einstein himself provided active support for the refugees in Princeton. He received countless requests for expert opinions and letters of recommendation. Immigration into the United States had been subject to a quota since World War I, which originally permitted

51,227 immigrants from Germany annually, but this number was halved to 25,957 after the stock market crash in 1929. Those who succeeded in obtaining a visa were often faced with the difficulty of establishing themselves professionally. He ran "an agency for the persecuted and for intellectual eccentrics," Einstein wrote in August

Hermann Broch was able to escape to the United States only thanks to the affidavits – written guarantees that the undersigned would accept responsibility for the livelihood of an immigrant if needed – assumed for him by Albert Einstein and Thomas Mann. Einstein even made his house on 112 Mercer Street in Princeton available to the

Albert Einstein and Thomas Mann in Princeton 1938 (photo: Lotte Jacobi)

1938, and he could ensure "that business is booming." Next to Thomas Mann, Einstein was one of the most prominent émigrés; many expected that a recommendation from him would give them a chance. A favorite anecdote tells of four physicists applying for a free position in a hospital: each of them was able present a letter of recommendation from Einstein. But it should not be forgotten that his goodwill was evinced in more than just nonbinding recommendations and expert opinions. For instance, the Viennese essayist and writer

writer so often beset by financial difficulties in exile from mid-August until mid-September 1939. Broch was not the only one Einstein sponsored: Although he was hardly without means upon his arrival in the United States, having invested funds in Switzerland and the United States while he still could, and although he occupied a well-paid position at the Institute for Advanced Study, in 1938 Einstein found himself compelled to refrain from issuing further affidavits, as doing so would have endangered the guarantees he had already submitted. This conveys an idea of the extent to which he had taken on financial obligations for refugees.

Thanks to the Institute for Advanced Study and its members, Princeton also became an important place of scientific exile, although this was not one of the declared goals at the institute's founding. Among the prominent émigrés there was the mathematician Hermann Weyl, who held a chair at the institute for seventeen years starting 1933, and who had already concerned himself with the theory of relativity in his scientific work for years before encountering Einstein again in Princeton. Appointed in the autumn of 1935 to the new School of Humanistic Studies set up at the institute in the same year, was Erwin Panofsky, an art historian from Hamburg, who became acquainted with Albert Einstein there and came to appreciate the interdisciplinary discourse with him and other natural scientists at Princeton. From the autumn of 1938 until spring 1941 Thomas Mann, too, lived just a few streets away from Einstein in Princeton, where he had accepted a guest professorship at the institute. A friend of Mann's for many years was Erich von Kahler, philosopher and cultural historian from Prague, who also followed Mann's recommendation in moving to Princeton after deciding to emigrate in 1938. From 1942 to 1948, his friend Hermann Broch lived in Kahler's home in Princeton. Einstein cultivated contact with both of these men; however, the large house the Manns kept in Princeton was not his world, however. Yet, in their commitment to the expellees, Mann and Einstein were of one mind and supported each other. Like many of the expelled scientists and artists, both felt a responsibility to help other refugees and to take a stand on political events

The writer Hermann Broch (1886–1951), around 1950

in Europe and the U.S. The émigrés sensed what Thomas Mann designated a "obligation to participate in politics." In Princeton this created a form of public that was far removed from Flexner's vision of a shielded institute dedicated exclusively to science.

Even in Germany, many years before his flight, Einstein had expressed his views on political topics. In the autumn of 1914 he was one of only four signatories to the *Appeal to the Europeans*, which he had and the physiologist and pacifist Georg Friedrich Nicolai of Berlin had written to protest against the appeal *To the World of Culture*, in which 93 prominent signatories denied that Germany was responsible for the war and defended militarism. This protest was the beginning of Einstein's public engagement for political goals. He generally advocated them independent of the political events of the day, and so fundamentally that he is often considered today as more of a philosopher than as an advocate of politics as the art of the possible, more as a moralist and prophet than as a political pragmatist, as Fritz Stern writes. In the political concepts of the émigrés, we again encounter the fundamentalism and claim to universality with which Einstein had promoted world peace and the maintenance of a common world culture. In the autumn of 1939, a group of émigrés and American intellectuals formed in Princeton, with the Italian historian and later son-in-law of Thomas Mann, Giuseppe Antonio Borgese, and Hermann Broch at their center, to work on a joint book project. This book was supposed to counter Hitler's plans for world domination with a political concept. The product that emerged was the book *The City of Man. A Declaration on World Democracy,* in which the authors promoted a democratic world state. Broch had also hoped to convince Einstein to participate, but Einstein refused. Yet Einstein shared the political convictions of the Princeton Group, that peace and security could only be created and defended by an international organization. Back in the Germany of the 1920s he had championed this goal, and had high regard for Wilson's League of Nations initiative, despite the difficulties in its political implementation. In a radio address to a student assembly in Chicago in May 1946, Einstein – now under the

Their political and personal experiences had reduced their "trust instability, indeed, in the viability of civilized human society," as Einstein wrote in 1939, the year in which the *Declaration on World Democracy* was published. The political concepts of exile were a response to this expe-

Albert Einstein in front of his house in Princeton, in winter 1939 (photo: Eric Schaal)

impression of the atomic bombings of Hiroshima and Nagasaki – spoke about the necessity of creating a world government. It was not wishful thinking, removed from reality and blind to political necessities, which moved Einstein and the other émigrés to the seemingly visionary appeal to create a world government and a world state.

rience, and much of their emphasis on human rights to peace and freedom and the necessity of international cooperation is found again in the 1945 Charter of the United Nations, a future organization to securing peace which Roosevelt began planning in the same year, 1939, the year the war broke out.

Gerald Holton

The Woman in Einstein's Shadow

I shall attempt in these few pages to draw attention to a woman to whom every historian of modern physics is indebted, but whose role is now known in any detail only to a mere handful of specialists – a woman who for twenty-seven years spent more

Albert Einstein and Helen Dukas during a benefit concert in the Synagogue on Oranienburger Strasse, Berlin, on 29 January 1930 (photo: Erich Salomon)

time face-to-face with Albert Einstein than perhaps any other person: Helen Dukas, the self-effacing but ever loyal and extraordinarily effective secretary and helpmate of Einstein.

From her first day of employment in 1928 to Einstein's death in 1955, and indeed importantly for many years afterwards, she was the person who read, typed and often translated Einstein's correspondence, and saw to it that the vast correspondence and manuscripts were saved (in the face of Einstein's own typical disinterest and neglect of such matters). Without her scrupulous collector's passion and devotion to Einstein, we simply would now have only a mere fraction of the collected papers of Einstein that have already sparked so much important scholarship. In the Berlin years as well as in Princeton, she was also a member of the household, and a fierce protector of the family's

privacy. On Einstein's death, she became a Trustee of his Estate according to his Last Will. I have told elsewhere how I first met her in the first of some fifty or sixty visits during the decades. I had traveled to the Institute of Advanced Study in Princeton on 13 August 1959 with a recommendation from Philipp Frank to Helen Dukas, in the hope of my being able to consult some of the documents in the *Einstein Nachlass* while working on a paper for a conference. After Einstein's death she had been relegated to the large, room-sized vault in the basement of Fuld Hall. That's where I found her, the whole scene illuminated only by her rather insufficient desk lamp. She was sitting at her desk, bent over some papers; a large stack of file drawers loomed in the darkness beyond. I could not help but think of Juliet in the crypt, after the death of Romeo. She was born on 17 October 1896 in Freiburg in Breisgau, the fourth of seven children. According to a Memorial Essay by Abraham Pais, she had to interrupt her Lyceum education at age fifteen, after her mother had died, and took charge of running the household and bringing up the younger children. Later she became governess in the home of Raphael Straus in Munich, one of whose new nephews was Ernst Straus. Let me give you a taste of this remarkable woman's wit and tough realism. As it happened, in the 1940s, Ernst came to the Institute, to be one of Einstein's assistants. When Ernst introduced himself to Helen, she said, "Of course I already know you well: I was present at your circumcision". Or again: In 1965, in one of her letters to me (and we had a correspondence of well over one hundred letters altogether), she said the Russian historian-philosopher Kuznetsov "has sent me the English translation of his new Einstein biography. With my letter of thanks, I enclosed two pages of corrections; and I have found since some more." To the President of the Israel Academy of Sciences she wrote in 1979, "I looked a little into the catalogue (of an exhibit he had sent). The Zionist Congress of 1949 took place in Zurich, not Munich. I was there!"

Her self-assigned task was now to attend to the continuing correspondence and inquiries, also to find new documents, retype old, fading ones, or

handwritten ones, particularly those in Gothic script. Her sharp memory and her utter devotion and reliability became quickly obvious to me. She had been immediately helpful in my initial visit. But more importantly, I soon came to feel that for the sake of the profession of historians of science one must somehow capture her experience, her memory of the events and correspon-

dence in which she had been involved. In the absence of serious help, she had been trying to type out a catalog of the papers, correspondence, and manuscripts. John Wheeler soon put the matter clearly in a letter of 12 December 1961 to Robert Morrison of the Rockefeller Foundation. He recommended that I be given support for a serious project, namely to put the huge heap of correspondence in good order for use by scholars. Only a few had dared to use it so far. Wheeler wrote, "... the great mass of the material is unorganized. Miss Helen Dukas works at this only in a limited way and without assistance or guidance by anyone trained in the history of science. An enormous task requires doing it, and it goes ahead only at a niggling pace." The financial support I looked for (and which was granted) was also needed for a first microfilming of at least the scientific part of the collection. To quote here from one of Helen's letters to me of those days: "The work you have in mind for me fascinates me, but also fills me with apprehension." And in another letter, "I have been hoping for something like this to turn up."

I also had to plant in Helen's mind the idea of eventually allowing publication, so as to provide scientists, historians of science, and philosophers of science with the necessary material for future good work. Moreover, she had to be made to see the historical value of the riches all around her, and to bring into the vault what she called the "personal stuff," which she kept at home,

Helen Dukas with Albert Einstein in the study on Mercer Street in Princeton, 1940 (photo: Lucien Aigner)

and which really was needed to supplement the "scientific correspondence."

This effort succeeded by September 1968, when she had a number of file cabinets brought from

Helen Dukas, Albert Einstein and daughter Margot take the oath of U.S. citizenship in Trenton, New Jersey, 1 October 1940

the house at 112 Mercer Street into the vault, to be included and cataloged. She also made available documents that she had kept in a safe within that vault; it contained correspondence with Freud, Roosevelt, Romaine Rolland, Elsa's letters. So eventually there were orderly, cataloged folders on Gandhi, Paul Valéry, Bertrand Russell, Chaim Weizmann, the Queen of the Belgians, Tagore, Schweitzer, Thomas Mann, Bernard Shaw, as well as the light-hearted verses of Einstein – all those joining the files of Schrödinger, Pauli, Curie, Lorentz, Bohr, Born, Ehrenfest, Infeld, Hilbert, Bose, de Broglie, Bohm, Debye, Eddington, and so on, to Meitner, Minkowski, and so forth to Wenzl, Wien and Zeeman. By 1963 there were one hundred and thirty such file folders done and catalogued, some very bulky, with Ehrenfest's having no fewer than one hundred and sixty-five items. And from about 1976 on, the strong editorial staff

of the Princeton University Press project greatly expanded what we had started. By the time John Stachel finished in January 1980, he had 42,000 items in his big index, which he wrote me to have been initially based on what he called our "little index."

The Foundation money I had raised and which Harvard administered was primarily to provide her with a salary for her work (which, to her amusement, made her my research assistant), and it also gave her the companionship of physics graduate students at Princeton, selected by John Wheeler and myself. These students were hired to come for a few hours or days per week; they did excellent work in cataloging, and also brightened Helen's life. To help with the work at hand, I made periodic visits to the Institute myself, at least monthly, sometimes weekly, starting in the early 1960s, and staying for longer periods as Member of the Institute in 1964 and as Visitor in 1971. Let me confess that I came to respect and even love Helen – much like one of my favorite aunts – and I think a little of such feelings might have been true for her too. We trusted each other fully. After the first few years, whenever she was occasionally ill, she would permit me to work in the vault on my own or to supervise the students, having given me the code for opening the vault and the keys for the files and the safe within. During that whole decade of visits and collaboration and correspondence, I recall not a single time when we were out of sorts with each other. In 1964, she had the idea of giving me a present. It was precious indeed – the set of reprints of Einstein's published papers that had been kept near his desk, and on a few of which he had made corrections and additions. (These are noted in the published Collected Papers.) The set was bound in three volumes, and on the first page, Helen had written a dedication to me: "To my helper [...]"

Among the countless other direct and indirect results of these first years of putting together the Archive, I will just mention one: On 2 October 1962, Tom Kuhn wrote Helen and me for examples of how we were cataloging the Archive, so as to serve as a model for ordering the Niels Bohr Archive in Copenhagen. Einstein and Bohr, wherever they were then, still exchanging ideas, must have been amused by that news."

It is a pitiful irony that Helen, in part because of the delays caused by the Executor, Dr. Nathan, did not live to see the publication of the first volume of the projected thirty-volume series of Einstein's Collected Papers. She died on 10 February 1982, at age eighty-six, having been in full possession of her lucid faculties to the end. From the late 1960s on, I and others had urged her to fulfill her own plan of moving to Jerusalem, where she had relatives, and perhaps to continue from time to time to attend to the papers she had so lovingly dealt

with through most of her life. In a letter of 11 February 1971, Yehuda Elkana of the Hebrew University offered her a room at the Library where the Einstein Archive was to be transferred upon the

Helen Dukas with Albert Einstein, early 1950s

decision of the Trustees (as in fact it was about a decade later), and also offered to produce living accommodations for her. She could have added much to illuminate the newly acquired materials in the Archive. But it was not to be.

Yet, she lives now, having been an "outsider" for the history of science, inside the Collected Papers.

This essay is based on chapter 2 of the forthcoming volume: Gerald Holton, *Victory and Vexations in Science: Einstein, Bohr, Heisenberg and Others*, Harvard University Press, 2005

Don Salisbury

Albert Einstein and Peter Bergmann

Peter Gabriel Bergmann was Einstein's young assistant from 1936 through 1941 at the Princeton Institute for Advanced Study. He and Einstein were the first to suggest, in an article published in 1938, that a fifth spacetime dimension was real, yet unobservable due to its small finite size. The fifth dimension was originally proposed in attempts toward the unification of gravity and elec-

Albert Einstein with Peter Bergmann and Valentin (Valya) in Princeton, 1940 (photo: Lucien Aigner)

tromagnetism by Kaluza and by Klein. This idea is a centerpiece of contemporary brane theory. Bergmann wrote the first English language introductory textbook in general relativity in 1942. It contains a preface by Einstein. The research program in general relativity initiated by Bergmann at Syracuse University in 1947 was the first in the United States, and for the following two decades virtually every leading relativist in the world came to Syracuse to explore with Professor Bergmann the modern frontiers of general relativity. The primary focus of Bergmann's research throughout his career was the creation and interpretation of a quantum theory of gravity. Given his expertise in both classical general relativity and quantum theory he was uniquely suited for this task, as attested by Albert Einstein himself in his

support of Bergmann's application for research funding in 1954, in which he says "Dr. Bergmann is one of the few who are completely at home in both theories. He is an imaginative theorist, very enterprising and undeniably gifted with an intuitive feeling for the essential ..." His Syracuse school and P. A. M. Dirac developed, initially independently, a mathematical technique for casting general relativity into a form susceptible to rebuilding as a quantum theory. A variant of their methods is the standard approach to almost all modern quantum theories that unify the fundamental interactions of nature.

Bergmann was born is Charlottenburg, Berlin, in 1915. His parents were divorced early in his life. Young Peter moved with his pediatrician mother, Dr. Emmy Bergmann, to Freiburg in 1921. His father, Dr. Max Bergmann, was from 1921 through 1933 the director of the *Kaiser Wilhelm Institut für Lederforschung* (now the *Max Bergmann Zentrum für Biomaterialien*) in Dresden. In 1935 his father assumed a position at Rockefeller University in New York City where he made significant advances in molecular biology. His mother and sister emigrated to Israel in 1935. His aunt was murdered by the Nazis, probably in Auschwitz. As a Jew it became evident to the young Peter as in 1933 he awaited acceptance of his application for admission that he would not be able to pursue his original plans to study theoretical physics in Berlin. Meanwhile, unbeknownst to him, his mother wrote to Albert Einstein seeking to persuade him to take on her son as a doctoral student. It is possible that Frau Bergmann felt emboldened to appeal to Einstein because of her acquaintance with Helen Dukas, who served as Einstein's secre-

Albert Einstein and Peter Bergmann

tary from 1928 until his death. The Bergmanns took over the Dukas home in Freiburg in 1933, when the Dukas' were moving to Berlin. Einstein replied through underground post from his temporary refuge in Belgium. He suggested that Bergmann pursue his doctorate in Zurich. He offered the possibility of bringing him to Princeton following the completion of his studies, asking only that in the following year Bergmann send a sample of original work so that he could form a judgement of Bergmann's scientific merits. Apparently Frau Bergmann forwarded the Zurich suggestion to her son, but did not communicate the source. Bergmann decided instead to enroll at Charles University in Prague, primarily since it was by far the least expensive of foreign German-speaking universities. He studied there under Phillip Franck. Shortly before he received his doctoral degree in

1936 he wrote to Einstein seeking a research position with him, and he did so completely unaware of his mother's previous letter. He included copies of his first two scientific articles. Nor was he aware that Franck, who had taken over Einstein's position in Prague, was a longtime friend who maintained a scientific correspondence with Einstein. Franck's recommendation was probably pivotal in Einstein's decision to invite Bergmann to join him as his assistant.

Bergmann later recalled first meeting the "big man" in 1936 as he was deeply engrossed in conversation with his other two young assistants Leopold Infeld and and Banesh Hoffman. Einstein's younger colleague John von Neumann was given the responsibility of acquainting Bergmann with the Institute. So began five years of intensive research "with and under Einstein." The day usually began late morning at Einstein's house on Mercer Street, where Einstein's assistants would join him for the walk to the Institute. After many hours at the blackboard the group would accompany Einstein back to his house. Often conversations would continue evenings over the house's party-line telephone – much to the consternation of the landlady. Discussions continued over the summer through written correspondence, often several times a week.

At the time of Bergmann's arrival in 1936, Einstein, Infeld and Hoffmann (ETH) were occupied with the relation between particles and fields in the general theory of relativity. Bergmann later recalled his astonishment that it was not known until twenty years after the birth of the general theory that the dynamics of the gravitational field fixed the motion of its material point sources. Today variations of the ETH result are routinely applied in computer calculations which have as their aim to predict and interpret gravitational waves which may soon be detected by an international system of laser interferometric gravitational observatories. Einstein's original suggestion to Bergmann was that he adapt methods that had been applied to the purely gravitational situation, and apply these methods to electrically-charged particle sources. Einstein's hope was that the particle itself would again prove to be superfluous, and that embarrassing singularities would thereby be eliminated from the theory describing the interaction of charged, massive particles. Unfortunately

Albert Einstein in his study in his Princeton home, 1939 (photo: Eric Schaal)

it soon became apparent that if a theory describing a classical spinning electron existed, it would have a ratio of mass and electric charge that would be incorrect by several orders of magnitude. The project was abandoned; this was Einstein's final effort to derive properties of fundamental particles from classical field equations. Einstein and Bergmann then turned to an effort that would occupy Einstein for the rest of his life, the search for a unified classical theory of gravity and electromagnetism. They were joined a year later by Valya Bargmann. The resulting papers published in 1938 (with Einstein) and in 1942 (with Einstein and Bargmann) contain two remarkable innovative ideas, both of which, judging from the existing correspondence, were Bergmann's inspiration. One finds here the first suggestion that a fifth

spacetime dimension was real but too small to be observable. This is a central component of current higher-dimensional unified theories. The authors also recognized for the first time that the well known freedom to transform potentials in electromagnetism is a consequence and expression of the freedom to undertake coordinate transformations in the five-dimensional spacetime. This is another common central feature of modern higher-dimensional geometrical unified theories.

Upon arrival in Syracuse in 1947, Bergmann immediately launched an effort to quantize general relativity. Since time duration is dynamically determined in Einstein's theory of gravity, it was not immediately apparent how the quantization procedures that were then being successfully applied to the other fundamental interactions of nature could be applied to gravity. The so-called canonical quantization program requires the reformulation of a theory as a problem in time evolution. But the general freedom to alter spatial and time coordinates manifests itself in highly non-linear equations of motion, some of which do not even involve accelerations. The first steps in disentangling the related conceptual and technical questions was reported by Bergmann in 1949 in a publication entitled *Non-Linear Field Theories*. A series of publications dealing with the reformulation of general relativity as a problem in classical time evolution followed in the next decade, with numerous collaborators. P. A. M. Dirac independently formulated and actually pushed further a parallel program, and the two equivalent procedures are now known as the Dirac–Bergmann approach. Indeed, the reformulation of general relativity as an algorithm for evolving the gravitational field in time, given the initial values of the field and rates of change of the field at an initial time, now makes it possible to find increasingly accurate numerical solutions of Einstein's equations.

In preparation for a study of the nature of observables in the initial-value formulation of general relativity, in 1956 Bergmann published his first paper discussing the nature of observable quantities in quantum electrodynamics. Later, Arthur Komar and he reformulated in the new formalism an idea concerning observables in general relativity that had first been proposed by Albert Einstein as he struggled to come to terms with the freedom to alter space and time coordinates without altering the physical content of general relativity. The idea was to let either the gravitational field or material sources fix a preferred coordinazation. In the early 1970s Bergmann and Komar turned their attention to the residual general-relativistic symmetries that are manifest in the initial-value formulation of general relativity. This symmetry, coupled with the idea they resurrected from Einstein, offers an approach toward the construction of an ultimate quantum theory of gravity being explored by many researchers around the world today.

Peter Bergmann was revered by the thirty-two students whose doctoral theses he guided, a comparable number of postdoctoral researchers who spent up to two years at Syracuse, and innumerable colleagues and friends. All prized his incisive intellect, his infectious modesty, his commitment to dialogue and social justice, and his personal warmth. He was instrumental in fostering international cooperation, playing a central role in the formation of the *General Relativity and Gravitation Society*, and seeing to it that relativists from all over the world, including behind the iron curtain, could participate in its functions. In 1963 he also helped to organize what became the Texas Symposium on Relativistic Astrophysics. In September 2002, along with John Wheeler he was awarded the first Einstein Prize by the American Physical Society. He learned of the prize shortly before his death on 19 October 2002.

Jörg Zaun

Josef Scharl and Albert Einstein:
The Story of a Friendship

"He was never amenable to feeble compromise, neither as an artist nor as a person; it would have been impossible for him to make his external existence more comfortable through such sacrifices of honesty. A great artist by nature, he followed only his inner voice, which allowed him unswervingly for the first time as part of the "Munich New Secession" exhibition.

At the invitation of the photographer Lotte Jacobi, Josef Scharl went to Berlin for several months in 1927. He had met Lotte Jacobi in Munich, where she studied photography and film, as one of the

Josef Scharl around 1927 (photo: Lotte Jacobi)

to find the sure path to heightened mastery and maturity." With these words Albert Einstein honored Josef Scharl in an obituary, following his death in December 1954. The physicist had first met the painter, twenty years his junior, in Berlin in 1927, but it was not until they were both in exile in the United States that their deep friendship emerged.

Josef Scharl was born in Munich in 1896, the son of a gardener, and initially apprenticed as a scene painter after finishing school. In 1915 the 19-year-old was drafted into military service; in the following year he was wounded near Verdun. After the end of World War I, Scharl enrolled at the Munich Art Academy, where he soon made contract with and found recognition in a circle of young artists. In 1923 he showed his own work

few women at the Art Academy. Together with her sister Ruth, also a photographer, Lotte saw to it that the young painter from Munich made many acquaintances in Berlin – among them, the already world-famous Albert Einstein and his wife Elsa. This meeting inspired Scharl's first oil portrait of Einstein. Einstein had not posed for it, though: Scharl painted it from memory, as he did the later oil portraits of Einstein in the U.S. Originally Scharl wanted to present Einstein the painting as a gift, but Elsa sent back the courier with the message that her husband, fortunately, did not look like the man in the painting.

In 1930, the city of Nuremberg awarded Scharl an Albrecht Dürer Foundation scholarship of 500 Reichsmarks. In return, the artist was expected to cede one of his works to the municipal gallery.

Josef Scharl in his studio in Munich, around 1933. The pictures that can be recognized are *The Beast*, 1933 and *Self Portrait*, around 1933 (missing)

Initially Scharl selected a landscape, but then substituted the Einstein portrait instead. With the scholarship Scharl was finally able to fulfill a long-held dream: to travel to Paris to study. Gradually Scharl's recognition as an artist began to grow, and he was able to show his first solo exhibition in Berlin in January 1933, shortly before the Nazis seized power. The presentation in the Neumann-Nierendorf Gallery was well received

Nuremberg, and again in the 1936 exhibition in Berlin. Scharl was bitterly disappointed that his pictures were no longer included in the famous Munich exhibition of degenerate art in 1937. The Einstein portrait was later removed from the art collection of the city of Nuremberg and all records of the picture were destroyed. However long after the end of World War II, it turned up again in a private collection.

Left:
Albert Einstein's affidavit for Josef Scharl's application for naturalization, 1941

Right:
Josef Scharl with his wife Leni and his son Alois around 1924

by the Berlin press. Yet Scharl was soon to learn what the new rulers held of his art. In April 1933, the first exhibition of undesirable art took place in the Municipal Gallery of Nuremberg, entitled "Chamber of Horrors"; Scharl's Einstein portrait among the works displayed. In 1935 the picture was shown in the *Entartete Kunst* exhibition on the occasion of the Nazi party convention in

Josef Scharl was forbidden to paint, but this ban did not keep him from working. However, it became increasingly difficult for him to earn his living as an artist. In 1938 he finally decided to emigrate to the United States. He left his wife Leni and his son (born in 1922), in Munich, in the hope that they could follow him later, when he had established himself as an artist in the U.S. He traveled to New York, where many of his old acquaintances had also landed; among them, the two photographers Lotte and Ruth Jacobi. Emigration saved Scharl's freedom, but his success as an artist, which had gradually begun to take shape in the last years before 1933, was something he could not take into exile with him. In the lean years, he worked as a book illustrator to earn a living.

albeit fleetingly, and Einstein was well known to be willing to take up the cause of political refugees. Einstein also declared himself willing to sponsor Scharl, and from the contact thus reinstated, a deep friendship between the physicist and the painter soon developed. Because Scharl was not permitted to leave New York during World War II, the friendship initially took the form of a lively exchange of letters. From 1946 on Scharl regularly visited the scholar in Princeton, resulting in the creation of numerous oil portraits and countless drawings. Scharl appears to have been fascinated by Einstein's hands as well, which he captured in dozens of sketches. Einstein apparently never purchased any of Scharl's paintings or drawings, but he provided considerable financial support for several of the painter's exhibitions, always remaining incognito.

Josef Scharl with Albert Einstein in Princeton, around 1950

For several years after the end of World War II, Scharl was greatly concerned about his son, who had been captured by the Soviets as a German soldier in the last year of the war. He was not released until after many petitions, which Einstein also actively supported, and returned home in 1950. However, Scharl had since become a stranger

In 1941 Scharl applied for citizenship in the United States. This application had to be sponsored by an American citizen. Scharl turned to Albert Einstein for this; exactly why he asked Einstein is not certain, but two had met before,

to his wife and his old country, and consequently he neither returned to Germany nor brought his wife to New York. During a stay in Switzerland in 1952, he did not even cross the border to visit Germany.

Scharl was a welcome guest in the Einstein residence in Princeton. He was an exceptionally good storyteller with a gift for comedy. His visits were much appreciated by Einstein's sister, Maja, his stepdaughter, Margot and his secretary, Helen Dukas. Einstein and Scharl were bound by a deep respect for each other's work and, similar political pay court to it not only when it is rebelling, but also in the intervals, when it is not up in arms. That may well be a kind of capitulation, but it is worth it."

Scharl's last documented visit to Princeton took place in February 1954. When Scharl died in December 1954, Einstein was already too weak

Portrait of Albert Einstein, 1927 Oil on canvas, 70.5 x 57cm Private collection, Berlin (photo: Jochen Littkemann)

Portrait of Albert Einstein, 1952 Oil on canvas, 48.5 x 39.5 cm Gillhausen collection, Hamburg

views – but also a shared ailment: both had a very sensitive stomach. In a letter from 1949 Einstein gives Scharl the friendly tip: "I have learned one thing in this matter: You have to to attend the funeral. The obituary he wrote for his friend was read by the physician and painter Felix Fuchs: "Everything about him was genuine, original and pure."

Tibor Frank

Closely Associated:
Leo Szilard and Albert Einstein

The nuclear physicist and biologist Leo Szilard (1898–1964) came to know Albert Einstein during Szilard's first year in Berlin: 1920 or right after. Einstein thought highly of the young Hungarian, whom he considered a "fine, clever man who would otherwise not lend himself to illusions.

Albert Einstein and Leo Szilard draft the famous letter to Roosevelt (re-enactment)

Perhaps, like many people of similar nature, he is prone to overestimating the role of the rational in human life."

Szilard was born in Budapest, Hungary into a middle-class Jewish-Hungarian family. After the Hungarian Bolshevik revolution of 1919 he felt threatened and decided to leave his native country at the very end of December 1919. This was a momentous year for young Szilard: he changed his country, his religion, his language, and his field of study.

In 1922 Szilard completed his doctoral degree under Max von Laue (1879–1960) at the University of Berlin. Both of his dissertations dealt with the foundations of thermodynamics. In his doctoral dissertation of 1922 (*Über die thermodynamischen Schwankungserscheinungen*) Szilard showed that statistical fluctuations, which hitherto had been taken as proof of the reality of atoms, could be included within the framework of phenomenological thermodynamics without making any reference to atoms. He arrived at his results in a highly characteristic way. He remembered much later that he had spent several agonizing months working hard on a problem in thermodynamics and had come to believe that he had no chance of solving it. He then took a vacation for a month, determined just to loaf, but while relaxing an unrelated idea in statistical mechanics occurred to him, and he solved the problem before the end of the vacation.

At first Einstein, much of whose past work had been based upon the analysis of statistical fluctuations, did not believe Szilard's solution when Szilard handed him his manuscript; but the next day von Laue telephoned the young man, saying that his work had been accepted as a doctoral thesis. Von Laue praised Szilard's dissertation as an "independent" and "a major achievement", primarily for the pioneering use of Einstein's analysis of energy fluctuations, and judged it to be outstanding ("Eximium"). The dissertation was published three years later.

In the late 1920s, Szilard worked with Einstein on a number of innovative approaches to cooling, making good use and taking some advantage of his celebrated co-worker by patenting his inventions under both their names. These included a refrigerator (*Kältemaschine*) and an apparatus for transporting liquid metal, especially for condensation of gases and vapors in refrigerators (*Vorrichtung zur Bewegung von flüssigem Metall, insbesondere zur Verdichtung von Gasen und Dämpfen in Kältemaschinen*) in 1927, an electromagnetic apparatus for producing an oscillatory motion (*Elektromagnetische Vorrichtung zur Erzeugung einer oszillierenden Bewegung*) in 1928, a compressor (*Kompressor*) in 1929, and a pump, chiefly for refrigerators (*Pumpe, vorzugsweise für Kältemaschinen*) in 1930. (For some reason, Szilard also patented basically the same inventions a little later under his own name alone in Germany.) The new principle that Szilard and Einstein conceived promised to lead to a new type of refrigerator that was more efficient and less dangerous to operate than existing ones. One of their main ideas was

to produce cooling by causing alcohol to be absorbed by water. However, this idea did not prove to be of lasting industrial influence. More important was their invention of what became known as the Einstein-Szilard electromagnetic pump, in which "a traveling electromagnetic field causes a liquid metal to move." The prestigious A.E.G. became interested and involved in developing a prototype of their pump for refrigeration purposes.

Although Szilard and Einstein patented their refrigerator in Great Britain, the United States, and Hungary, and although Szilard later tried to promote interest in it in the United States, it was never marketed, mainly because of the devastating economic effects of the Great Depression. Einstein also declined to lend his name to advertisements for the refrigerator, which materialized only as an A.E.G. prototype. Shortly before his death, Szilard was asked about this refrigerator. He replied: "Oh that? That went into the atomic bomb."

Two weeks after the Nazi party's success in the Reichstag elections of 14 September 1930, Szilard assessed its significance in a letter to Einstein: "From week to week I detect new symptoms, if my nose doesn't deceive me, that peaceful [political] development in Europe in the next ten years is not to be counted on ... Indeed, I don't know if it will be possible to build our refrigerator in Europe."

In 1931 Szilard asked Einstein for a letter in support of his application for a U.S. visa, drafting a few sentences for Einstein's signature. Einstein changed Szilard's modest wording from "he is well known to me ("*persönlich gut bekannt*") for many years of joint work" to "he is closely associated" (*"eng verbunden"*) with me ...," adding that he had "a direct interest" in Szilard's journey to America. In October 1931 Einstein also supported the granting of a non-quota immigrant visa to Szilard. Szilard finally left Berlin after the Reichstag fire at the very end of March 1933.

Jewish groups in Europe considered raising funds to establish a new university somewhere in Europe to be staffed by a refugee faculty (*Flüchtlingsuniversität*), an idea conceived by Einstein; but his long-time colleague and friend Leo Szilard was able to convince him "that this would not be an easy task," and that instead he should "concentrate on one promising effort." Einstein took Szilard's advice and lent his support to the idea of an Academic Assistance Council, which was established in Great Britain with Nobel Laureate Lord Rutherford (1871–1937) as president and Sir William (later Lord) Beveridge (1879–1963) and Charles S. Gibson (1883–1950) as secretaries. A prime mover behind the founding of the Academic Assistance Council, Szilard functioned like an entire team of people in the organization, rescuing hundreds of lives and important academic careers. In 1939 Szilard was in a unique position to call on Einstein, the most famous Nobel Laureate in the world, to alert President Franklin D. Roosevelt (1882–1945) to recent developments in nuclear physics. Pointing out the need for funds to carry out research on nuclear-chain reaction, which could lead to the construction of extremely powerful bombs, the first, undated draft of Einstein's famous letter to Roosevelt of 2 August 1939, mentioned future Nobel Laureate Eugene Wigner (1902–1995) as the prime mover behind it. However, from Szilard's correspondence with Einstein

Leo Szilard, photo around 1960

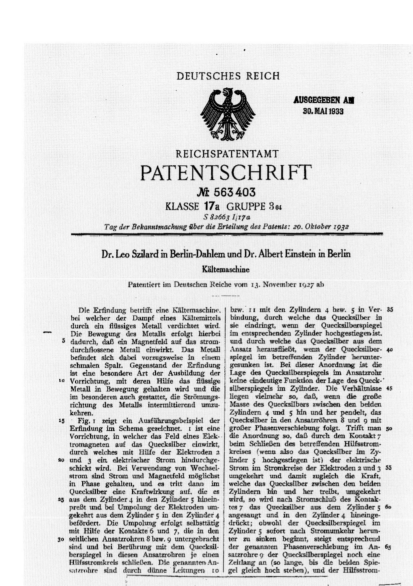

DEUTSCHES REICH

AUSGEGEBEN AM 30. MAI 1933

REICHSPATENTAMT

PATENTSCHRIFT

№ 563403

KLASSE **17a** GRUPPE 3₀₄

S 82663 I/17a

Tag der Bekanntmachung über die Erteilung des Patents: 20. Oktober 1932

Dr. Leo Szilard in Berlin-Dahlem und Dr. Albert Einstein in Berlin

Kältemaschine

Patentiert im Deutschen Reiche vom 13. November 1927 ab

Patent specification by Leo Szilard and Albert Einstein for a cooling machine, patented in the German Empire on 13 November 1927, announced on 20 October 1932

rians thus may have learned from Fermi's failure and seen that Einstein was the only possible scientist who was likely to gain the ear of the president. Even so, it took two and a half months before the economist Alexander Sachs of the Lehman Corporation, who was known to have close ties to Roosevelt, felt free to deliver Einstein's warning by hand to him, a telling example of the status and image of scientists in America at this time. Refugees from Hitler's Germany, however, knew full well that the Nazis would use every means at their disposal to further their war aims, and the Hungarian refugee scientists in particular, with their experiences during and after the Hungarian revolution and counter-revolution of 1918-1920 had ample reason for concern. Szilard also attached a letter of support from Einstein to his 16 August 1939 letter to the famous American aviator Charles A. Lindbergh (1902 – 1974), trying to enlist his aid in securing

through October 1939 we know that Szilard worked closely with fellow-Hungarian Wigner in this effort. This amounted indeed to a Hungarian initiative, since it was Edward Teller (1908–2003) who lent his driving abilities to the effort by taking Szilard to Einstein's home.

These Hungarian refugees proved to be more effective than the recent Italian Nobel Laureate Enrico Fermi (1901–1954), who tried to approach the U.S. Navy Department in March 1939 for funds, but was dismissed as a "crazy wop." The Hunga-

governmental funds for purchasing the uranium needed to sustain a nuclear-chain reaction.

By the spring of 1945 Szilard had convinced himself that the bomb should not be used. He knew that the defeat of Germany was imminent, which obviated the original motivation for creating the Manhattan Project, and that dropping the bomb on Japan would harm the peace process. "Ever eager for a cause," Szilard decided to alert President Roosevelt to the dangers of a nuclear-arms race in a memorandum, and again asked

Einstein for a letter of support for it, which Einstein promptly provided. In a letter to the President dated 25 March 1945, Einstein pointed out that Szilard was "greatly concerned about the lack of adequate contact between scientists who are doing this work and those members of your Cabinet who are responsible for formulating policy." Szilard attached a memorandum to Einstein's letter, in which he called for the establishment of an international system of controls in the field of atomic bombs in an effort to avoid an arms race. He apparently succeeded in scheduling an appointment with President Roosevelt, allegedly for 12 April the very date on which the President died. Szilard then modified his memorandum and sent it along with Einstein's letter to President Harry S. Truman (1884–1972) on 25 May 1945, who in turn passed it on to his future Secretary of State James F. Byrnes (1879–1972). Szilard's arguments were completely rejected by Byrnes when the physicist visited the future Secretary in his South Carolina home later in May accompanied by U.S. chemist and Nobel Laureate Harold C. Urey (1893–1981).

In an age of unprecedented disasters and threats, Leo Szilard played the essential role of a man of conscience. He was not only able to identify major issues, he had a gift for dramatizing and marshalling broad support for them. In an effort to give greater weight to his many initiatives, he made ample use of his close relationship with Albert Einstein, who supported his former disciple and close associate for some thirty-five years, from the beginning of their friendship to the very end of his own life. Their cooperation in war and peace for the great causes of the twentieth century reveals their shared social sensitivity, joint sense of global responsibility, and their recognition that the goal of science is always to serve humanity.

Eugene Paul
Wigner, 1992

Mark Walker

Albert Einstein, Carl Friedrich von Weizsäcker, and the Atomic Bomb

In 1939 it was clear to many scientists (like many other people) that war was coming. The émigrés Leo Szilard and Edward Teller were just two of the many scientists who recognized that nuclear fission might become a new weapon. Now in the

Albert Einstein in front of his house in Princeton, 1939 (photo: Eric Schaal)

United States, they made a direct appeal to President Roosevelt that he recognize the potential threat of German nuclear weapons and support American work in this direction.

In order to get FDR's attention, they enlisted the help of their more famous émigré colleague, Albert Einstein. Szilard and Teller wrote a letter; Einstein signed it on 2 August 1939. Of particular interest is what Szilard and Teller say about the possibility of a German atomic bomb:

"I understand that Germany has actually stopped the sale of uranium from the Czechoslovakian mines which she has taken over. That she should

have taken such early action might perhaps be understood on the ground that the son of the German Under-Secretary of State, von Weizsäcker, is attached to the Kaiser-Wilhelm-Institut in Berlin where some of the American work on uranium is now being repeated."

This letter is interesting in many respects. First of all, on 2 August 1939 the German uranium project had not yet been started, significant experimental work on nuclear fission had not yet begun at the KWI for Physics, and von Weizsäcker was only involved to the extent that he had discussed the consequences of nuclear fission with Otto Hahn and other colleagues. However, this changed once war began. Von Weizsäcker now began working on the theory of nuclear fission chain reactions in nuclear reactors (what the Germans called "uranium machines"). In December of 1939 Werner Heisenberg submitted a secret report to the *Heereswaffenamt* on the theory of chain reactions, including pointing out that pure uranium isotope 235 would be a nuclear explosive of unimaginable power. But this would require isotope separation of uranium, which appeared to be a daunting task.

In the summer of 1940 von Weizsäcker drew upon American publications and came to the conclusion that there was another, much easier way to make nuclear explosives. Uranium 238, the most common isotope, inside the nuclear reactor would absorb neutrons produced by fission and transmute, first into element 93, and finally into element 94, which is now called plutonium. This new element 94 could be separated chemically from the uranium, and would be an explosive as potent as uranium 235. Von Weizsäcker drafted a report on element 94 in nuclear reactors, and sent it on to the *Heereswaffenamt*.

Recently discovered documents show that von Weizsäcker in fact went a step further and drafted a secret patent application in 1941 for using plutonium in nuclear reactors and as a weapon:

Albert Einstein, Carl Friedrich von Weizsäcker, and the Atomic Bomb

"Patent Claims

1. Process for generation of energy from U238 characterized in that U238 is bombarded by two thermal neutrons, whereby at first an 'element 94'

Carl Friedrich von Weizsäcker, photo from 1949

of mass 239 is created by b-decays, in which the second neutron then induces nuclear fission that is accompanied by release of an enormous energy as well as creation of further neutrons and nuclei.

2. Process for generation of energy from U238 according to claim 1, characterized in that the neutrons, necessary to convert significant amounts of U238 into 'element 94,' are generated in a 'uranium machine' [reactor].

3. Process for generation of energy from U238 resp. element 94 according to claim 1 characterized in that the 'element 94' created by neutron enrichment is separated from the remaining uranium by known chemical methods and is then available for use in pure or suitably chosen concentration.

4. Process for generation of energy and production of neutrons from hardly fissile heavy nuclei (such as U238, Thorium, Lead, Bismuth) characterized in that the element 94 obtained as in claim 3 is added to these elements in a suitable amount so that they can undergo an auto-catalytic fission process by neutrons, during which energy is generated and new nuclei are created.

5. Process for explosive generation of energy and neutrons from fission of the element 94, charac-

terized in that the element 94 produced as in claim 3 is brought to one place, for example in a bomb, in such an amount that an overwhelming majority of the neutrons released by fission are consumed in exciting further fission and do not leave the substance.

6. Process for production of very small handy machines to generate nuclear energy and neutrons, in part by claims 3 and 4 and based on the knowledge about a 'uranium machine' distinguished in that only such an amount of 'element 94' is accumulated at one place (possibly mixed with suitable neutron-retarding and/or -absorbing additive elements) that explosions are avoided and the energy release proceeds in a continuous fashion. For the attached patent application to the Reich Patent Office as inventor is to be named Mr. Dozent Dr. C.F. von Weizsäcker"

Later that summer, this patent application was revised in two ways: it was now under the names

Left: Franklin D. Roosevelt during a radio address in September 1941

of several physicists at the Kaiser Wilhelm Institute of Physics; the references to nuclear weapons had been removed. Thus these patent applications demonstrate a clear understanding of how nuclear weapons would work, and an eventual ambivalence about the weapons themselves.

Right: Edward Teller, photo from the 1960s

Erdmut Wizisla

An Excellent Play for the Spoiled Contemporaries: Einstein writes to Brecht about Galileo

Brecht had been warned: "Einstein is nothing for drama," Leopold Infeld, Einstein's former employee, had declared in 1955 "partner; with whom do you intend to have him talk?" An in fact, Brecht's last fragment, *The Life of Einstein*, presents a monologue: eight clumsy lines of verse, in which Einstein tries to explain the change in his attitude to war. The collection of materials in Brecht's estate is considerable, however, comprising a reference library, several files of articles and considerations about the conception of the play, which are anything but superficial.

Brecht, approximately 1939

It was Einstein's death that inspired Brecht to take on the project. Among the materials is a handwritten report about the physicist's last hours, traced back to information provided by Einstein's secretary Helen Dukas. Brecht, whose heart was always

Brecht scribbled the potential closing of the play: "then his aorta burst." The founder of the theory of relativity had died thirteen days before. Brecht had been interested in Einstein for a long time. In 1930 he heard the scholar lecture on cau-

Mr. Bertolt Brecht, Paris, 4 May 1939

Dear Mr. Brecht:

I very much enjoyed your Galileo. You seem to have a profound grasp of not only Galileo's personality, but also of the importance of his presence in the development of the history of ideas and thus of history in general. Your portrayal also provides deep insight into the research questions Galileo faced and into the way pre-Galilean science approached experience. You managed to create a dramatic context that is extraordinarily captivating and must interest us as well, especially due through its strong relation to the political issues of our day. Hopefully our spoiled contemporaries will also be able to appreciate what an excellent play you have set before us. With friendly greetings

Your [Albert Einstein]

sality at the Marxist Workers' School in Berlin (MASCH). In the 1930s Brecht attempted to apply this concept to art. For him, Einstein's thought was the standard of the scientific age. Like Henry Ford and Lenin, the physicist represented the progressive stance that Brecht hoped for in the audience of his epic theater. He introduced himself to the New York theater scene in 1935 with the apodictic touch "I am the Einstein of the new stage form." At the same time, for Brecht Einstein embodied the failure of modern sci-

"somewhat wobbly," reacted with empathy to the description of the circumstances of his death. Einstein had rejected an operation on a burst aortal tumor. "Such bad taste – prolonging life artificially," Helen Dukas related his refusal. "I will go when *I* want, elegantly!" On an admission ticket for the grandstand of the 1 May demonstration in 1955,

ence. Aghast, he registered the vote of the former pacifist for the development of the atomic bomb. On the other hand, there was no perception that Einstein was interested in Brecht. All that was known was the physicist's rather sweeping estimation written to Alice K. Orlan on 27 December 1946, c/o Brecht, Santa Monica, "My regards to

Bert Brecht as well, whose art I sincerely revere most of those German-language authors I know of today." The letter Einstein wrote to Brecht years before, on 4 May 1939, provides justification for the reverence. The letter, a typed carbon copy without a signature, is preserved in Einstein's estate and today is preserved in the Albert Einstein archive of the Jewish National and University Library in Jerusalem. For decades the document could be located through the card catalog of the archive; since 2003 it is also available on "Einstein Archives Online," the joint database of the Jerusalem Archive and the editing office of the Einstein Papers Project in Pasadena (www.alberteinstein.info).

There is no need to doubt that the letter was sent, yet it remains an open question whether Brecht received it. It was addressed to Paris, but Brecht had been living in Sweden since April 1939. It is possible that the letter was brought or forwarded to Brecht. There is no indication of this in his estate in the Bertolt Brecht Archive. The occasion of the letter was the delivery of the first version of *The Life of Galileo* in March 1939. Einstein was

Report of Einstein's death from Helen Dukas (outline). From Brecht's estate

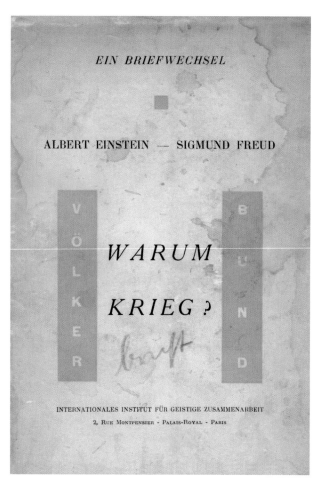

Albert Einstein /
Sigmund Freud:
Why War?
Paris 1933.
From Brecht's
library

version," through his cunning he is ultimately victorious over the Curia. The astronomer's self-recrimination first appears in the second (American) version, which was written after Hiroshima. This was the first time Brecht spoke of Galileo's "crime," which could be regarded as the original sin of the modern sciences. Galileo may have enriched astronomy, but he made it into a specialized science and thus robbed it of meaning for the aspiring bourgeoisie. "The atomic bomb, as both a technical and a social phenomenon, is the classical final product of his scientific achievement and his social failure."

Einstein composed his letter to Brecht in a state of at least presumed innocence. Yet the future had already begun. Three months later, on 2 August 1939, Einstein signed the fateful letter to President Roosevelt, in which he reported that it was technically possible to construct atomic bombs, and called upon the American administration – with reference to the cessation of German uranium sales – to provide funds for the intensification of atomic research. After Hiroshima and Nagasaki, Brecht's tone was nearly always spiteful when he spoke of Einstein. "Einstein issues the demand that the atomic bomb may not be delivered into the hands of the other powers, especially not Russia," he noted in his journal on 28 October 1945. "The 'world government' Einstein calls for appears to have been conceived in the image of Standard Oil, with enterprisers and underlings. – The brilliant specialized brain, implanted in a poor violinist and eternal prep-school student with a weakness for generalizations about politics."

one of the recipients of the wax-matrix copies of the play produced by Helene Weigel and Margarete Steffin. Einstein's thanks for the play turned out to be more formal than expected from the otherwise quite eloquent scientist. His assessment is concentrated and analytical nonetheless, if only for its emphasis on Galileo's personality, the astronomer's importance in history and the timeliness of the play. The reference to the category of experience reminds of Galileo's approach of placing empirical and sensory perception above all speculation – a conviction Einstein had shared from an early date.

Einstein's position remained open as to the question Brecht had entitled "Prize or Damnation of Galileo." One reason for this is that while in the first version Galileo recants, known as the "Danish

Brecht looked with fascination upon what he regarded as the schizoid existence of the scientist who does not reflect on the consequences of his knowledge, the risk of abuse. "The SS leader Heydrich," he writes in a later diary entry, "was an 'outstanding Bach specialist'; Einstein plays quartets and is a humanist, and somewhere there are atomic bomb factories that work day and night." The leitmotif of Brecht's notes is the lack of a scientific ethic in a matter that threatens humanity. "The goal of the scientist is 'pure' research," he

between pacifism and the contribution to the development of a weapon that killed the masses. "news of the bomb on Hiroshima reaches Princeton," Brecht noted on the grandstand ticket. "the population looks with fear to the great Einstein – the champion of peace." Einstein could have known "that his triumph had been transformed into defeat, because he, too, cannot withdraw the

Bertolt Brecht's grandstand ticket for the demonstration on 1 May 1955, with notes

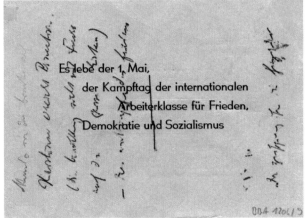

writes, "the product of research is less pure. The equation $E = mc^2$ is thought eternally, bound to nothing. Thus others can make the link: the city of Hiroshima has suddenly become quite ephemeral." The point is the life-threatening discrepancy between producing and conceiving, of which Günther Anders spoke in *The Obsolescence of Man*. Or, in Brecht's words: "progress in the knowledge of nature / coupled with a standstill in the knowledge of society / is fatal."
Brecht's play could have drawn from the antagonism between genius and political irresponsibility,

great equation when its deadliness has been demonstrated." Brecht's judgement is harsh and not free of defamation; its nadir is the formulation of the "notorious Einstein." His relentless look at the Galileo of the 20th century sharpens the perception of the problem.
What might Albert Einstein have thought about the American version of *Galileo*? He declared his interest in the new version in December 1947, after the performance in New York, again via Alice K. Orlan. But it is not known whether Brecht, who was already in Zurich, ever sent him a copy at all.

John Stachel

Einstein and the American Left

Perhaps the first cause actively supported by Einstein after he came to America that was specifically identified with the American left was his support of Loyalist Spain during the Spanish Civil War (1936–1939), at a time when the Roosevelt Administration promoted a policy of neutrality. This really should not have been a left-wing issue since it involved support of the legally-elected

1930s up to the time of the German-Soviet non-aggression pact (1939) and again after the German attack on the Soviet Union in 1941, was the Communist Party and allied "fronts" as they were often called. Non-communist socialist groups and non-Stalinist communist groups did exist and had a certain influence in some places (notably in the trade union movement); but even they were large-

Combatants of the International Brigades in the Spanish Civil War, 1936

Republican government of Spain against a military insurrection led by General Francisco Franco and openly supported by the fascist governments of Italy and Germany. But the support of the Catholic Church for Franco on the one hand, and the open support of the Republicans by the Soviet Union on the other, soon transformed it into a right-left issue in the United States.

It must be recognized that, for good or ill, the dominant force on the American left during the

ly forced to define themselves in relation to the issues raised by the dominant Communist Party. So, for example, military participation by Americans on the Republican side in the Spanish Civil War took place under the aegis of the Abraham Lincoln Brigade, a Communist-led organization, and support for Republican Spain inevitably meant alliance with Communist front groups. But even before the Spanish Civil War – indeed almost from the time of his arrival in the United States

over Japan followed by attempts to use "atomic diplomacy" to intimidate the Soviet Union and the growing intensity of the Cold War, Einstein was forced into more and more open conflicts with United States government policy on many issues. This opposition often brought him into alliance with left-wing, pro-Communist and even pro-Soviet groups in the United States. But he always maintained his independent stance, never afraid to challenge erstwhile allies when he disagreed with them over another issue. But he never allowed such differences to weaken his principled stand for civil liberties, publicly protesting the indictment and imprisonment of the U.S. Communist Party leaders under the notorious Smith Act. He came to advocate world government as the only certain method of avoiding a nuclear war be-

Albert Einstein with Charlie Chaplin at the premiere of City Lights in Hollywood, 30 January 1931. Einstein and Chaplin were united in their similar political views. Chaplin was forced to leave the United States in 1952 as a consequence of McCarthyism. He lived in Switzerland until his death in 1977. Not until 1972 was Chaplin "rehabilitated" through the award of an Oscar for lifetime achievement.
(photo: Emil Hilb)

Kion vi faras por evitition? (What are you doing to prevent this?) Poster by the People's Front government against the support of rebellious Franco troops by the fascist states of Germany and Italy, 1937

until his death, Einstein was the subject of an ongoing investigation by the American Federal Bureau of Investigation (F.B.I.), which attempted to prove that he was a "communist spy."

During World War II, Einstein's general anti-fascist position and support for all opposition to the Axis powers, including that of the Soviet Union and the Communist parties, put him into the mainstream of American politics, even though right-wing opposition to this mainstream concensus continued. That Einstein was no tool of Soviet or Communist interests is proved by his support for the war efforts of Britain and France, and his opposition to isolationist policies in the United States during the period of the German-Soviet non-agression pact (1939–1941).

After the end of World War II, accompanied as it was by the explosion of the first atomic bombs

tween the great powers with its potentially catastrophic consequences for the very existence of mankind. This brought him into open conflict with Soviet foreign policy; as did his advocacy of com-

-3-

Der leitende Gedanke allen politischen Handelns müsste deshalb sein: Was können wir tun, um ein friedliches, ~~und~~ im Rahmen des Möglichen befriedigendes Zusammenleben der Nationen herbeizuführen? Erstes Problem ist die Beseitigung gegenseitiger Furcht und gegenseitigen Misstrauens. Feierlicher Verzicht auf gegenseitige Gewaltanwendung (nicht nur Verzicht auf Verwendung von Mitteln der Massen-Vernichtung) ist zweifellos nötig. Solcher Verzicht kann aber nur dann wirksam sein, wenn er mit der Einführung einer übernationalen richterlichen und executiven Instanz verbunden ist, welcher die Entscheidung der mit der Sicherheit der Nationen unmittelbar verknüpften Probleme übertragen wird. Schon eine Erklärung der Nationen, an der Realisierung einer solchen "beschränkten Welt-Regierung" loyal mitzuarbeiten, würde die imminente Kriegsgefahr bedeutend herabsetzen.

Letzen Endes beruht jedes friedliche Zusammenleben der Menschen in erster Linie auf gegenseitigem Vertrauen und erst in zweiter Linie auf Institutionen wie Gericht und Polizei; dies gilt ebenso für Nationen wie für Eimzel-Individuen. Das Vertrauen aber gründet sich auf eine loyale Beziehung des give and take.

Wie verhält es sich aber mit der internationalen Kontrolle? Nun, ~~ich denke etwa so wie zwischen den Staaten New York und New Jersey!~~ sie mag als sekundäre polizeiliche Massregel nützlich sein. Man wird aber gut daran thun, ihre Wichtigkeit nicht zu überschätzen. Der Vergleich mit der "Prohibition" stimmt nachdenklich.

A. Einstein Archive 28-870

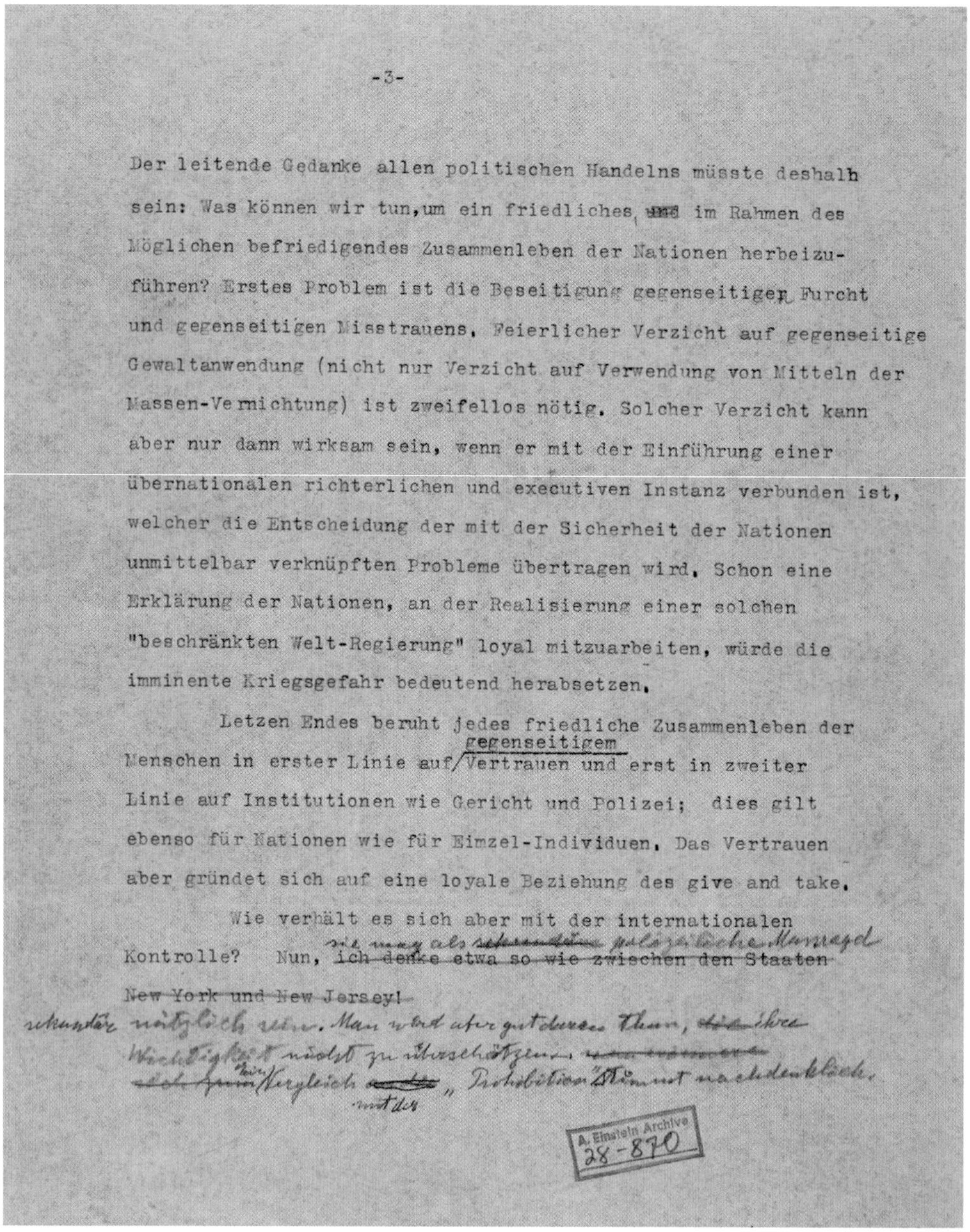

Einstein manuscript on thoughts for a world government, 13 February 1950 © The Hebrew University of Jerusalem, Albert Einstein Archives

plete national disarmament in favor of a world government at a time when the Soviet Union was striving to develop first atomic fission and then nuclear fusion weapons.

While he had indirectly contributed to the U.S. atomic bomb program through his famous 1939 letter to President Roosevelt warning of the dangers of a German A-bomb, he resolutely opposed the development of the H-bomb after the War.

It was primarily on issues of the cold war and domestic policy that he joined forces with the American left. He saw the cold war as an excuse for the promotion of militarism within the United States and for attacks on the civil liberties of Americans. This led him to support many groups that advocated American-Soviet friendship and the most vigorous defense of American civil liberties. During the 1950s McCarthy era, he even

Joseph Raymond
McCarthy (1909–
1957), Chairman
of the Committee
against "Un-Ame-
rican Activities," in
conversation with
his main investiga-
tor, the former FBI
man F. P. Carr,
around 1950

advocated civil disobedience, in the form of refusal to respond to Congressional or other government inquiries into personal political beliefs and
activities, and he earned the wrath of such "respectable" mainstream journals as the New York
Times for his stand. Again, his stand here differentiated him from the Communist position, which
advocated refusal to answer such questions on
the grounds of the U.S. Constitution's provision
against self-incrimination (Fifth Amendment) as
a means of avoiding imprisonment. Einstein felt
this to be undignified, and advocated refusal to
answer on the grounds of the Consitution's guarantee of the right of free speech (First Amendment), even if this meant going to jail.

In his later years, Einstein came to advocate openly a socialist reorganization of society. He felt that
capitalism encouraged the individual to forget his
or her social ties and obligations and to develop
an egoistic individualism, and that a socialist reorganization of the economy was ultimately the only
way to avoid such social atomism. Yet, faced with
the example of the Soviet Union, he was aware of
the dangers of the demand for blind social obedi-
ence and conformity of thought that could result
in tyranny by a small elite. It was in a form of
socialism that navigated in the narrow democratic
space between social atomism and social tyranny
that he saw the future hope of mankind.

Helge Kragh

Einstein as a Historian of Science

Einstein was not a historian of science – he was a scientist who *made* history. Although he never articulated his views on the history of science, he had a deep interest in the subject and often framed his philosophical reflections on science in

Leopold Infeld

the form of historical narratives. These were not accounts in agreement with the standards of professional, critical history, but they were historical accounts nonetheless. In a conversation he had with I. Bernhard Cohen (1914–2003), a historian of science, a few weeks before his death in 1955, he expressed his interest in the history of science, not as documentary history but as an attempt to reconstruct the development of thought processes in order to follow them logically from the past to the present.

It is well known that young Einstein was deeply influenced by the works of the Austrian physicist-philosopher Ernst Mach (1838–1916), one of the fathers of positivism. Einstein was thoroughly familiar with Mach's celebrated *The Science of Mechanics*, a critical and historical study of key

concepts such as space, time, mass and substance. "This book*,*" Einstein wrote in his autobiographical essay of 1946, "exercised a profound influence upon me." He similarly praised the book in a long and sympathetic obituary he wrote upon Mach's death in 1916. Although it was Mach as a philosopher of science who appealed to him, rather than Mach as a historian, in Mach's thinking history and philosophy of science were so closely intertwined that the two aspects can scarcely be separated. Moreover, Einstein's way of using the history of science had much in common with Mach's. Many of Einstein's philosophical reflections were organized along historical lines, as he was not content to show what science *is*, but also wanted to show how it *develops* over time. The temporal or dynamical dimension was an integral part of his view of science and naturally invited historical reflection. For example, in an article of 1930 on the concepts of space and field he adopted a historical framework in explaining how prescientific thoughts about space had evolved into the modern notion as found in the general theory of relativity. According to Einstein's way of thinking, one can identify in history one or more principles that can be used to order historical events in such a way that their intrinsic significance stands out. He likened the task of the conceptually oriented historian of science with that of the archaeologist and believed that a method of "intuitive archaeology" would lead to a clearer, more logical picture of the past. Such a kind of history he admitted was a *construction*, not a factual history based on the right chronology and causal sequence of events. In Einstein's view, the recovery of the significance of historical events was more important than their chronology and contexts. Like Mach, his primary interest in the history of science was to use it exemplarically to illuminate philosophical points.

Many of the same themes can be found in other papers and addresses by Einstein, especially in the 1930s and 1940s, when he wrote numerous memorial articles on great physicists such as Kepler, Newton, Maxwell and Planck. In 1927, upon the bicentenary of Newton's death, he wrote an article which, characteristically, was not so much about Newton as it was a semihistorical survey of the conceptual development from Newton to contemporary physics. His most extensive work of a historical form was *The Evolution of Physics*, a popular book he wrote in 1938 jointly with his

than that; it also included Einstein's views on the historical development of physics and its relevance for the philosophy of physics. For this reason it should not be dismissed as merely a popular book on physics.

What were the two authors' view on history? They freely admitted that their account was a constructed history which included only events "which seemed to us most characteristic and significant" and that it was written "without

Planets and worldview, allegorical representation. Painted tabletop by Martin Schaffner, created in Ulm in 1533 for Asymus Stedelin

collaborator, the Polish physicist Leopold Infeld. Although the idea of writing the book came from Infeld, its structure and material came from Einstein. The book was highly successful and a masterly example of popular science, yet it was more

bothering too much about chronological order." What interested Einstein was the emergence and fate of new, fundamental concepts in physics, from the law of inertia to the general covariance associated with general relativity. In accordance with his general philosophy of science he stressed by means of historical examples that such conceptual advances did not arise inductively from experiments and phenomena, but were the results of great scientists' creative imagination, their ability

(He was, however, willing to make an exception for thermodynamics, which he believed would never be overthrown.)

Some of the historical ideas employed by Einstein, both in *The Evolution of Physics* and at other occa-

System of the world according to Ptolemy (2nd century B. C.), Copernicus (1543) and Tycho Brahe (1588). Copper engraving, around 1696

to emancipate the intellect from illegitimate restrictions. At the end of the book he wrote: "Science is not just a collection of laws, a catalogue of unrelated facts. It is a creation of the human mind, with its freely invented ideas and concepts." He believed that this view received support from history. Einstein further used the history of science to argue that science is a never-ending process: there are no eternal theories of physics, he concluded. Even the crowning achievement of his own creativity, the general theory of relativity, would one day be replaced by a better theory.

sions, have an affinity with those expounded by later historian-philosophers, most notably by Thomas Kuhn (1922–1996) in his *The Structure of Scientific Revolutions* from 1962. Thus, Einstein pictured the history of physics since Galileo as a brief sequence of conceptual revolutions in which new physical ideas were born in the struggle with old ideas; the continued development, on the other hand, was evolutionary or what Kuhn called paradigm-governed. However, Einstein's revolutions differed from Kuhn's in retaining a close connection with past science. In Kuhn's view, revolutions

Newton's
apotheosis.
Copper engraving,
around 1738

a comparison, we could say that creating a new theory is not like destroying an old barn and erecting a skyscraper in its place. It is rather like climbing a mountain, gaining new and wider views, discovering unexpected connections between our starting-point and its rich environment. But the point from which we started out still exists and can be seen, although it appears smaller and forms a tiny part of our broad view gained by the mastery of the obstacles on our adventurous way up."

This is a most un-Kuhnian point of view. Of course, many of Einstein's semi-historical accounts were concerned with his own fundamental research, in particular his invention of the theory of relativity which he saw as the culmination of a long historical process with roots back to Galileo and Newton. In his capacity as an expositor of the historical development of science, he was an interpreter as well as actor. In the Herbert Spencer Lecture, delivered at Oxford in 1933, he said: "If you want to find out anything from the theoretical physicists about the methods they use, I advise you to stick to one principle: don't listen to their words, fix your attention on their deeds." He was fully aware that the principle applied to himself as well and considered himself a poor source of information concerning the genesis of his own ideas. The historian, he believed, was in a better position to get insight in to the thought processes of a scientist than the scientist himself.

mark separations between incommensurable world views, whereas Einstein stressed continuity and correspondence to older theories. In describing the transition from the mechanical worldview to field theory, he expressed his view as follows: "The new theory shows the merits as well as the limitations of the old theory and allows us to regain our old concepts from a higher level. This is true not only for the theory of electric fluids and field, but for all changes in physical theories, however revolutionary they may seem. [...] To use

Jürgen Ehlers

Einstein's General Theory of Relativity in Contemporary Physics

Physics theories are seldom the work of a single scientist. The theories of electrodynamics, the special theory of relativity and quantum theory: all were advanced with the participation of several physicists. By contrast, the general theory

Albert Einstein and Walter Sydney Adams at Mount Wilson Observatory, 29 January 1931

Pages 362–363 Components from the German Electron Synchrotron DESY (photo: Peter Ginter)

of relativity, the modern theory of gravitation, was devised between 1907 and 1915 by Albert Einstein alone.

The general theory of relativity differs from the physics theories which proceeded it, that is the theories of mechanics, electrodynamics and of special relativity, and the quantum theory advanced after it, in a fundamental regard: in it the structure of space and time is not assumed to be given permanently, once and for all, but rather as a changeable field subject to natural laws, similar

to the electrodynamic field. While spatial distances' relationships to each other and to periods of time had been regarded previously as immune to any influence exerted by events in space and time, according to Einstein's general theory of relativity, the spatial-temporal model is distorted by every piece of matter and everything that has energy and an impulse. Inversely, this metric field, together with other interactions, determines the structure and the motion of material bodies and the propagation of light. Einstein's gravitational field equation at the core of the theory expresses a mutual relationship between the metric of spacetime and of matter; in this case the electromagnetic field is also considered to be matter. Isaac Newton (1687) attributed gravitation interactions to forces between bodies, which exerted instantaneous effects across any distance. He regarded space and time as the unchangeable stage of natural events. According to Einstein, however, the pliant spacetime mediates these mutual effects. To visualize this, one must imagine spheres rolling around on an elastic membrane spanned horizontally. Each sphere dents the membrane in its immediate vicinity, and when two spheres come close to each other, the dents moving with them influence their paths. In this analogy, one has to imagine that the space surrounding the membrane does not exist, for the membrane is only intended as a model of space, or of spacetime, whose dents represent the moving bodies contained "in" the spacetime continuum.

As early as 1915, Einstein was able to explain by means of his general theory of relativity the motion of the planet Mercury. Since 1859 it had been known that the planet's observed path deviated from that yielded by Newton's calculations; Einstein's theory predicted new, observable effects, some of which were not verified with measure-

field (such as those induced by the motion of stars) should propagate at the speed of light in the form of waves, as qualitatively suggested by the membrane model described above. The as yet elusive direct proof of such gravitational waves is a current research emphasis of many collaborative working groups involving physicists from the

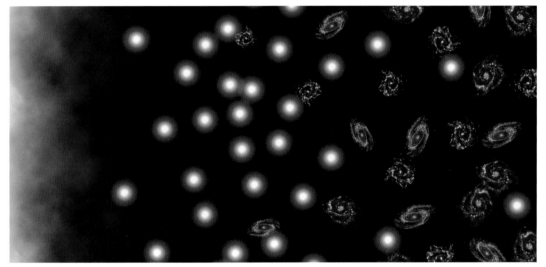

The development of our universe after the big bang: When the universe was still young, it regularly generated massive galaxies, like small bees from a bee colony. Over time the universe bore ever fewer "offspring," and newborn galaxies (represented as white circles) matured into older ones like our Milky Way (represented as spirals)

ments of satisfactory accuracy until after his death (1955) – to date, all of them successfully. These include the influence of the rotating disk, the oscillations of light, and the velocities of light through gravitational fields, as well as the deflection of light rays through the gravitational fields of stars and galaxies, an effect which has been used to determine the distribution of matter in outer space (the gravitation lens effect) since 1979. An experiment has been underway since April 2004 to verify an effect predicted by Josef Lense and Hans Thirring back in 1918, also known as "frame dragging," which is a tiny motion exerted by the rotating Earth on the axes of rotation of gyroscopes carried on board a satellite. Many of the relativistic effects mentioned above are used in the technology of global positioning systems (GPS) for navigation.

Back in 1916 Einstein already recognized that, according to his theory, changes in the gravitational

U.S., Japan, France, Italy, Australia, Great Britain and Germany. In Ruthe, near Hannover, Germany, a British-German team jointly operates a gravitation wave detector.

An indirect proof for gravitational waves, for which the radio astronomers Hulse and Taylor received the Nobel Prize in 1993, resulted from their observations of a double galaxy whose period of revolution of approximately eight hours became around 70 microseconds shorter each year as a consequence of the emission of gravitational waves predicted by the general theory of relativity. The direct proof for gravitational waves, and further, the analysis of the waves emitted by different kinds of "transmitters," would open up a new "window" on the observation of such processes in outer space, about which astrophysicists have no other means of acquiring information. This is true for double stars that rotate around each other ever more closely on spiral orbits and ultimately fuse together – in some cases to form a black hole, oscillating and rotating stars, supernovae and other exotic processes.

Probably the most peculiar conclusion from the general theory of relativity is when great masses collapse, and when this occurs in the instable final stage of massive stars, and perhaps also in

The general theory of relativity was the first to result in a consistent model of the structure and development of outer space, which also fit together quantitatively with a number of empirical observations. According to this theory, the stars and galaxies formed over the course of around fourteen billion years in a kind of condensation

Artistic rendition of our home galaxy, the Milky Way

the formation of galaxies, the consequence can be the formation of what are known as black holes. A black hole is an area separated from the outside world by a closed surface of finite area, which matter and radiation can pass through in only one direction: from the outside to the inside. The properties of these eerie creations were not determined theoretically until after Einstein's death, starting in 1960. In the past thirty years astronomical observations have yielded indications about the actual existence of black holes. In particular, it is fairly certain that the center of the Milky Way galaxy, to which the solar system belongs, holds such a "hole" with the mass of three million suns.

process, from an initially extremely hot, dense, rapidly expanding gas. This standard cosmological model can be reconciled not only with the empirically observed expansion of the galaxies, but also with the properties of the thermal radiation remaining from the hot initial stage, which has been measured with increasing precision in recent years, and other observable facts. However, the model only "works" under the assumption that there is much more as yet unknown, not directly observable matter, and moreover a repellent anti-gravitational force counteracting gravity, perhaps caused by what is known as "dark energy." The open issues this implies will probably puzzle physicists and astronomers well into the future.

Schematic drawing of an extremely massive black hole with an "accretion disk" (yellow), outer dust ring (blue) and two tightly bundled particle rays, called jets

In the general theory of relativity, matter is portrayed only on the macro level, without any consideration of the fine structure described by quantum theory. Inversely, quantum theory does not cover gravitation. Many physicists are working to develop a theory that includes those features of quantum theory and the general theory of relativity which have been confirmed by experience, and excludes those features which present obstacles to their unification. The problem has been a research object since 1930, but has yet to yield any significant successes. Such a unification presumably can have success only if fundamental assumptions of both the general theory of relativity and quantum theory are altered. In order to be able to describe gravity and the other interactions in a uniform manner, the theory aspired to also would have to represent the smallest dimensions of the spacetime structure in a new mathematical way. The two primary efforts in this direction are called "string theory" and "loop quantum" gravitation. In 1975 Stephen Hawking gave an indication of the properties of quantum gravitation with the surprising discovery that black holes – deviating from the "classical" general theory of relativity – should disintegrate when they encounter thermal radiation. However, no connection has been found yet between this theoretically interesting, possible effect of quantum gravitation and facts obtained from empirical experience.

Ulf von Rauchhaupt

Lots Happening in Spacetime

Maybe the star Eta Carinae is no more. Sure, last night we could see it burning brightly in the southern sky. But this light spent more than 8000 years getting to us. It shows us a sun in the throes of death, which could flare up into a supernova at any moment. Maybe that's already happened. If so, then the explosion's flash of light is accompanied by a bunch of gravitational waves as they chase through the galaxy. These are vibrations excited by the flare in the fabric of space and time itself (see "ripples in spacetime"). Only in recent years has it become technically possible to detect

Straight lines in a curved world

Non-physicists consider the theory of general relativity more difficult than anything ever conceived by man. Evidence of this is the myth that at no time were there more than three people who really understood it. In reality there have always been a lot more than that – and for them Einstein's theory was not necessarily the greatest subtlety physics has to offer, but certainly its most beautiful creation. Presumably these two conceptions have to do with one and the same feature of this theory: it describes its subject, gravitation, as a geometrical property of space and time. Basically

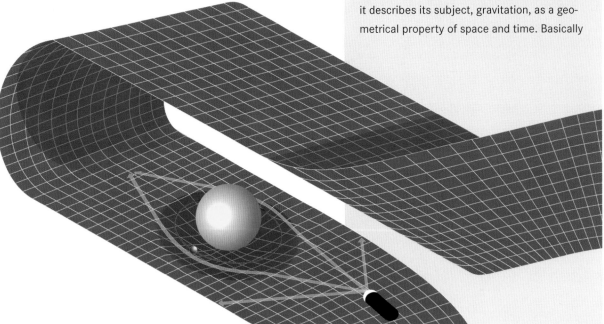

such waves. So now we expect and hope for a strong space quake. Several detectors are already set up. They have hardly passed the testing phase, but they should be able to detect gravitational waves from a supernova if it occurs within our own galaxy.

That would be Albert Einstein's ultimate triumph. The existence of gravitational waves was one of the first predictions from the theory of gravity he had perfected in 1915 – and today it's the last of its confirmations that is still missing. But these

in Einstein's universe there is only one direction: straight ahead! Massive bodies or light rays move only on straight lines, but straight lines in a curved world. We cannot escape from this world, not even in thought, in order to describe them from the higher plane of an external background, uninfluenced by physical events. That there should be no such background, no omniscient "exterior," is a highly intellectual fascination for the connoisseur –

and for normal mortals not ordained into the higher realms of tensor analysis it is a reason for permanent brain-wracking.

For easier digestion one can take the elegant four-dimensional world and disassemble it, as we tried to do here: instead of spacetime we consider only space, and even that in only two of its three dimensions. So we obtain a section of the universe's spacetime represented as a surface. The theory does not insist that going along this surface one can return to the initial point. But it is a possibility. To visualize gravitational effects we can now bend this surface into the remaining third dimension, simulating rather than visualizing the actual relationships – after all, we are missing the distortion of time, which goes hand in hand with every deformation of space.

Still, we can get a feeling how gravitation can be understood as geometry. The curvature that causes a star's gravity (yellow ball) is represented as a bump on the surface, and an orbiting planet (blue ball) by the path of a roulette ball. We have to pretend that there is no friction, so the ball never loses its momentum. The deflection of light (blue lines) in a gravitational field also becomes reasonable. We can see that a gravitational field is even able to reunite initially divergent rays from a light source, like the lens of a magnifying glass. Einstein already recognized this lens effect of gravity, but thought it could not be observed. Fortunately it can. Exploiting the lens action of the gravitational fields of galaxies and stars, and lately even of distant planets, is today a very active branch of astronomy.

Holes in Space

These may be the best known of the exotic predictions of general relativity: regions where collapsing matter curves space so strongly that the curvature is amplified without bound – or rather, until new, as yet entire unknown laws take over. Einstein found this idea unbearable, and so he later attempted to prove

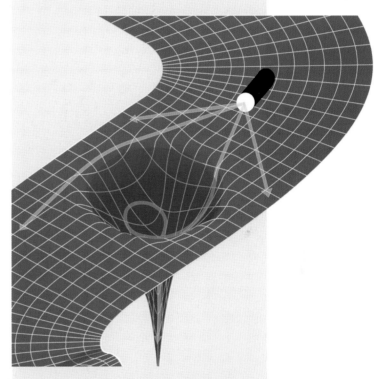

that there cannot be any such monstrosities. The attempt was not successful. Today we know such objects by the name of "black holes," because in their immediate vicinity gravity is so strong that not even light (blue) can escape from them. Although for this very reason black holes cannot be seen, there is today overwhelming evidence that the cosmos is full of them.

Ripples in Spacetime

In principle it's easy to generate gravitational waves. You take a lump of matter and accelerate it. It's more convenient to let two space objects orbit around each other, as in this picture. Because there are countless orbiting bodies in the universe, and every so often exploding stars accelerate a lot of mass, gravitational waves are plentiful. Fundamentally, detecting them is also not difficult: you simply look at a yardstick. The moment a wave comes along, space expands and contracts periodically, and the stick likewise. So why don't we

need the collision of two neutron stars, at least.

What would be even better – except for a close-by supernova – would be the collision of two black holes. But even such gravitational waves would expand and contract a yardstick reaching from here to the Sun only by the diameter of one atom. By now we have devices (not yardsticks, but arrangements of enormous lasers) that can actually measure such tiny changes in length.

The three most advanced detectors are the two American LIGO detectors and the German GEO600. When they will see the first gravita-

already have a vibrant science of gravitational wave astronomy? That's because spacetime is an extremely stiff sort of thing. The waves' propagation speed shows how stiff. By comparison air is as soft as a pillow, and sound in air barely manages a speed of 340 meters per second. In wood or metal the vibrations move already ten times as fast. But vibrations of spacetime propagate at the fastest possible speed, 300 000 kilometers per second, that is, at the speed of light. To excite something as stiff as that into sensible vibrations is not so easy. In fact, our Earth, for example, radiates just 200 watts in gravitational waves as it obits the Sun. For radiative power that can be detected today or in the near future over astronomical distances we

tional waves depends on the frequency of black hole collisions in the universe, and that is known only approximately. "With some luck we'll have a signal by 2005 or 2006" says Bernard Schutz of the Max Planck Institute for Gravitational Physics in Golm near Potsdam. "If we're very unlucky we may have to wait 10 years." By then at the latest the age of gravitational wave astronomy will have begun.

wandering folds of spacetime are only one of many consequences, among them some highly peculiar ones, that can be derived from Einstein's equations. We have collected here some of the most spectacular ones. Strange as they may seem, all of them are consequences of a theory that has by now been confirmed in countless tests. Another name for the theory is general relativity, in contrast to the theory of special relativity published ten years earlier by Einstein. The latter was itself an epoch-making breakthrough not least because in one fell swoop it did away with the traditional concept of space and time as an absolute stage on which all physical action is played out. According to special relativity theory the rate of a clock or the length of a ruler always depends on how fast the observer is moving with respect to these measuring instruments.

The theory of special relativity is a prime example of a modern scientific theory. In a conceptually cogent way it explains previously mysterious experimental results and integrates them into existing knowledge. At first sight it would seem that the general theory of relativity was developed for different reasons. At the time Isaac Newton's theory of gravity prevailed. It was in full agreement with all features of gravity that could then be observed – neglecting for the moment a tiny anomaly in the orbit of the planet Mercury, an insufficient reason to second-guess the great Newton. Therefore it seems as if it was more of a philsophical desire for an encompassing theory of nature that impelled Einstein toward his gravity project. Research in the history of science paints a different picture. When he was asked in 1907 to write a review article on special relativity, Einstein saw himself confronted with a problem: In Newton's theory, changes in the gravitational action of one body can immediately be felt in the whole universe. But this contradicts special relativity, which forbids anything, including fluctuations of gravity, to spread faster than light. So Einstein had

to dream up a new theory of gravity. He realized that this forced him to generalize his theory of relativity, because he could simulate gravity by accelerated motion. Accelerations generate forces – just think of the force that pushes the occupants of a space vehicle back into their seats when the rockets start up. The insight of Einstein's genius was this, that the inertial force of accelerated motion and the gravitational force are ultimately of the same nature – so astronauts inside a light tight capsule cannot tell the rockets' ignition from a sudden increase of Earth's gravity. Transforming this idea into a mathematical formalism cost Einstein eight years of arduous work. But when the theory was perfected and observations confirmed the first predictions of the theory, physics was shaken in its foundations. Space and time – until then the framework of every physical description, even if made relative – had themselves become an object of physics, a peculiar entity called spacetime that can be squeezed and stretched like rubber.

For according to Einstein, gravity is just a consequence of the curvature of space. It curls up not just where there is matter, but also where there is pressure and where energy is stored. Even the totally massless light rays therefore feel gravity. But not only that: they also cause it. Above a burning light bulb space is minutely more curved, and an object is a tiny bit heavier than when the light is turned off. But then the energy of gravity itself is also a source of gravity. Two masses that attract each other have together more gravity than would result from the addition of the two separate gravitational fields. The curvature they generate itself produces more curvature. This property of gravity to act back on itself finds its mathematical expression in the so-called "nonlinearity." It is just this nonlinearity that makes for many of the weird effects in general relativity theory. This goes so far that gravitational waves can theoretically be so strongly focused that their

Bridges through Space and Time

Nothing travels faster than light. This result from Einstein's special relativity could one day become an existential problem for our adventurous humanity. For light speed is snail's pace, by interstellar measure; the journey even to the nearest stars would take years. But hypothetically there is an alternative: Relativity theory admits the existence of so-called wormholes: tunnels in spacetime that can be a short cut for travel to far distant regions of

special cases, but no one yet knows whether the throat of a wormhole is such a case, and if so, whether this suffices to stabilize the tunnel. So most physicists see no reason to believe in the existence of wormholes. And many turn up their noses as soon as wormholes are mentioned – and not only because they don't appreciate science fiction. It's also because in certain cases a wormhole could also be used as a time machine, allowing travel back to one's own past (but not to times prior to the

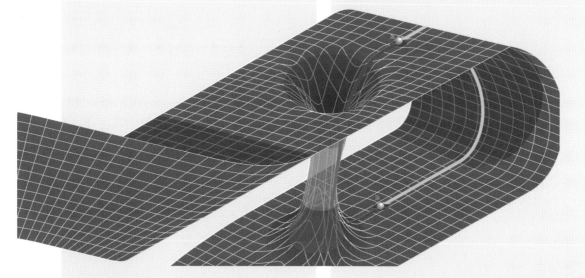

the cosmos. Two plants A and B, separated by many light years of normal space (yellow), would be only a few million kilometers apart via a suitably placed wormhole (red). There is a problem with that: it's not like a black hole that forms when a massive star collapses, we know no natural process for wormhole formation – not to speak of one that could be technologically controlled. In addition, such a construction would be highly unstable and would recollapse immediately after it was formed – except if we can stuff its throat with exotic material of negative energy density (blue). Such material does exist in

time machine's construction). That's also allowed according to general relativity. Supposing it actually works, and given the fact that one cannot change one's own past without getting involved in logical contradictions, some researchers derive a profound assertion that is today coming back into fashion: the world would be completely predetermined, and free will would be impossible.

concentrated curvature makes a black hole out of pure energy (see "Holes in Space"). True, the conditions would have to be so special that this sort of process can practically be excluded in the macrocosm – not everything allowed by Einstein's theory also has to occur in reality.

Still, scientists speculate that precisely such effects are common in that unexplored realm of the very smallest, where Einstein's theory probably has to be modified in order to fit together with quantum theory. In that realm, mini black holes can continually form only to evaporate again, and wormholes can open and close (see "Bridges through Space and Time"). Under extreme enlargement the all-pervading rubbery spacetime could look like seething foam. Here the strangest consequences of Einstein's theory could become reality. In front of our very nose.

Tevian Dray

The Nature of Time in Relativity

Perhaps the most fundamental change wrought by Einstein was to our understanding of time. With the advent of special relativity, space and time lost their separate meanings, becoming unified as spacetime. According to relativity, the universe simply does not have an absolute notion of time. Time was no longer universal, but rather observer-dependent. As with the ancient question of where one should stand to weigh the Earth, one must al-

ways specify just whose time one is talking about. As the famous twin "paradox" demonstrates, a round-trip journey involving travel at a significant fraction of the speed of light takes less time according to those traveling than according to those left behind. This can be interpreted geometrically using the relativistic interval to measure "distance" in spacetime: the spacetime path which seems "longest" according to our naïve notion of distance turns out to take the least time. An interstellar trader going back and forth between different star systems would therefore age significantly less than her clients. In this sense, she can travel into the future – but never the past.

It is in fact remarkable that one can distinguish future from past at all. After all, space and time have been unified, and there is no such unambiguous distinction for space. Put differently, one can turn around smoothly to face in the opposite direction; there is no way to identify the precise instant when one has stopped facing one way, and started facing the other. Such is not the case with time, where the distinction between aging and getting younger is unambiguous: All relativis-

14.3.1879 – 14.3.1979*

* Ermöglicht wurde dieses ungewöhnliche Bilddokument durch die von Albert Einstein in der Theorie begründete und hiermit in der Praxis bewiesene Zeitdilatation. (Näheres dazu in den Werken zur Relativitätstheorie, erschienen im Verlag Vieweg, Braunschweig, Wiesbaden.)

Albert Einstein tastes the birthday cake for his 100th. Poster by P. Troschow, 1979

tic observers will agree on the difference, even when disagreeing about the rate at which aging occurs.

There is however nothing in relativity which indicates which direction in time should be toward the future, and which should be toward the past. This distinction is embodied in the Second Law of Thermodynamics, which says roughly that things eventually fall apart, but the origin of this "arrow of time" remains a fundamental, unanswered question.

wards in time. There is no way out of this argument so long as relativity is assumed to be a good approximation in ordinary space, which it appears to be. To paraphrase science fiction author Arthur C. Clarke (1917–), many books normally classified as science fiction should really be classified as fantasy for precisely this reason.

In short, traveling faster than light is inextricably associated with travel backwards in time. There could be another way to travel into the past, though, depending on the shape of the universe. It is an open question whether the universe is open or closed, that is, whether you will return to your starting point if you travel far enough in a given direction. This won't help you travel into the past, but suppose you could circumnavigate the universe by traveling (forwards) in time, rather than in a spatial direction. You're already traveling in time, simply by getting older! So the question is whether there are closed timelike curves, loops in time which return to their starting point.

Model of a black hole surrounded by matter

What if one could travel faster than the speed of light? There are two scenarios: "warp speed", in which one travels through ordinary space faster than the speed of light, and "hyperspace", in which one travels through some other realm. It is hard to determine how much the trader ages in "warp speed" scenarios, since there are no ground rules. Consider therefore the "wormhole" version of the hyperspace scenario, in which the trader jumps "instantaneously" from system to system. The big question is, instantaneously according to whom? A technology which can transport the trader from one location to another "faster than light", no matter how, could be used starting from some other inertial frame (e. g. a spaceship traveling at relativistic velocity) to transport the trader back-

Such an ability to travel into the past would complicate things considerably. Leaving aside the question of what happens if you were to kill your own ancestor, this would make it impossible to distinguish cause and effect, rendering useless much of science as we know it. Science seems to describe nature pretty well, so it seems more reasonable to assume that travel into the past is indeed fantasy. Cambridge physicist Stephen Hawking (1942–) elevated this to the "Chronology Protection Conjecture", which says that the laws of physics conspire to prevent time travel, at least on a macroscopic scale. It is worth emphasizing that there is no theoretical way to decide this question; relativity allows this possibility, but does not require it.

But Einstein's impact on our notion of time goes even further. General relativity predicts the birth, and perhaps the death, of the universe itself. Very simple assumptions, roughly that gravity is always attractive, combined with the current observation that the universe is expanding, lead inexorably to the prediction that the universe originated in a big bang. Spacetime is the universe; "before" the big bang there was no notion of time at all! A similar argument leads to the prediction

then, regardless of the precise matter distribution, it must recollapse.

However, general relativity breaks down at the predicted matter densities of a big bang or big crunch. In order to understand the physics of the early universe, one must go beyond relativity. Einstein spent his later life trying unsuccessfully to reconcile general relativity and quantum mechanics, and although we have had some tantalizing clues since then, no such unified theory yet exists. What might such a theory have to say about time?

Quantum theory involves probability, and therefore uncertainty. It is not possible to locate an object precisely; even asking whether it has a precise location makes no sense. Similar restrictions apply to determine just when something takes place. The scale at which such effects matter is determined by the fundamental constants of nature, in this case Planck's constant, Newton's gravitational constant, and the speed of light. These constants combine to give fundamental length and time scales, and it seems unlikely that a unified theory will permit one to resolve information on smaller scales than this.

Expansion of the universe after the big bang

that the universe could recollapse into a big crunch. Whether this happens depends on whether the density of matter in the universe is enough to stop the current expansion; if it does,

theory will permit one to resolve information on smaller scales than this.

One must therefore ask whether space and time are continuous or discrete. We are accustomed to

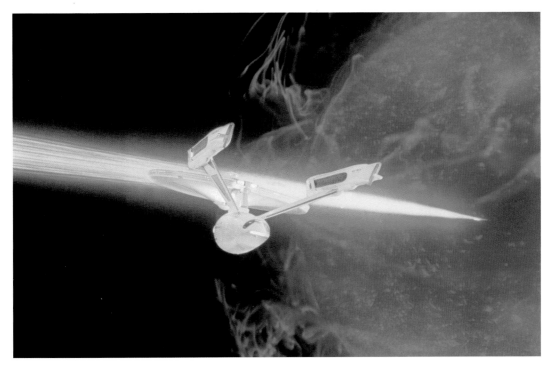

Star Trek V:
The Final Frontier,
1989

thinking of matter as being made up of discrete parts, whether atoms, protons and neutrons, or quarks. Surfaces which appear smooth to us are really complicated crystal-like structures. Does time pass the same way?

With discrete ticks rather than a smooth flow? Perhaps string theory is the answer, perhaps something else. Either way, we will likely have to rethink our notion of time yet again. Time will tell.

Axel Jessner

Pulsars: Einstein's Cosmic Clocks

One of Einstein's central motifs in his special and general theory of relativity concerns the behavior of clocks under different conditions: What is the reading of clocks when they are in uniform or accelerated motion, or in a gravitational field? Alas, in most cases the effects are very small, and we need extremely precise clocks to detect them. For all of human history up to the introduction of

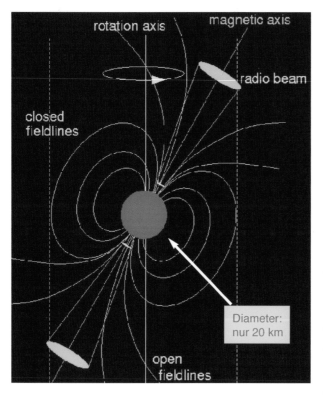

Scheme of the pulsar magnetosphere (diagram: M. Kramer)

the quartz clock it was astronomy that provided the most accurate time standards via the stable rotation of a large inert mass, the Earth. First in sundials and later by means of exact observation of stars it was possible to achieve precision of 0.01 s per day. By comparison, today's atomic clocks are accurate up to 10^{-9} s per day. This sufficed to verify the difference in reading predicted by relativity between clocks on earth and those in orbit about the Earth. But since the mass involved is comparatively small, the effects are rather tiny. For a proper test of the theory of relativity one would ideally like an extremely precise clock in

an accurately determined orbit near a large concentration of mass.

In July 1974 Russel Allan Hulse and Joseph H. Taylor discovered just such an object: PSR B1913+16, a pulsar that completes, together with a second neutron star, a tight (866 000 km), strongly elliptical orbit about the common center of mass in only 7.75 hours. The two masses approach each other at the periastron regularly once per revolution, only to then retreat from each other again.

Neutron stars are the nuclei of massive stars that remain after a supernova explosion. Within a sphere of diameter ca. 20 km there is a concentration of 1.4 solar masses. The mean density of $7 \cdot 10^{17}$ kgm^{-3} is here more than twice that of nuclear matter. One therefore assumes that the overwhelming part of the mass is in the form of neutrons. The escape velocity from the surface is as high as 30% of the speed of light – we are dealing with an object in whose surroundings relativistic effects play an important role. Neutron stars rotate, like all celestial bodies; the fastest ones we know of even turn about 600 times per second. Moreover, neutron stars possess an extremely strong magnetic field of $10^4 - 10^9$ Tesla. The rotation of the magnetic field induces enormous electric potentials of ca. 10^{13} volts, which can accelerate electrically charged particles above the magnetic poles to high energies. This current of particles emits strongly focused radio waves in a range of frequencies from less than 30 MHz to over 90 GHz. When this radio beacon sweeps momentarily over the Earth during its rotation we receive regular radio pulses from the neutron star, and one speaks in this case of a pulsar. Because of the large mass and comparatively low losses of rotational energy by radiation and outflow of particles the rotation is extremely stable, and the arrival of the radio pulses is accurately predictable. The most regular pulsars are found to pulse with a similar precision of 10^{-9} s per day, while the arrival time of single pulses can be determined with accuracy up to 10^{-6} s per day. Since their 1968 discovery in Cambridge (UK) by Jocelyn Bell and Antony Hewish, more than 1500 pulsars have been found and are being investigat-

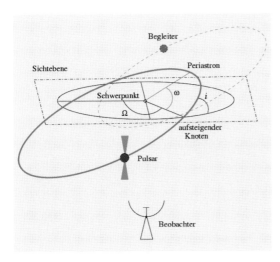

ed by the world's largest radio telescopes. In Germany this is being carried out at the 100 m radio telescope in Effelsberg.

Most pulsars are single objects, so far we know one with a planetary system (the first such system discovered around a star other than the Sun), and a few are part of a binary star system. From the motion of the partners of a binary star system astronomers can deduce the orbits by means of Kepler's laws. These elliptical orbits are in turn described by seven parameters:

The orbital period P_b (time for one revolution), the length a of the orbit's semimajor axis, the orbit's eccentricity e ($0 < e < 1$, with $e = 0$ for circular motion), the angle of inclination i between the elliptical orbit and the ecliptic, the angle Ω of the ascending node (intersection of the orbital plane with the ecliptic), the time of the two stars' closest approach (periastron τ), and the angle ω between the periastron and the ascending node (see figure).

The speed of a pulsar on such an elliptical orbit varies in time, and that can easily be seen in the Doppler shift of the radio pulses. But neither the distance to the pulsar system can be accurately determined, nor can the size a of the orbit be directly measured, as it can for visual binary stars. Therefore we lack independent data on the two stellar masses (M_1, M_2) and the inclination angle. Astronomers using the rules of classical mechanics can directly determine only five parameters $(P_b, \varepsilon, \omega, \tau, \Omega)$ in this case. Instead of the axis a and the inclination i one can calculate only the product $a \cdot \sin i$ and the "mass function" $f = (M_2 \cdot \sin i)^3 : (M_1 + M_2)^2$ from the variation of the pulsar's orbital speed.

It is just at this juncture that general relativity first comes to the rescue: One of the first of Einstein's successful predictions was the relativistic advance of the perihelion of Mercury's orbit. This effect is very small (46" of arc per century) in the solar system, but in the system PSR B1913+16 the equivalent periastron motion is ten million times larger: $\omega = 4.2°$ per year. This quantity does not depend on the orbital inclination i, it is a simple function $\alpha (M_1 + M_2)^{2/3}$ of the system's total mass. A second effect, already verified on Earth, is the slowing of clock rates in a gravitational field. The pulsar on its elliptical orbit regularly approaches its companion, and then the "pulsar clock" finds itself in the vicinity of a large mass and slows down accordingly. The magnitude γ of this effect is well measurable in the system PSR B1913+16, having a value of 4.3 ms. It is also an inclination-independent function of the two masses $\alpha M_2(M_1 + 2M_2)(M_1 + M_2)^{-4/3}$.

Through these two relativistic effects it is possible to calculate the masses separately, with an accuracy of $2 \cdot 10^{-3}$ ($M_1 = 1.442 \cdot M_{sun}$, $M_2 = 1.386 \cdot M_{sun}$), and then the mass function f can be used to find the inclination as well, yielding 54°. Thus the binary star system is completely determined! If still other independent parameters can be measured,

Diagram of the elliptical paths of a binary star (diagram: Axel Jessner)

the system is already overdetermined – as for a system of k equations with n unknowns, where there is a simultaneous solution for $k > n$ only if all equations are in agreement. For the theory of relativity this means that the equations derived from it must predict the additional measurements exactly.

A third relativistic effect, also known from the solar system, is the "Shapiro time delay." If the line of sight to a planet grazes the Sun, then the curva-

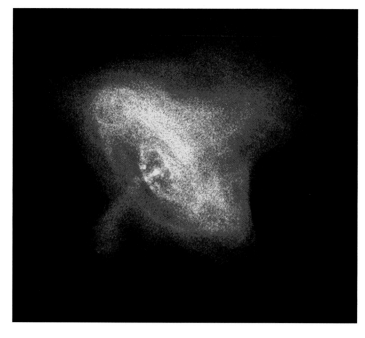

Chandra X-Ray, detailed observatory view of the Crab Nebula. In the center is a pulsar, which makes the structure glow. Taken on 1 September, 1999

ture of space caused by the Sun's mass leads to a small increase in the path length of the light, compared to the case that the Sun's mass is not present. In the solar system this path difference amounts to at most 120 μs of travel time, as Shapiro determined it in 1964 by means of radar echoes from the planet Mercury. From the shape of the time delay curves for pulsars one can even determine two independent parameters, r (range) and strength s. Of these r depends only on the mass M_2 of the companion, and s depends only on the inclination i. In the system PSR B1913+16 it was only possible to determine r and s with relatively large uncertainties. But within these uncer-

tainties the measured values are in good agreement with the predictions of general relativity. When two celestial bodies are in orbit, there is an effect on the spacetime continuum: gravitational waves are generated. This radiation carries away rotational energy from the system, the orbital path slowly becomes tighter, and the revolution times shorter. This effect also depends on the orbit's eccentricity, the system's total mass, and the product of the two stellar masses. In the solar system this is not measurable, but in PSR B1913+16 the orbital period decreases by 100 μs per year, so the orbit shrinks by about 1 cm per day. This measured change in period of $\dot{P}_b = (-2.427 \pm 0.026) \cdot 10^{-12}$ is to be compared with a prediction of general relativity of $\dot{P}_b = (-2.40216 \pm 0.00021) \cdot 10^{-12}$, a brilliant confirmation of the theory of relativity and at the same time proof of emission of gravitational waves! For this Hulse and Taylor were awarded the Nobel Prize in 1993. In the meantime a few comparable systems were found, and PSR B1913+16 is no longer the best test system for Einstein's theory, for it is overdetermined by only one parameter. An even tighter binary system – PSR 0737-3039 – was found in 2003. It has two pulsars that are even visible from the Earth. Here a revolution lasts less than 2.5 hours, the orbital semimajor axis at 450 000 km amounts to only little more than that of the moon's orbit. The inclination of 89° means that we look at this system's orbit edge-on, so that the radio waves of either pulsar pass close to the partner at certain phases of its orbit. It is then absorbed by the charged particles that populate the partner's magnetic field. In this system all orbit parameters are already fixed by classical celestial mechanics. In addition the relativistic effects described above are larger and more easily measured because of the tighter orbit. A diagram with the possible masses M_1 and M_2 as coordinates can be drawn (see figure).

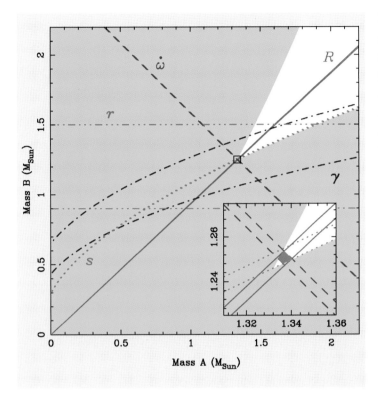

In this diagram the relations between M_1 and M_2 implied by the various measured parameters are shown as curves, or as bands in case of significant measurement uncertainties. In the case of PSR 0737-3039 there is an intersection or overlap of four curves in a tiny region – therefore a twofold overdetermination of the system. This object has been observed for only a year, so we can expect even more accurate confirmations of general relativity in the future. However, we ought not take too much time about it, because another relativistic effect, geodetic precession, leads to a rapid change of the rotation axis in such systems, and to a loss radio wave beaming toward the Earth. This change has also been observed already, and therefore we will be able to see the Hulse-Taylor pulsar PSR B1913+16 only until the year 2025.

Radio observations of pulsars provide us with the only avenue to date for precise measurements in regions where the theory of general relativity has very strong influence. Einstein's thought experiments are becoming reality – a pity that Einstein could not witness this triumph of his deductions!

Mass – mass diagram for the system PSR 0737–3039 (Lorimer & Kramer 2005). *R* is given by the mass ratio determined by celestial mechanics; the white area represents the total of combinations permitted by the mass functions determined by celestial mechanics; the other parameters are explained in the main text. The inner diagram is an enlarged section of the interface of the three best-known curves

Gerhard Börner

Expansion: From Redshift to Dark Matter

Einstein and Hubble

At the beginning of the 20th century physicists and astronomers experienced fundamental changes in their conception of the world. Besides quantum mechanics and Einstein's relativity

there were the discoveries of the American astronomer Edwin P. Hubble that had dramatic consequences. Hubble found that the objects known to astronomers at the time as "nebulae," such as M31, are in reality gigantic systems of stars, like our Milky Way, and that almost every such system or galaxy exhibits a shift of its spec-

This disproved the standard picture of an unchanging, infinitely extended distribution of stars. Strangely enough this picture was solidified at the end of the 19th century, disregarding its blatant conflict with the darkness of the nighttime sky. Johannes Kepler had already remarked in 1610 that this fact is inconsistent with a universe that is uniformly filled with stars of average brightness and size. If this were so, then every straight line originating from our location would eventually and with unit probability meet some star, and the whole sky would have to appear as bright as the surface of the Sun. (In today's standard cosmological model the stars evolved only some time after the big bang, and their distribution developed in such a way that a ray from us would hit a star only with negligible probability. So the night sky is dark because the stars originated at a finite time in the past.)

Even Albert Einstein at first adhered to this general view, and therefore he tried in 1918 to derive a

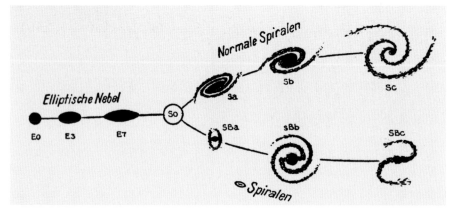

Edwin P. Hubble, 1932

Classification of the galaxies according to Hubble: Elliptical galaxies, spiral nebula, barred spiral galaxies, drawing 1938

tral lines, compared to those measured in terrestrial laboratories. This "redshift" z (where $1+z$ is the ratio of the received to the emitted spectral wavelength) is larger the greater the distance to the galaxy. The explanation of the redshift as a Doppler effect finally led to the conclusion that the galaxies fly away from us with a velocity proportional to their distance.

static world model from the equations of general relativity. To do so he had to introduce a repulsive force that balances at cosmic distances the gravitational attraction of any masses present in the model. He succeeded by supplementing the equations with an additional constant, the so-called cosmological constant L. But when Alexander Friedman found expanding and collapsing cosmological solutions of the general relativity equa-

Albert Einstein among famous U.S. physicists on the occasion of a conference at the Carnegie Institute of Mount Wilson Observatory: (from left to right) Milton L. Humason, Edwin P. Hubble, Charles St. John, Albert A. Michelson, Albert Einstein, W. Wallace Campbell and Walter Sydney Adams, 14 January 1931

tions in 1922, and when Hubble's discoveries became public, Einstein rejected the cosmological constant term in his equations. He later characterized his adoption of this quantity as the act of a jackass.

The Cosmological Models

Actually there were no grounds for such a self-critical remark, because his static solution and the subsequent expanding models of Friedman and Lemaître are still the foundation of modern cosmology. In these simple, highly symmetrical models the expansion is conceived as an outflow of an idealized, uniformly distributed fluid, whose density and pressure vary only in time but not in space. Any two particles of this fluid are separated by a distance that changes in time in proportion to a function $R(t)$. To gain a mental picture of these models we omit one spatial dimension and consider only two space dimensions and time. At any given time we can visualize space as the surface of a sphere (for positive curvature), as a saddle surface (for negative curvature), or as an infinite plane (for zero curvature, Euclidean space). Just these three different space forms are all one can have. Let's look at a spherical surface a little more closely, one made of elastic material like a rubber balloon. We'll mark the galaxies as points on this surface. As the balloon is being inflated these points move away from each other. This ex-

Albert Einstein in conversation with the Belgian priest and astronomer Georges Lemaître, Pasadena, 12 January 1933

pansion can be experienced from any one of these points, and will always look about the same. The distances between galaxies increase as the size of the balloon increases, even though their locations (latitude and longitude) on the spherical surface do not change. The distances change because the elastic material stretches. You can take that as a rather good way to visualize the relationships as described by Einstein's theory: the distances increase because of a change of the fabric of space-time, not because of any motion of the galaxies themselves. In this way we, as observers in one

of the galaxies, have the impression that the other galaxies move away from us. The radius of the balloon increases in proportion to the function $R(t)$ – call it the expansion factor – and in the same way the distances on the spherical surface change proportional to this function. The redshift ($1+z$) is simply the ratio of the present expansion factor to that at the moment of emission of this radiation.

If we turn time backwards, the balloon shrinks; the marked points approach each other ever closer. The cosmological models lead to an initial state in which all distances approach zero, and the density becomes infinite. This "big bang" cannot be further described by the physics we know. Here Einstein's theory reaches its limit of applicability. In the balloon picture we reach the big bang when the radius of the balloon approaches zero. But there is no point on its surface that marks the spot where the big bang happened. All points of the surface are always present, even arbitrarily close to the bang and for an arbitrarily small balloon. In this interpretation it seems as if the center of the sphere were the point where the big bang occurred, but this point does not belong to our world, which consists exclusively of the balloon surface.

This simple picture received decisive support by the discovery of a distribution of cosmic radiation in the microwave region. These radio waves were discovered in 1964 by Arno Penzias and Robert Wilson at a wavelength of 7 cm. Numerous observations in balloons and satellites have to date confirmed the cosmic origin of the radiation and its

The Andromeda
Galaxy,
photographed on
15 December 2003

thermal character. The temperature of this cosmic microwave background (CMB) amounts to only 2.7 Kelvin (about –270 °C). However, it was witness to a hot early phase of the cosmos, when matter and radiation were closely coupled to each other. The radiation cooled because of the expansion and decoupled from the matter as the first atoms were forming. From that time of "recombination" the CMB radiation was able to propagate unhindered and to cool off by expansion to the temperature we measure today.

Observations and Measurements

In the last five years astronomers succeeded in measuring the Hubble-relation quite accurately. They were able to exploit the fact that the radiation power emitted by a certain type of stellar explosion, the type Ia supernovae, is relatively con-

stant. Their brightness and hence their distance can be determined rather accurately from the maxima and the decay of their radiation power curve. The result for the typical time of the expansion turned out to be about 14 billion years. In addition it was found that the expansion is currently accelerating. A repulsive force, acting over cosmic dimensions, must be responsible for this acceleration, namely the above-mentioned cosmological constant.

This surprising result is supported by measurements of the CMB achieved by the NASA satellite WMAP. It provided accurate maps of the sky, where temperature variations could be seen as more and less cold patches. The differences in temperature are minute, only a hundred thousandth of the mean temperature ($DT/T = 10-5$), but they paint a picture of the small variations in density at the time of recombination, about

Scale of the
redshift

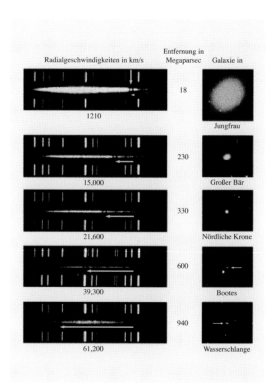

400 000 years after the big bang, which served
as seeds of the structures we see today, such as
galaxies and clusters of galaxies. All cosmically
relevant quantities had an influence on the form
and distribution of these small patches. A statistical analysis of the sky maps allows evaluation
of many parameters to within a few percent:
The curvature of space is zero, that is, the geometry is the everyday, Euclidean one. But this means
that the sum of all mass and energy densities
must attain a certain critical value, and that is
possible only if there is a cosmological constant
amounting to about 70 % of the critical value.
This fits perfectly with the measurements on
supernovae. We do not yet know whether this
cosmological constant, often called "dark energy,"
is really exactly constant, or whether it perhaps
represents a slowly changing energy of some
field that appears to us as a cosmological constant. Much about this is still a puzzle. Why does
the dark energy dominate just now, at the present
epoch? Why is it so small, a value about 100

orders of magnitude less than what one can estimate from quantum field theory for the energy
density of the vacuum?

At any rate it is the dark energy that is responsible for the presently observed acceleration of
the cosmic expansion.

Another remarkable fact is that the remaining
30%, due to matter, comes overwhelmingly from
as yet unknown forms of dark matter. Known,
baryonic matter accounts for only 5%. Physicists
hope to identify particles of dark matter in a few
years, thereby diminishing the extent of the
mystery.

The Cosmic Observer:
Redshift and Time Development

Astronomers in such a universe face the following
situation: Light signals from distant galaxies and
from the CMB arrive here today, but they were
emitted a long time ago. Therefore we observe the
galaxies not as they are today, but as how they
appeared in a past epoch. Astronomical observations present us with a cross section through the
history of the cosmos, and its present state can
be extracted only by means of a suitable model.
In the balloon model we can mark the observable
region as a circle. Objects within the circle can
be observed, the circle itself is our horizon, and
beyond it there are regions that we cannot observe. But our horizon grows at the speed of light,
proportional to time. The balloon, in turn, also
expands – in the simple cosmological models this
depends on the density of matter and energy.
As long as matter and radiation is the dominant
component of the cosmic substrate, the balloon
expands more slowly than the horizon, and we
can observe more and more of the universe. ($R(t)$
varies as $t^{2/3}$ for matter, and as $t^{1/2}$ for radiation,
whereas the radius of the horizon increases as t)
If dark energy determines the expansion, then
the balloon expands more rapidly than our hori-

zon, and some galaxies gradually escape from our field of view. For the same reason our view loses its reach as we follow the time development backwards into the past: in a matter and radiation-dominated cosmos the horizon shrinks much

A tiny region, part of a causally connected domain within one horizon, can then become our entire observed universe. This is how the model solves the horizon problem, but then we have to resign ourselves to the idea that only a tiny fraction of the total cosmos is accessible to our scrutiny.

Three-dimensional view of the distribution of galaxies in our universe; at the center, the Milky Way

faster than the universe itself contracts. This leads to the strange result that our horizon contains less and less of the whole world as we approach the big bang. This holds in the same way for every other point, that is, the cosmos frizzles up into many small, separate regions that no longer interact, because not even light signals can be exchanged. Immediately after the big bang each point stands alone, so to speak, unrelated to its neighbors – a peculiar consequence of the big bang. The inflationary universe model avoids this by changing the way the universe expands at the earliest epoch: again one takes the energy of a field as temporarily dominant, quite analogous to the dark energy, leading to an explosive inflation. The field energy changes into radiation after a very short time, about 10^{-35} seconds after the bang, and then the model continues its development like a normal cosmology.

The Five Phases of Cosmological Development

Immediately after the big bang there was a short phase when fields fluctuated with high energy, then inflation blew up a small region into the seed for the observable cosmos. After about 10^{-35} seconds the inflationary period was at its end and radiation dominated the expansion. About 400 000 years after the big bang matter took over command, and today as in the beginning a "dark energy" is responsible for the accelerating expansion. If this continues all neighboring galaxies will escape from our horizon. In the end everything will be torn apart.

The dark energy that Einstein first considered in 1918 determines the expansion of today. Equally important are its implications for the foundations of elementary particle physics. It is evidence of the deep inner connections among the natural sciences that astronomical observations of the largest possible object, the universe, can contribute to our understanding of a phenomenon that is closely linked to the smallest constituents of our world.

Erhard Scholz

The Standard Model of Contemporary Cosmology

In professional circles and beyond the opinion has prevailed since the 1960s that the universe is expanding and originated in the so-called *big bang*. In this model the general theory of relativity determines the geometry of space and time in accordance with fundamental assumptions about directional independence of the structure of space ("isotropy"), and with the empirically known characteristics from astronomy and astrophysics (cosmological redshift, microwave background

Arthur Stanley
Eddington

etc). The postulated cosmic space and time structure can be mathematically specified by four numerical quantities (space-time coordinates); this system constitutes a (semi-) Riemannian manifold, as it is called. Remarkably, all current models of space-time admit a cosmic time parameter t: for every value of t there is a purely spatial section R_t in the model, a kind of instantaneous space, whose points can be considered as marked empirically by galaxies, quasars, star clusters, stars, etc. The implied cosmic "simultaneity" is valid only for

the special observers who follow the mean motion of the galaxies. Therefore this cosmic time does not contradict special relativity's critique of "absolute" time, divorced from relationships in space. The total spacetime of the cosmos can be viewed as a continous family of such spatial sections. The relations between sections characterize the different types of cosmological model.

It is simplest to assume that all spatial sections are geometrically alike and that the special observers are at rest relative to each other. For obvious reasons one calls these cosmologies static models. The next simplest hypothesis would be one where sections are derived from each other by uniform enlargement or reduction *(a similarity mapping* by a t-dependent factor $a(t))$. In that case both the distinguished observers and the typical galaxies approach or recede from each other, according as the space sections contract or expand. Einstein chose to assume constant space sections in the first relativistic cosmological model he proposed. In this Einstein Universe, as it was later named after him, he supposed that all space sections had the geometry of a three-dimensional sphere of constant radius, that is, of the surface of a four-dimensional ball. According to the theory of general relativity this type of cosmological model could exist only if on the average it were uniformly filled with matter, a cosmological medium like an ideal gas or an ideal fluid of "constant mass-energy density."

Moreover, the gravitational forces exerted by the distant masses would have to be balanced by inward-directed forces of the cosmic medium. The "cosmic fluid" would have to resist without tearing (or collapsing) when stressed by tension rather than compression, like certain structural girders in buildings. This behavior, expressed mathematically as a negative pressure, seemed physically impossible to Einstein. He tried to evade the problem by introducing a hypothetical "cosmological term" in his gravitational equations.

But this introduced other problems, as noted immediately by Arthur Stanley Eddington. However, his contemporary Tullio Levi-Civita considered a cosmological medium resistant to tension a quite realistic possibility.

At present a different standard model of cosmology is widely accepted. Its fundamental idea is the expansion of spatial sections, each of which separately carries a geometry that is nearly the same as that of Euclidean space ("almost flat" spatial

ment's doctrine replaces it by the assumption that the so-called quantum vacuum, a quantum mechanically interpreted "activity of the void," produces negative pressure. This is now interpreted dynamically, as a cause for the expansion of the spatial sections, and it has time-dependent magnitude. Here one works in is a class of models that depend on two parameters, the relative energy density of matter and radiation, usually denoted by Ωm, and the fractional contribution to the en-

Abel 2218 galaxy cluster, Hubble telescope photograph

sections). If we think back in cosmic time parameter by about 14 billion years, then the "world" (or, more accurately, the geometry of the spacetime model) shrinks to a single point. Near that point in time – if it existed – conditions must have prevailed that were rather extreme. One speaks of an "initial singularity" of the standard model, or in more colorful language, of a "big bang".

In the standard model Einstein's cosmological term reappears in a different guise. The establish-

ergy density by the "vacuum" Ω_Λ, (here the subscript m refers to matter, and the notation Λ was originally introduced by Einstein for his "cosmological term", and taken over in the new context). Both of these are relative quantities, specified by dimensionless numbers.

In connection with the part Ω_Λ one also speaks of a dark energy that is supposed to cause expansion in the spatial sections of the model – even accelerated expansion, as is frequently emphasized for the

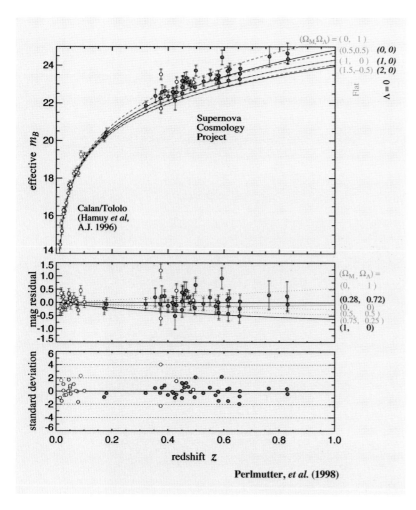

$(\Omega_M, \Omega_\Lambda) = (0, 1)$
$(0.5, 0.5)$ *(0, 0)*
$(1, 0)$ *(1, 0)*
$(1.5, -0.5)$ *(2, 0)*
Flat
$\Lambda = 0$

Supernova
Cosmology
Project

Calan/Tololo
(Hamuy *et al*,
A.J. 1996)

$(\Omega_M, \Omega_\Lambda) =$
$(0, 1)$
(0.28, 0.72)
$(0, 0)$
$(0.5, 0.5)$
$(0.75, 0.25)$
(1, 0)

Perlmutter, *et al.* (1998)

Relative magnitude m_B of supernovae Ia as a function of red shift z compared to the standard model predictions for different values of the parameters Ω_m and Ω_Λ, from S. Perlmutter et al.

length. On its way towards us, visible light shifts its color toward the red (cosmological redshift, conventionally denoted by the letter z). The greater the distance of the light source, the greater the redshift in the spectrum. For small distances and shifts the relation is linear, so the two increase in proportion. This was found and measured in the late 1920s by Edwin Hubble (Hubble's law). One way to explain this is to assume that the galaxies' mean positions are fixed at points of expanding spatial sections. They then recede from each other as the cosmic time parameter advances, with greater speeds for those already separated by greater distances. This results in a redshift of cosmic light sources as it is observed, analogous to the decreasing pitch of a receding fire engine's siren, the so-called "Doppler Effect."

2. As distance increases the relation between distance and redshift is no longer linear (in proportion). But firm data about distances of observed objects in the universe are available only to a limited extent, because at greater distances such data always depend on the theory used to evaluate the observations. In this connection it is very helpful that astronomers and astrophysicists found a class of supernovae, very bright combustion processes in stars, that all have nearly the same luminosity due to physical reasons concerning their interiors. The supernovae that can be most accurately assessed are denoted as "class Ia", SN Ia for short. Towards the end of the 1990s two different projects measured very accurately the dependence in supernovae Ia between the observed magnitude m (a measure of the relative brightness at the observer) and the corresponding redshift z. These measurements provide very good means to determine the parameters of a cosmolo-

current model parameters. However, the attribute "dark" is ambiguous with unintended self-irony. On the one hand the quantum vacuum is not directly visible, on the other hand this time dependence of the "dark energy" fraction compared to the matter content is so peculiar, that it introduces an obscure feature into the model. Even more irritation arises when one attempts to calculate the amount of "dark energy" from established quantum field theory. Any such attempt leads to a totally bizarre result, too large by a factor 10^{120}. The four most important empirical arguments given for the standard model are the following.

1. Radiation from distant cosmological sources such as galaxies, quasars, etc. gets to us, the observers, with a systematically enlarged wave-

gical model and test whether a class of models can represent reality. In this way the free parameters in the standard cosmologies were determined with relatively great accuracy (see p. 390). On the basis of these measurements and other indications

2.73 °K, as calculated by Max Planck on the basis of his quantum hypothesis (Planck radiation), and it is largely independent of direction (isotropic). Only in angular ranges of less than a degree can tiny fluctuations of radiation temperature be ob-

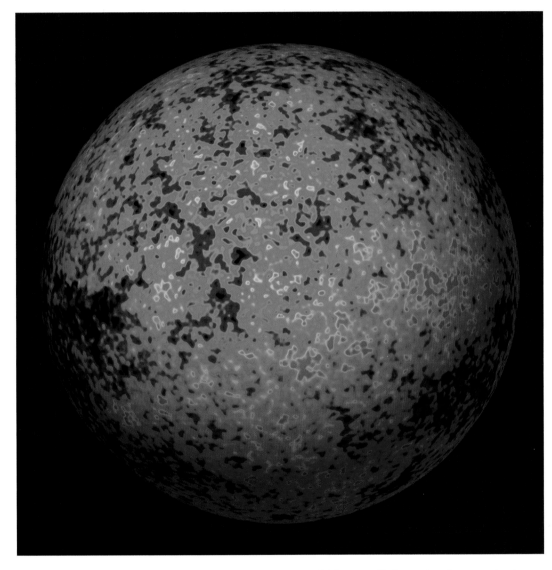

"Map" of the cosmic microwave background, projected on a spherical surface

the values are today taken to be $\Omega_m \approx 0.3$ and $\Omega_\Lambda \approx 0.7$ for a realistic model within the standard class.

3. The entire celestial globe is filled with (weak) radiation in the microwave range. This corresponds very closely to the thermal equilibrium radiation of a so-called "black body" of temperature

served. This so-called cosmic microwave background (CMB) was discovered in the 1960s. Over the years its properties were measured with increasing accuracy, in particular the tiny direction-dependent fluctuations, the so-called "anisotropy." The currently still running satellite project WMAP (Wilkinson Microwave Anisotropy Probe) is an example. In the standard picture this cosmic microwave background is interpreted as the redshifted relic of a high-temperature state of the universe a few hundred thousand years after the big bang. The anisotropies are interpreted as signals that shed light on the "early history" of

"With unbelievable presence of mind God had snapped a picture of the big bang, which He still found rather impressive", from Bernd Pfarr: *Eines Tages war Zeus das Blitze-schleudern leid*, Frankfurt/Main 2001

cosmic evolution, somewhat like a snapshot of the universe shortly after the big bang.

4. Through observations of the spectrum of light arriving here from stars and galaxies, astronomers and astrophysicists were able to estimate the distribution of the chemical elements in the cosmic material. They found a large component of the periodic table's light elements, of hydrogen in particular. Stars convert this in several cycles of nuclear burning to heavier elements, so one might think that there cannot really be so many light elements left in the universe. That is, they are in excess, compared to first, naïve expectations. One speaks of a cosmic abundance of light elements. In the late 1940s George Gamov and others framed the hypothesis that atomic nuclei formed only in some very early phase of "cosmic evolution". By today's estimate this must have happened about 2–3 minutes after the absolute

beginning of time, the big bang. From the conditions in such information of nuclei, in more distinguished words at "primordial nucleosynthesis", one could include an abundance of light elements in good agreement with present observational values. However, Fred Hoyle and other physicists supported the hypothesis that the distribution of elements is to be understood by transformations within the cosmos as we see it today. They first focused on the study of the different nuclear burning cycles inside of stars. With increasing understanding of these stellar processes their hypothesis proved very fruitful and of great explanatory value. Only for the abundance of light elements does the standard model take recourse to a refined version of Gamov's primordial nucleosynthesis.

A word of caution is appropriate: assuming that nuclei are synthesized primordially leads to a rather low upper limit for the resulting density of matter (usually denoted as "baryon density", Ω_b) $\Omega_b \approx 0.1$. This value is much less than the total matter density as derived from other data ($\Omega_m \approx 0.3$, see above). In the standard cosmological model one therefore introduces the ad-hoc hypothesis that this gap is closed by a totally unknown and exotic type of "dark matter."

The development of new observational techniques (extension of the frequency range of astronomical signals into the radio, UV, and X-ray region, refinement of observations in the visible, satellite observations etc.) and the new objects thereby amenable to observation (quasars, radio and X-ray sources etc.) have led to many new observations that must be taken into account by any cosmological model before it can claim to represent reality.

From the mathematical point of view the standard model is suitable to reconcile theoretically the different types of data from observational cosmology. However, a certain amount of healthy skepticism concerning the oft-cited "claim to reality" of these models is by now more than appropriate. The search for candidates for this (non-baryonic) "dark matter," pursued with great enthusiasm by elementary particle physicists, theoreticians among them, has not met with success. By the criteria for validity of physical theories as conventionally practiced elsewhere the hypothesis of primordial nucleosynthesis really ought now to be regarded as empirically refuted. But only a few physicists will draw this direct conclusion. The dynamical properties of the "dark energy" that follow from the assumptions of the standard model continue to be unexplained physically and will probably remain so. One may therefore have doubts that these and other obscurities can be cleared up within the confines of this model. Nevertheless an important majority of the scientists working today in cosmology continues to be firmly convinced that the standard model's picture of an expanding and accelerating universe is confirmed by the latest results about redshift of supernovae, anisotropy of the microwave background etc. The more thoughtful ones regard the difficulties as anomalies that call the present certainty of the picture into question. But to dateonly a minority of the scientific community has drawn the consequence to work on alternative explanatory schemes. It will be interesting to see whether this leads to a viable alternative, and what direction the future development will take.

Erhard Scholz

Einstein-Weyl Models of Cosmology

The hypothesis of an "expanding universe" is not the only possible explanation of the cosmological redshift. The foundation for accurate research of the cosmological redshift was laid in the 1920s and 1930s by the astronomical measurements of Edwin Hubble, who considered the mutual "recession speed" of galaxies as only a convenient language to describe his observational data. He did not take a "real" expansion of the universe for granted. His colleague and contemporary Fritz Zwicky supposed that the cosmological redshift was caused by a gravitational interaction of photons, not yet understood in detail, which withdraws energy from the latter over cosmological time intervals. By Einstein's interpretation of Planck's quantum principle, $E = h\nu$, this corresponds to a decrease in frequency, that is, an increase in the photon's internal time measure.

An extension of Riemannian differential geometry suggested by Hermann Weyl assumes that the geometrical (and hence physical) distance and time scales are introduced by conventions that may vary from place to place – similar to the different length "gauges" such as the rod, pole, and perch of former times. As these are transferred from one place and time to another they are subject to a possibly path-dependent gauge change (Weyl's gauge geometry). Weyl described this mathematically by introducing a so-called "length connection" (or gauge connection), and by allowing changes in scale, that is, gauge transformations in the strict sense of the word.

Hermann Weyl (1885–1955), photograph 1950

This sort of geometry is ideally suited to represent physical relationships that involve a change of the internal scale when different times and places are interrelated, as by an exchange of light signals. Exactly this is the case for the cosmic redshift. Unlike what Weyl suggested in 1918, Einstein's general relativity theory (GRT) can indeed be formulated without drastic changes in the context of Weyl's gauge geometry. This can be done within a conservative extension of Einstein's theory. Here "conservative" means that GRT does not only appear as a limiting case of the extended theory, as is typical in physics when theories are extended, but that it is even a special case of the extension, so that under certain conditions it remains valid not only approximately but strictly. Initially one gains merely additional conceptual-symbolic scope by being able to choose different gauges. But in the next step this also leads to an extended variety of possible physical interpretations of the mathematical structure. This is so even if Weyl's gauge geometry hardly differs from the Riemannian geometry that is normally used in general relativity theory. In the latter case on speaks of "integrable Weyl geometries." They form the basis of the following models.

Analyzing the geometry of the well-known relativistic cosmological models (so-called "Robertson-Walker manifolds") from this point of view allows us by choice of gauge to transform between a kinematical interpretation of the redshift, as a Doppler shift, and a non-kinematical one (such as loss of photon energy caused by gravity). Einstein was able to regard gravitational fields and accelerated systems of observers as equivalent (under certain conditions) with the help of Riemannian differential geometry. Similarly in the present case the assumption of expanding spatial sections in cosmological models and the assumption of a non-kinematical (gravitational) decrease in photon energy prove to be mathematically equivalent under certain conditions. In this sense Weyl's geometry makes it possible to extend the equivalence principle to the cosmic redshift. This results in new points of view, because unusual gauges from the point of view of Riemannian geometry may acquire physical meaning.

In one of the possible gauges the simplest models that can be constructed by this scheme look like an Einstein universe (or like other static models of classical relativistic cosmology). But in addition they are now endowed with a Weylian "length connection" (transport of lengths in Weyl's sense) that implements the redshift without necessarily involving a real expansion of space. Therefore this will here be called the Hubble connection of the cosmological model. Assuming even the simplest case, a constant Hubble connection already leads to cosmological models that shall for obvious reasons be called Einstein-Weyl models. They can be specified by the accurately known Hubble constant H of astronomy and by the curvature κ of spatial sections. Mathematically the only essentially free parameter of the model is the ratio $\zeta = {}^{k}/_{H^2}$ (it determines the model "up to isomorphism"). For purposes of physical interpretation of the model there are good reasons to identify this geometrical parameter with the relative density of matter energy in the universe. This is usually denoted by Ω_m.

Calculation of the redshift z and of the decrease of the relative luminosity of cosmic sources, as expressed by their astronomic magnitude m, shows an essential dependence of the derived relation $m(z)$ between these quantities on the parameter ζ. One notices that the very accurate data on m and z measured toward the end of the 1990s for supernovae Ia agree very well with the prediction of these geometrically very simple models. Put more technically, this class of models fits the data, for the value $\zeta \approx 1$, about as well as the standard cosmological model with the parameter $\Omega_m \approx 0.3$ and

relative vacuum energy density $\Omega_\Lambda \approx 0.7$ (see figure below). The differences in goodness-of-fit (by the method of least squares) are not significant. In view of the obscurities occurring in the standard model – from the standpoint of theory and history of science these are robust anomalies of the presently prevailing theory – this may be a first occasion to consider whether the assumptions of this model can possibly correspond to reality.

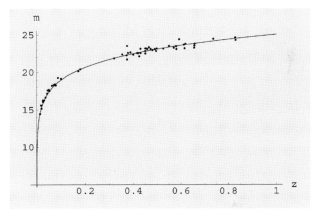

Relation between relative luminosity m and redshift z of supernovae Ia, compared with the Einstein-Weyl model, $\zeta = 1$, Hubble constant $H_0=70$, for absolute luminosity $M = -19.3$. SN_{Ia} data (dots) (Perlmutter 1999)

Should this be so, general relativity would have to be slightly modified by the Hubble connection. The modification would be so slight that its effects on experiments or observations on the solar system would be below measurement error. Only at the level of cosmological scales would the decrease of photon energy (presumably caused by gravity) have consequences for geometry and physics. But there the consequences would be substantial.

Most decisive is the absence of the "big bang" in these models. From this point of view the Bang appears as an artifact of the standard model, even if there are excellent reasons for it within that context. Hence it may possibly be pointless to study many of the questions that turn upon the big bang, taken literally. In the opinions of the majority of scientists this appears at present not very attractive, or worse. On the other hand, a part of the standard assumptions should in any

Albert Einstein with Charles Edward St. John at Mount Wilson Observatory in California, winter 1931

Number of uniformly distributed objects in Einstein-Weyl universe with $\zeta = 0.3$ (longest dashes), $\zeta = 0.6$ (long dashes), $\zeta = 1$ (medium dashes), $\zeta = 1.5$ (continuous), and $\zeta = 2$ (shortest dashes), compared to the quasar counts of the Sloan Digital Sky Survey (dots); with constant bin interval $\Delta z = 0.076$

which however also still lacks a physical explanation.

The Einstein-Weyl models also provide a reason for the appearance of the cosmic microwave background, CMB. Here it is nothing to do with the effects of a big bang, rather it appears as the equilibrium state of the (quantized) electromagnetic field on the three dimensional sphere of the Einstein universe. The mathematician and mathematical physicist I. E. Segal proved in 1983 in a somewhat different context the existence of such a state of equilibrium.

The small perturbations of the CMB's directional dependence, its so-called anisotropies, correspond in this view to perturbations of the geometry of the Einstein-Weyl universe, caused by irregularities of the mass distribution, such as galactic clusters and super cluster. The angular sizes of super clusters in a wide segment surrounding

case be viewed as problematical or even obsolete. The so-called "primordial nucleosynthesis" in particular can no longer be made to agree with the empirical data on mass densities in the universe. Attempts to find unusual ("exotic") dark matter is by now no longer regarded as convincing from the standpoint of observational cosmology. This argues for a return to the view of Zwicky (who first assumed a perceptible amount of dark matter in the universe), that we deal here with usual baryonic, but non-luminous, matter in interstellar and intergalactic space. Furthermore the Einstein-Weyl models encompass no expansion (certainly not the accelerated kind), whose explanation the present standard theory hopes to find in a mysterious "dynamic dark energy." Instead the model indicates only a constant tensile stress of the cosmic medium,

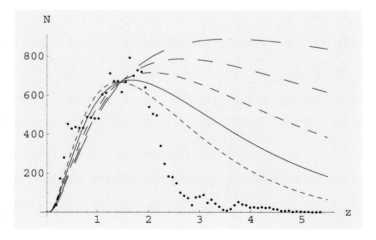

the equator of the spatial sphere of the Einstein universe correspond rather exactly to the angular size of the CMB anisotropies. It also appears possible that the background radiation's photons lose energy when they pass through regions of super clusters (by the so-called "Sunyaev-Zel'dovich effect" in these regions). The most recent studies have shown that at least a part of the anisotropies correlate so well with the positions of galactic clusters that the interpretation of the anisotropies as properties of the "snapshot of the big bang" is empirically no longer tenable with its present claim to exclusivity. We should come to terms with the possibility that the snapshot interpretation may as a whole turn out to be a fiction, as such correlations continue to be investigated in the realm of higher redshifts.

In conclusion let us mention a remarkable observation concerning quasar frequencies, which the standard cosmological model can make plausible at best only by ad hoc assumptions on matter evolution. In the Einstein-Weyl model it is a simply predictable effect of the space-time geometry. If the volumes covered by an observer's backward light cone in the spatial sphere of the Einstein-Weyl model are calculated as a function of red shift, the results are curves as in figure 2 (see p. 396) for different values of the parameter ζ. For $z \leq 2$ they agree surprisingly well with the most recent results on quasar counts over large angular ranges of the celestial sphere. The rapid decrease of the observed number of quasars beyond $z = 2$

appears due to selection effects, not least because of the drastic decrease of sensitivity of the CCD detectors in this frequency range. Thus in the Einstein-Weyl scheme the quasar data imply an approximately uniform distribution of quasars when averaged over sufficiently large cosmic regions. Moreover, if ζ is interpreted physically as Ω_m they indicate an even higher mean mass density in the universe $(\Omega_m = \zeta \approx 1.5)$ than established so far by dynamic mass determination $(\Omega_m \approx 0.3)$. This is by no means empirically unreasonable; after all, quasar data aggregate observations over far greater ranges of volume than can be accounted for in the estimates of dynamic mass density. From the Einstein-Weyl models' point of view this favors the assumption of a higher fraction of known baryonic matter than may be assumed in the standard model. If one listens to astronomers whose vision is not limited by the framework of the standard model, then we could simply be dealing with molecular hydrogen, which is difficult to detect astronomically. Its frequent occurrence (abundance) could be explained by energy-absorbing fission of intermediate-mass nuclei by high-energy cosmic radiation (so-called "spallation"). The "primordial nucleosynthesis" might then be replaced by a kind of cosmic recycling of baryonic matter.

Günther Hasinger

Black Holes: The Beginning as Well as the End?

At the Russian front in December 1915, only a few weeks after Einstein had published the general theory of relativity, the astrophysicist Karl Schwarzschild found the first exact solution of Einstein's field equations. He calculated the curved spacetime in the vacuum exterior to an electrically neutral, non-rotating, spherically symmetric star. To Einstein he sent his preliminary, "exterior" solution, which later was given the name "Schwarzschild-Geometry." Einstein congratulated him on the simplicity of his solution and in Schwarzschild's name he lectured on these results at a session of the Berlin Academy of Sciences in January 1916. According to this theory each mass determines a critical distance, the so-called "Schwarzschild radius," within which the curvature of space becomes so large that even light cannot escape from inside. For the Sun, for example, this radius is 3 kilometers, for the Earth it is 1 centimeter. A few weeks later Schwarzschild sent a second proper solution, where he calculated the spacetime curvature inside the Schwarzschild radius as well. In the "interior" solution, space and time interchange their roles in a bizarre way, leading inexorably to a singularity, where the curvature and the density become infinitely large. A few weeks later, afflicted with a serious skin disease, Karl Schwarzschild had to return to Potsdam and died in May 1916. During his life Albert Einstein rejected the construct of the "Schwarzschild singularity" as illogical and unphysical and thus rejected the predictions of his own theory. Previously, towards the end of the 18th century, using Newton's theory, the British parson Rev. John Mitchell and the French mathematician and astronomer Pierre-Simon Laplace had already calculated conditions for stars to be so compact that not even light can escape from their surface. Even after Schwarzschild's solution, it still took several decades of scientific dispute among such great physicists as Eddington, Chandrasekhar, Oppenheimer and Landau before the reality of compact objects dominated by relativistic effects was accepted. Before the Second World War, Baade and Zwicky had already predicted the existence of neutron stars. These stars were to have about 1.5 – 2 solar masses, but a radius of only 10 – 20 kilometers, and were to be stabilized by quantum mechanical forces. Only after the 2nd World War's technical advances in radio and rocket technology could be applied to scientific research did the 1960s bring three important discoveries in rapid succession, which established "relativistic astrophysics": Compact X-ray sources and X-ray background (see below) in 1962, "quasars" – highly energetic centers of active galaxies – in 1963, and "pulsars" – rapidly rotating neutron stars – in 1968.

In 1968 John A. Wheeler introduced the concept of "black holes". According to the models of stellar evolution, neutron stars and black holes are produced at the end of a very massive star's life,

Brought about by friction effects in what was known as the "accretion disk", matter begins to glow brightly in X-rays shortly before falling into a black hole

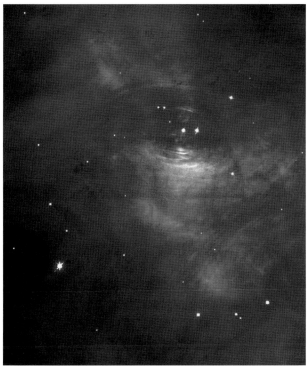

Crab Nebula, with black hole at the center

Light radiation by tightly bundled particle rays, known as "jets," from the surroundings of a black hole

masses. In 1970 the compact object in the X-ray binary star system Cygnus X-1 was characterized as the best candidate for a stellar black hole. Extremely massive black holes are also suspected in quasars, which often are a thousand times brighter than the center of their galaxies. Why does a black hole radiate at all – should it not absorb all light? In reality we see not the black hole, but matter that is being swallowed by this maelstrom. In this process the matter reaches velocities close to the speed of light, and it is heated by friction in the so-called "accretion disk" (Fig. 1) to such high temperatures that it begins to radiate very strongly, for example in X-ray light. It is as if we see the "last gasp" of the matter shortly before it falls into the black hole. Frequently, in addition to the strong radiation, the surroundings of the black hole also emit matter, which is accelerated to relativistic speeds in sharply focused beams of particles, the so-called jets, emitting light in a wide range of the electromagnetic spectrum, from radio to gamma radiation (see above).

Although several good candidates for black holes were identified in the 1970s and 1980s, the idea of such weird qravitational traps continued to be regarded as very exotic and with great skepticism. But in the last decade there was a change of paradigm. By monitoring in the infrared at major telescopes with extreme resolution, German and American research groups were able to "capture" a black hole of about 3.5 million solar masses in the center of our Milky Way. Thus the first determination of the velocities and orbits of separate stars about the galactic center became possible (see p. 400), thereby proving beyond doubt that

when the fusion furnace at its center ceases and the central "heap of ashes" can no longer withstand its own gravity. Black holes are thought to form from stars with more than 25-30 solar

the dark mass must be a black hole. There are also indications that a black hole with over 100 million solar masses exists in our twin galaxy, the Andromeda Nebula. Successively more sensitive observations in recent years have pointed to black holes at the center of almost all nearby galaxies, with masses up to three million times that of the Sun. Surprisingly, the mass of the black hole depends directly on the brightness of

nate in the early universe, only a few billion years after the big bang; whereas the smaller "garden variety" black holes, such as that in our galaxy, were formed only significantly later. Gigantic galaxy collisions are often involved, where two Milky Ways fuse, including presumably their central black holes. Recently it was a particular success to spy two active black holes "in flagranti" at such a wedding of galaxies (see below).

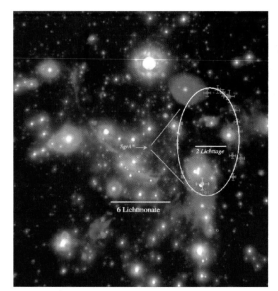

Orbital paths and speeds of individual stars around the galactic center of our Milky Way prove the existence of a black hole

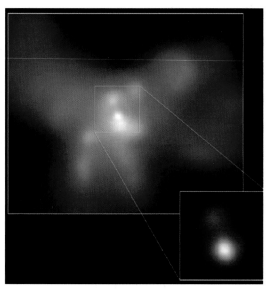

When two Milky Ways merge, the black holes presumably merge as well

the galaxy or on the velocities of the stars at its center. So there must be a presently mysterious mechanism that provides for simultaneous growth of galaxy and black hole, hand in hand as it were.

Our whole sky is filled with a diffuse X-ray light, the X-ray background, discovered in 1962. More sensitive and accurate X-ray telescopes made it possible in recent years to resolve this background radiation into hundreds of millions of separate points of light, which upon closer study turned out to be distant active galaxies, with growing black holes at their centers. Accordingly the X-ray background originates during the eat-and-grow phase of the total population of massive black holes in the universe. It turns out that the biggest black holes, the so-called "quasars," origi-

But what came first, the black hole or the galaxy? Put together the puzzle pieces found in recent years and you get a fairly well-rounded picture: about 400,000 years after the hot big bang dark matter dominated the processes in the universe. It clumps together under its own gravity to form an invisible cosmic network of filaments and clusters. It drags along normal matter, which will later make galaxies. After the big bang the universe had to wait another 200 to 500 million years until it became cool enough for the first stars to form. According to newer insights the first star that formed at that time in the deepest quantum well of dark matter must have been very massive, so that after a short time it ended its life in a gigantic supernova explosion, and presumably left behind the first black hole in the universe.

This black hole will spend the rest of its life near one of the greatest concentrations of mass in the universe, and like a spider in its web only has to wait for something edible to come by. The feeding is probably best for the black holes during a galaxy collision as described above, where they can grow exponentially in a few million years to billions of solar masses, thus forming the first distant quasars. In this picture the black holes originate as "galaxy seeds", that is, in fact much earlier than most stars, and then they grow together with their galaxy.

The oldest object in the local universe is probably the black hole in the galaxy M87 at the center of the Virgo cluster (see p. 399 above), whose attraction extends to our Milky Way as well. But what is the future fate of black holes? Massive black holes can in principle live almost indefinitely. Nonetheless they radiate a tiny amount of energy, as Stephen Hawking found out, and therefore their lifetime is finite. A black hole of solar mass can exist for about 10^{67} years, considerably longer than all other forms of matter and structures in the universe. If the recently discovered dark energy really drives universal expansion forever, then black holes can grow to hundreds of billions of solar masses, and live about 10^{100} years – a truly unimaginable time. The first compact objects to enter the stage of the universe will then also be the last to leave.

Bruno Bertotti

The Cassini Experiment: Investigating the Nature of Gravity

Why do bodies fall? What governs the amazing harmony of the orbits of planets? What is the deep nature of gravity? The exploration of these questions has developed in three stages, from Newton to Einstein's theory of general relativity and, lastly, to the present search for deeper and different views. They show a steady progression in the use of unintuitive and abstract concepts, for which mathematical tools are essential. An important highlight of this long endeavor is the experiment carried out in 2002 with the Cassini spacecraft, at that time on its way towards the planet Saturn.

In Isaac Newton's *Principia* (1686) the gravitational force produced by a mass is not attributed to its essence, or to a direct pulling action. In the space around a massive body, even if empty of matter, Newton maintains, there is a mathematical trace of its presence, consisting in just a single number at every point. This is called, technically, a scalar field. This number can be measured: it is the energy required to move a mass of one gram from that point away to large distances. Newton established the rules with which this field quantity is produced by the bodies around and showed that, in admirable and accurate agreement with observations, it determines the orbits of celestial bodies. This field is an abstract construct and cannot be perceived directly by the senses; but, in Newton's view, it fully embodies the very nature of gravitation and its concrete realization.

The second stage began in 1915, the year in which Albert Einstein, in a gigantic conceptual revolution, put forward a radically different view of gravity, in which its field is not, so to speak, imprinted on space, but is inherent to space itself, and determines its geometrical properties. Space, or, rather, the union of space and time called spacetime, is in fact the analog in four (three for space and one for time) dimensions of a curved, ordinary surface. The well-known theorems of geometry, in particular, Pythagoras' theorem, have a

Cassini space shuttle

different, more complex form. In Einstein's view, all the information about the gravitation produced by the masses is embodied in the geometrical structure of the spacetime around them, in particular, its curvature. It turns out that the determination of this structure requires assigning at every event (that is to say, at every instant and every point) *ten quantities*; together they constitute a mathematical construct called a *tensor field*. With the theory of general relativity Einstein showed that in the particular conditions of the planetary system, in fact only one of these quantities is by and large sufficient: this is just Newton's scalar field; thus outstanding success

of celestial mechanics is fully recovered. But in this theory there is much more than in Newton's conception. Besides its radically new view of the physical world, the observational consequences of general relativity are different. In the solar system and in the laboratory it predicts minute, but significant discrepancies, which have been the target of extensive experimental programs.

Isaac
Newton

The main ones are:

1. The reddening of light coming from a region where the gravitational field is stronger.

2. The deflection of light rays by a massive body. It is interesting to note that this effect arises also in a Newtonian framework, in which gravity is described by a singe scalar field; *but the deflection angle is exactly half Einstein's value.*

3. An anomaly in planetary motion, consisting in the motion of the *pericentre*, particularly important for Mercury.

In addition, general relativity requires that the fall of a body – the acceleration it experiences due to a massive body nearby – be independent of its nature and mass. This very peculiar property, which goes under the name of equivalence principle, is unique to gravity, as compared to other forces, like electricity; it has been confirmed with great accuracy in laboratory experiments; the three discrepancies listed above have also been

confirmed to a very good accuracy. It is quite remarkable that, so far, no evidence against Einstein's predictions has emerged, while several alternative theories have been disproved: general relativity, put forward ninety years ago, has an extraordinary longevity.

The second discrepancy was the object of Cassini's experiment. When a beam of light or radio wave grazes the Sun, it is bent inwards by the gravity of the latter. The deflection angle is very small indeed: it is about equal to the angle at which a cent (1 cm in diameter) is seen at 1 km. As a consequence, the stars which appear in the sky near the Sun are slightly displaced outward.

Albert Einstein,
around 1921

Due to the strong competing solar radiation, it is practically impossible to measure this deflection optically, unless the Sun is blocked out. This occurs during an eclipse; in fact, the first (rough) confirmation of Einstein's prediction was provided during the 1919 eclipse. It is much better to use radio waves. Over the years, with the improvements of techniques and instrumentation, the efforts to advance the measurement of the relativistic deflection angle have made big progress; no discrepancy with the prediction was demonstrated and the last experiment before Cassini, performed in 1995, had an accuracy of about 0.1% of the angle itself. Note that this accuracy corresponds to the angle at which a 1 cm object is seen at the distance of 1000 km. It should also be noted that for the other dynamical test of general relativity (No. 3 in the above list), no substantial increase in accuracy beyond the level of 0.1% has been attained.

In June 2002 Cassini and the Earth were almost aligned with the Sun, with the latter in the middle: an optimal condition for the experiment. Rather than the angular displacement of a source in the sky, Cassini measured the effect that the deflection has on the (sharp and stable) frequency of a microwave radio beam sent from the ground station to the spacecraft and retransmitted back. The deflection changes the angles between the beam and the direction of motion of the two end points and, therefore, affects the frequency received on the ground. The beam in its up and down passes traverses the solar corona around the Sun, where strong disturbances, due to solar flares and coronal mass ejections, greatly disturb electromagnetic propagation. The experiment has been made possible by the special radio system available to Cassini (with high and multiple frequencies), which successfully compensates these disturbances. At that time, the spacecraft, on its cruise to Saturn, was in very quiet and steady conditions, ideal for the elimination of other disturbances. Cassini did not find a violation of Einstein's prediction, but the accuracy of the measurement was 0.002%, about fifty times better than before.

Gravity Probe B (GPB), launched by the NASA space mission in April, 2004 will certainly contribute to testing the dynamical aspects of relativistic corrections. In 2008 *Microscope*, a space mission of the French CNES, will carry out an important test of the equivalence principle by measuring, to a much better accuracy than before, the difference in the accelerations of two masses free to move inside the probe. *GAIA*, the astrometric space mission of the European Space Agency, to be flown probably in 2013, is expected to measure the gravitational deflection caused by the Sun on the light of stars with an accuracy much better than Cassini's. Other, more advanced projects are under consideration and study. These major developments could mark entering in the third stage of the quest for the nature of gravity, in which Einstein's general relativity may be transcended by an even more complex and abstract theory, and violations – small, but radical – may finally be detected. Indeed, no theory can be regarded as final: scientific progress consists just in going beyond – theoretically and experimentally – an accepted scheme, opening the way to new developments.

What is being questioned in this third stage is whether gravity is entirely described by geometrical properties of four-dimensional spacetime – Einstein's view. On broad terms, our understanding of elementary particles and fields, especially at very high energies, seems to require additional dimensions and new geometrical entities, beyond the traditional concepts of point, line, plane, etc; moreover, the quantum theory of fields is essential. In this view spacetime is only a very particular manifestation, evident only at low energies, of a much wider framework. Deeper and more complex structures are normally inaccessible to observation, except very near the big bang, during the initial phases of the cosmological expansion. Many of these recent developments, in particular *string theory*, have one common feature: a scalar field in spacetime which contributes to the long-range force between neutral bodies and, therefore, acts just like gravitation. This field may have had a paramount role in primordial cosmology; but, as the Universe expanded and acquired its present aspect, with galaxies, stars and chemical elements, its effects have dwindled to almost nothing. But even a tiny remnant would at the present epoch contribute to gravity in a way not reducible to purely geometric properties and lead to violations of all Einstein's predictions: different bodies would not fall with a unique acceleration and all the relativistic dynamic corrections would be slightly different than expected. Regrettably, these new conceptions have not yet been really developed into precise theories, capable of provid-

by a very small amount. This is a striking example of how a very minute discrepancy could throw light on the fundamental nature of gravity; determining its sign would either disprove or confirm the broader scenario under consideration, based

ing specific observational tests; but on one point they agree: the deflection of light due to a massive body (Item 2 in the above list) should be less than Einstein's prediction. On an elementary level, this can be easily understood. Since in a Newtonian scalar scheme the corresponding angle is half the relativistic prediction, the addition of a scalar field should decrease the predicted value, albeit

upon a scalar field of cosmological origin. In this scenario gravitational dynamics in the laboratory, in the solar system and in the pulsars are affected by what happened near the big bang; vice versa, accurate experiments at the present epoch may provide information about that very early age of the Universe.

Left:
Cassini experiment to test the Einsteinian theory of gravitation

Right:
GAIA: astrometric space mission of the European Space Organization

Bernard Schutz

Gravitational Waves

Gravitational radiation is the last great prediction of Einstein's general theory of relativity that has not yet been verified by direct experiment or observation. This challenge to modern physics may not last much longer: large-scale gravitational wave observatories are nearing completion in several countries, including Germany. The long-awaited event in which physicists will register

for the first time the passage of a wave of gravity may happen at any time in the next few years. Exotic as waves of gravity may seem, they are natural within Einstein's view of physics. His special theory of relativity made it clear than no influence can travel faster than light. When gravity changes, therefore, the changes must spread out through space at a finite speed. According to Einstein, this speed is in fact exactly the speed of light.

For example, imagine two stars orbiting around one another in a binary star system, observed

through a telescope by an astronomer on Earth. Their orbits take them first closer to and then further from the Earth, so the astronomer will not only see the motion of the stars, but will also register the corresponding changes in their gravitational influence at exactly the same time! Unfortunately, it is a lot easier to see the binary stars than to feel their gravity. The extreme weakness of gravitational waves has prevented their direct detection so far. Physicists have been trying to catch them since Joseph Weber began building the first primitive detector in America around 1960. Only now, after more than four decades, are the instruments nearing the required sensitivity. In fact, the detectors now being put into operation are the most sensitive scientific instruments ever built.

In general relativity, gravity is ascribed to a distortion, or curving, of space and time. Waves of

Right:
Artist's drawing of waves rippling out from the motion of a binary star system

Left above:
At the core of the gravitational wave detector GEO600 all devices are handled under clean room conditions (photo: Wolfgang Filser/Max Planck Society

Left below:
Ultra-high quality optics are suspended by fused silica wires for the gravitational wave detector GEO600 (photo: Wolfgang Filser/Max Planck Society

gravity are therefore waves of geometry, changing the distances between objects in their path. They affect only distances perpendicular to their direction of motion, just as a water wave pushes a floating cork up and down as it passes underneath. In general, waves in stiff materials have higher wave speeds than those in soft materials. Sound

Even imperceptible waves of gravity can therefore carry huge energies through space, and the loss of this energy can have a big effect on the sources of the wave. This has led to scientists' strongest indirect evidence for gravitational waves. In 1974 the radio astronomers Joseph Taylor and Russell Hulse discovered a binary sys-

waves travel faster in metals than in air because metals are much stiffer. Since gravitational waves travel at the fastest speed of all, it follows that space-time is the stiffest wave medium of all! This stiffness explains why waves of gravity are so weak: it takes a lot of energy to make even a small distortion in a stiff material. Deforming space itself is much harder than deforming even the stiffest of ordinary materials.

tem containing a pulsar, which sends out pulses as precisely as the ticks of an atomic clock. The loss of energy to gravitational waves should make the orbit shrink, drawing the stars closer together. By tracking the orbit of this pulsar over three decades, Taylor has shown that the orbit is shrinking at exactly the rate predicted by Einstein. For their discovery Hulse and Taylor shared the 1993 Nobel Prize for Physics.

The gravitation-wave detector GEO600, developed by the Max Planck Institute for Gravitational Physics (Albert Einstein Institute) at the University of Hannover

The Hulse-Taylor waves are too weak for present instruments to see, and too low in frequency. Astronomers today are looking for waves from neutron stars and black holes. The first directly detected waves may come from two black holes merging together. This would be a double first: not only the first gravitational wave but also the first direct observation of a black hole. Other expected sources are merging neutron stars or spinning neutron stars (pulsars). Only objects with such extreme gravitational fields can produce sufficient

The strongest wave that astronomers expect to pass through this apparatus in the next five years will probably change the separations of the mirrors by no more than one part in 10^{-21}, or only one ten-thousandth of the diameter of a proton! And this disturbance will last only a fraction of a second. Incredible as this task sounds, the physicists building GEO600 have already demonstrated that it has the sensitivity to do this.

But if GEO600 senses such a disturbance, how can the scientists be sure it was a gravitational wave,

Artist's conception of the three-part LISA spacecraft in a space-time containing gravitational waves from a black hole merger. Not drawn to scale: the true distance between the spacecraft would be 5,000,000 km

Part of the vacuum system of GEO600, holding some of the mirrors at the junction of the two arms

energy in gravitational waves to be seen by current detectors.

To see how small the distortions created by even these strong sources are, consider the GEO600 detector, built in a field south of Hannover, as a cooperation among the Max Planck Society, the University of Hannover, and two British universities (Glasgow and Cardiff). It consists basically of two pairs of mirrors, each pair separated by 600 m in perpendicular directions. The mirrors hang from supports so that they are free to move horizontally if a gravitational wave passes. Laser light reflecting up and down the two arms in an almost perfect vacuum is used to compare the lengths of the arms for the slight distortions that a gravitational wave would induce.

and not, say, a seismic wave that broke through the vibration-isolation system around the mirrors? The answer is to compare gravitational wave detectors around the world. The GEO600 team works in close partnership with the largest gravitational wave project in the world, LIGO in the USA. LIGO has built two instruments of a similar design but with arm-lengths of 4 km. The larger size of these instruments makes them even more sensitive than GEO600, yet even these instruments may need to wait several years for the first gravitational wave. Only after LIGO is upgraded, in partnership with GEO600, after 2010, will the sensitivity improve to the point where detections should become routine. Another large (3 km) detector is being built outside Pisa, Italy, by the VIRGO consortium, and a further detector is being planned in Japan.When all these instruments operate, gravitational wave astronomy will become a reality.

But all these instruments register waves only at frequencies above 5 Hz or more. Lower frequen-

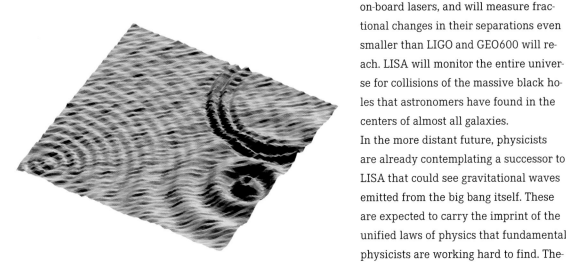

cies will be detectable only from space, and the European Space Agency (ESA) and the American agency NASA are planning a joint mission called

LISA to open up lower frequencies. LISA will be launched in 2013 and will go into orbit around the Sun. Its three spacecraft, arrayed in a triangle with sides that will be an incredible 5 000 000 km long, will monitor their mutual separations using on-board lasers, and will measure fractional changes in their separations even smaller than LIGO and GEO600 will reach. LISA will monitor the entire universe for collisions of the massive black holes that astronomers have found in the centers of almost all galaxies.

In the more distant future, physicists are already contemplating a successor to LISA that could see gravitational waves emitted from the big bang itself. These are expected to carry the imprint of the unified laws of physics that fundamental physicists are working hard to find. These incredibly weak waves are the ultimate goal of gravitational wave detection: with them, scientists will see the big bang directly.

Representation of the geometry of a space containing many gravitational waves. The distortions shown are much larger than the waves actually expected

Gregor Schiemann

God Does Not Play Dice

Einstein's Still Topical Critique of Quantum Mechanics

There is hardly a remark by Einstein as famous as his statement that God does not play dice. Not only is it included in many portrayals of Einstein's life and work, it has even provided the title for a number of books on issues of modern physics and mathematics. The popularity of the sentence stands in notable contrast to the rather private context from which it originates. Einstein did not so much elucidate his comment in his writings on physics as he implemented it in his correspondence, and even more frequently in oral discussions with other physicists. The subject was the "statistical interpretation" of

Max Born, around 1931

the atomic theory established in the 1920s known as "quantum mechanics." There is much to suggest that Einstein directed this remark against a view represented by this interpretation and still held in physics today: the belief that in the world of the very small, there are no causes for the spatial-temporal occurrence of individual events. For a long time Einstein's critique was held to be reactionary in the face of the innovations and successes of quantum mechanics. Einstein, thus the general opinion, was a representative of an antiquated worldview, whose proximity to Spinoza's determinism made it irreconcilable with the worldview of modern physics. In the last decades, however, scholars have expressed misgivings about this opinion, which deserve to be taken

seriously. Einstein's views on modern quantum mechanics (the same quantum mechanics still valid today) have piqued renewed interest on the part of many scholars.

The shift in the attitudes of scholarship to Einstein's arguments with quantum mechanics throws another light on his God who does not play dice. The more recent works emphasize that Einstein's critique of statistical interpretation is not the expression of an untenable view of physics, but rather refers to future potential developments of atomic theory, which still remain to be achieved. This also gives new meaning to the sentence about God not playing dice.

Similar to the manner in which Einstein left this remark unexplained, however, his entire position on quantum mechanics has not remained unambiguous. His dice metaphor provides latitude for opposing points of view. On the one hand it can be linked with recent results of research; on the other it points unchangingly to the reactionary elements in Einstein's thought. I will turn first to these latter elements, and then look at the opposing interpretation, which picks up on recent results.

Isn't Chance at the Root of Natural Phenomena?

If one looks more closely at the subject of his criticism, it is not surprising that Einstein's critique of statistical interpretation seemed antiquated to his contemporaries. The interpretation he rejects picks up on perhaps the most revolutionary finding in the atomic physics of the past century: The individual events of atomic physics which have been measured – e.g. radioactive decay and the deflection of particle beams – can be predicted statistically, but not with precision. The point of time when a radioactive atom emits a certain particle, for instance, is chance in the mathematical sense; in principle, the point in time is arbitrary. Only the probabilities of emission can be calculated, and with a large number of particles this can result in a high precision of predicted measurement values.

Probabilities were introduced to physics long before quantum mechanics. They already played a key role in the atomic theories of 19th century classical physics, with which Einstein was extremely well acquainted. Statistical assumptions about the motions of the invisible atoms, distributed by chance, were used to explain measurable

this state? The dice metaphor stands for this consideration. If the conditions of the motions of a rolled dice were known well enough, it would be possible to predict how the dice would fall. This would then reveal how chance is produced according to causal laws. Formulated as a paradox, chance would lose any element of chance. But God already has this knowledge. Thus what humans see as a roll of the dice is, from a divine perspective, not chance at all.

Niels Bohr and Albert Einstein in Leiden, at the end of 1920s (photo: Paul Ehrenfest)

heat phenomena with classical theories. An increase in the temperature of a gas, for instance, was traced back to an increase in the average velocities of the gas atoms. It was believed that the knowledge about the motion of individual atoms, while not available at that time, would be obtained in the future. Why should it not be possible to determine the state of motion of an atom exactly and specify all causes that led to

The statistical interpretation of quantum mechanics asserts the inapplicability of such ideas to the field of the smallest dimensions. According to this interpretation, mathematically calculated probabilities are not an expression of ignorance about the state of atomic objects, but rather a characteristic of their state. All his life Einstein disagreed with this, because, in his view, one of the tasks of a theory is to give causes for the phenomena it

describes. One had to leave open the possibility of later providing a deterministic foundation for the formulation of any physical theory.

With this view Einstein falls back on the ideas of 19th century classical physics and on the world-view associated with it, of a strict system of natural laws, which is effective on a fundamental level and does not provide any room for coincidental events. In this sense his remark that God does not play dice refers to an outdated agenda for the deterministic explanation of nature.

Or Are Dice Not at the Root of Chance?

More recent history of science research has shown, however, that Einstein's critique is not exhausted in its backward-oriented, problematic elements. In 1986 Arthur Fine presented significant arguments to this end in his highly regarded book *The Shaky Game. Einstein, Realism and the Quantum Theory*. According to Fine's analyses, Einstein does not object to the mathematical formalism of quantum mechanics, but rather to its conception as a complete theory in need of no further elaboration.

Einstein links his rejection of statistical interpretation's claim to integrity with the conviction that microphysical phenomena require a new kind of theory. In his view, the basic conceptualization of quantum mechanics should not be improved through minor corrections, but rather replaced by another "point of departure." Late formulations from the 1940s and 1950s suggest that he believed atomic theory would not be applicable in the future because of the still existing structural analogies and contextual relationships to classical physics. With this Einstein wanted to turn the tables on the critique directed against him: not his search for a realistic and causal theory of micro-physics, but rather quantum mechanics in its present form would be far too bound up with a traditional conceptualization.

His previous rejection of what he called the "interference explanation" could also speak for a thrust in this direction. It goes back to Werner Heisenberg and is still quite influential even today. According to this interpretation, the acausal character of measurements in atomic physics is a result of the fact that the measurement process inevitably and uncontrollably interferes with the objects it is supposed to measure. What is dubious about this assumption is the tacit prerequisite that the objects had classically definable local and pulse characteristics before their interaction with the measurement apparatus. Accordingly, the acausal character would not appear until after the fact

Werner Heisenberg, around 1958 (photo: Fritz Eschen)

and (in contrast to "statistical interpretation") not belong to the nature of the objects. By rejecting the interference explanation, Einstein intuitively – thus one could perceive his critique – abandons the attempt to ground the assertion of microphysical processes' supposedly undeceivable acausality by linking it back to ideas of classical physics.

From this perspective, his comment that God does not play dice appears in another light. The metaphor of playing dice expresses the conviction that coincidences are brought forth by nature, which is itself causally composed, in analogy to classical physics. If the conditions of the movements of the dice could be recorded exactly, then it would be possible to recognize the causes from which the results of each roll of the dice necessarily must proceed. Similar considerations can be related to the interference explanation: if interference through measurement could be minimized, then the deterministic basic structure of nature would be revealed. But God does not play dice. If the observable atomic coincidences are based on anything, it cannot be of anything like a dice game, whose causes can be researched in principle. Today it remains unclear whether the contingency of atomic phenomena is part of their nature or whether it results from a process that is perhaps not coincidental. Einstein's comment and its effect have made a great contribution to keeping us aware that the solution of this problem is one of the tasks of future physics.

Thomas de Padova

The *Conseil Européen pour la Recherche Nucléaire* (CERN)

They are digging the shaft of Babel. A hole one hundred meters deep, amidst cornfields, villages and churches. The view across the wide, open plain rising from Lake Geneva to the French Jura Mountains suddenly plunges dawn deep along a concrete wall. A skyscraper would fit into this abyss.

Particle accelerator
1991

On the edge of the concrete hole, in front of the assembly hall, stands a group of five Polish workers with yellow helmets on their heads. They are on their lunch break and chatting in their native language. Polish is one of around thirty languages blending into each other at this location. Here, Russians and Americans, Croatians and Serbs, Pakistanis and Indians, Chinese and Taiwanese are working together to construct one of the most ambitious buildings in science, under the roof of the world's largest research center for particle physics, the *Conseil Européen pour la Recherche Nucléaire* – CERN, for short – founded over fifty years ago.

In these fifty years it has become the Mecca of physics and cosmology. Everyone who wants to know what holds the world together at its core makes a pilgrimage to this site. In its laboratories, scientists generate antimatter and conditions of extreme heat, the phenomena that must have existed immediately after the big bang. CERN even invented the World Wide Web, now used by millions, giving it over to the world community without demanding any licensing fees whatsoever to be remitted to research. It was designed by an Englishman, Tim Berners-Lee, to further improve internal communication among scientists. Researchers at CERN are long since at work on a new computer network: the Grid. The project eclipses all existing high-speed networks. The goal is to link large computing centers worldwide in order to observe what will occur deep below the earth three years from now. At that time, an underground firework of as yet unknown energy is to be set off in gigantic caverns. "The data generated by a single one of our experiments will amount to the total of all data exchanged in all telephone calls around the world at this minute," says Henrik Foeth.

The sixty-three year-old physicist in jeans and a striped T-shirt takes another look down into the empty shaft. Then he walks to the entrance of the 60-meter long assembly hall. Inside it looks like a shipyard. The colossal physical apparatus, which soon will fill the underground laboratory, is gradually taking shape. Workers are welding and milling; cranes are crawling across the ceiling. They move high above rings of red iron plates towering 15 meters high. Iron supports hold a cylindrical magnet twelve meters long.

Foeth has been at CERN since 1967 and has been building devices to prove the existence of physical particles his whole life. But none on this scale. It is to record what occurs in its interior, with pinpoint precision. From the summer of 2007 atomic nuclei will be shot toward each other with enormous force, after building up momentum in a 27-kilometer-long ring accelerator. The tiny particles will burst into individual parts, fly apart in all directions, and trigger a cascade of physical process. "The most exciting moment surely will be when we observe the first particle collisions," says Foeth.

Not even he knows exactly what will happen in these collisions. Like many of his colleagues, he is hoping for the discovery of new elementary particles, above all for the long-predicted Higgs particles. This finding would be a decisive key to answering the fundamental question of physics: "What is mass?" All other basic components of material are presumed to be closely related to this extremely heavy particle.

But whatever the upshot of the experiments: the physicists and engineers on Foeth's team are staking everything on following the spectacle to the minutest detail. For this they are building the 27-kilometer-long accelerator in experimentation halls at four locations. The shaft for the new particle accelerator, the "Large Hadron Collider," was dug in the 1980s. The tunnel once housed a similar, albeit weaker machine. It was shut down for good several years ago. Now CERN is modernizing the entire inventory of the tunnel. Six thousand magnets, many of them 15 meters long, must be joined into a new ring accelerator. They generate the strong magnetic fields that force the racing atomic nuclei onto a circular path, Foeth explains. "The magnetic fields are 100,000 times stronger than the magnetic field of the Earth."

With all of this CERN is approaching the limits of its capabilities. The costs of just the accelerator and laboratories amount to more than two billion Euros. In order to realize the goal, CERN has had to give up nearly all other research activities and cut hundreds of jobs.

In spite of it all, the atmosphere is relaxed away from the construction sites, in that strange mixture scientists are known for: dramatic research and a drab exterior. The central campus of CERN is a gray collection of industrial architecture, of flat-roofed buildings and warehouses with flaking paint. The physicists, engineers and computer ex-

perts who come here take up their work in humble offices in cheerless blocks on streets named after Einstein, Rutherford and Democritus. The liveliest place is the cafeteria. Small groups come and go constantly; mostly men aged twenty-five to forty-five, dressed casually or carelessly. Their heads full of physical ideas, they exchange their thoughts and compete with each other in an everyday ritual. "The most important thing here is conversation," says the forty-four-year-old Austrian Michael Doser, sipping his coffee in bleached-out black jeans and a gray T-shirt. "Anyone who works alone is in danger of getting on the wrong track. You fall in love with a hypothesis and suddenly see nothing but confirmations for it.

The Cafeteria's terrace: a place where researchers meet and exchange ideas

Then you tell a colleague about it and have to start all over."

Every group at CERN works on its own design. But behind these small teams is an overall plan for the new discovery machine, to which all current Cernians are dedicated: the around 2,500 permanent staff who provide the infrastructure

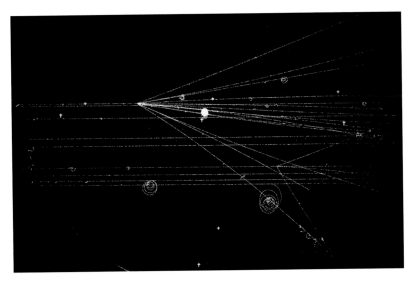

Above:
The interaction of a proton in the hydrogen bubble chamber at CERN

Right:
Taking measurements from bubble chamber exposures in the 1970s

for research, and the many thousands of scientists from universities all over the world, who arrive again and again for stays of a few weeks or months.

Will it be a new departure? Or is it the last stand for an international community of researchers that is collapsing under the weight of its own successes and the dimensions of its apparatus? "In Cern we have created the leading laboratory in the world," says Horst Wenninger, the former director of CERN. But in the meantime other areas such as genome research have benefited from stronger public interest. Even the competition within physics is increasing, especially from astro-particle physics and research on gravitation waves.

Nuclear and particle physics are generating less attention, the sixty-six-year-old tells us. In the 1930s and 1940s it was a key research area. In particular, the military was interested in master-

ing nuclear forces, in order to unleash them at targets at a later point in time, such as in atomic bombs.

After World War II the competition for the highest achievable energies continued with larger and larger machines. Europe joined the race by founding CERN on 29 September 1954. However, from the outset the initial twelve member states were not concerned with military objectives. CERN was the first international organization the Federal Republic of Germany was allowed to join as an equal member after the war. "At the beginning the Americans made all of the discoveries in particle physics," says Wenninger. Today American scientific institutions pay over five hundred million dollars to experiment at the "Large Hadron Collider."

Wenninger is standing in the back courtyard of the entry building, in front of one of the few technical gems displayed at CERN: a "bubble chamber" he helped build at the end of the 1960s. While he tells stories about the

founding years and the discoveries made at CERN, the steel housing with five portholes stares into the sky like a spaceship. Once there were five cameras mounted in front of the large portholes.

When two particles collided in the "bubble chamber," many new particles resulted. Each of these generated small bubbles of steam in a fluid along its trajectory. "The cameras recorded these traces," Wenninger says. "And an army of young women

stored on conventional CDs, the data to be recorded here starting in 2007 would yield a stack 20 kilometers high. If the scientists are not successful in fishing spectacular details out of the billions of collisions, the "Large Hadron Collider" could be

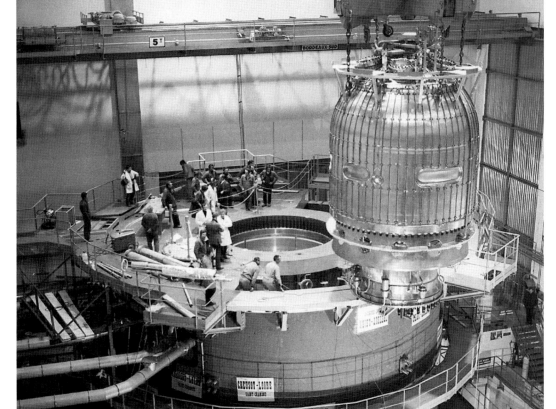

Bubble chamber installation at CERN, December 1971

examined the millions of pictures taken according to certain criteria."

Today electronic detectors record greater numbers of much more violent reactions. Were they to be

the world's last great accelerator project. But if they can open the window to a strange world even just a crack, CERN has no need to worry about its future.

David Cassidy

The Einstein Myths

Myths and legends are social phenomena. They are created and accepted by people living in specific social and cultural settings. As the settings change, so do the myths.

During the first half of the 20th century Einstein

Hannah Höch: *Schnitt mit dem Küchenmesser Dada durch die letzte Weimarer Bierbauch-Kultur-epoche Deutschlands*, collage, 1919, Albert Einstein's head above left

and modern physics were particularly susceptible to myth- and legend-making. The word "relativity" seemed to indicate the sudden loss of cherished values and traditions. The general public desperately welcomed ways to understand the bewildering new world in which it found itself. There was no shortage of myth-makers ready to provide mythological answers.

The years following World War I were indeed ones of great and frightening events: the horror of the brutal world war in a supposedly enlightened age; the threats to European liberalism posed by the Bolshevik Revolution in Russia; and in Germany, the defeat in war and the abdication of the monarchy in favor of social democracy. All of these gave the universal sense that people had suddenly lost control of their lives and destinies. They signaled in the words of Oswald Spengler (1880 – 1936), *Der Untergang des Abendlandes.*

Einstein's relativity theory entered this depressing state of affairs as a reinforcement for some and as a contradiction for others. Public incomprehension of this theory seemed indicative of the incomprehensible postwar world in which they lived. Einstein had published the theory of general relativity at the height of the world war. The British confirmation of his predictions about the bending of star light in 1919 made his theory an instant candidate for myth-making, and Einstein became an instant, legendary hero – a superhuman genius who had revealed the secret knowledge of the gods to humankind. Einstein himself, of course, repudiated all such attempts to canonize him or to apply his physical theories beyond physical phenomena. "I believe that the present fashion of applying the axioms of physical science to human life is not only a mistake but has something reprehensible about it."

It was not the technical theories themselves that were turned into myths for the public, but how the public perceived the meaning and implications of these theories as presented by writers who wrote for the general public. In most cases the popularizers were in fact physicists who exploited public anxieties by creating myths and legends about relativity theory and Einstein for their own personal advantage.

One extreme example of this is the case of Philipp Lenard (1862–1947) and Johannes Stark (1874–1957), two Nobel-Prize winning experimental physicists. They were among the earliest supporters of Einstein and his quantum hypothesis, but, like many others, they strongly opposed the rise of social democracy in Germany. They became actively political when they perceived a challenge

treme forms in other diverse cultures at that time. British physicists Sir James Jeans (1977–1946) and Sir Arthur Eddington (1882–1944) were among the most prominent and widely popular writers on Einstein and modern physics in England and America. But their portrayals were decidedly idealistic and myth-like. Eddington told his readers: "The stuff of the world is mind-stuff" and "Without

Albert Einstein giving his Nobel Prize acceptance speech in Göteborg, 11 July 1923

to their professional status as experimental physicists after the sensational confirmation of Einstein's theory in 1919. At the same time, other figures in the arts and literature were arguing that relativity theory meant that "everything is relative." Traditional meanings and mores seemed no longer valid. Exploiting this anxiety and the rising anti-Semitism in Germany, Lenard went on the attack. He proclaimed that all science is racially dependent and that relativity must be rejected because of its non-Aryan and theoretical origins. A similar mythology of relativity theory in exploitation of public anxiety occurred in less ex-

the mind there is but formless chaos." Jeans conjectured that the ultimate understanding of nature will result in "the total disappearance of matter and mechanism, mind reigning supreme and alone."

As characteristic of myth, their accounts of modern physics – relativity and quantum mechanics – provided reassurance to the Anglo-American public regarding its main science-related anxiety – that scientific determinism and Darwinian evolution denied individualism and free will and therefore undermined traditional ethics and religion. Relativity theory still upheld the determinism of

Caricature from
Karl Prühäuser,
(*Die Brennessel*,
München,
15 April 1931)

Größen der Zeit

Karl Prühäuser

Albert Einstein
der relative Indianerhäuptling

These are all characteristics of the classic hero, and Einstein himself tended by his actions and persona to reinforce them – although inadvertently. Einstein gave the appearance of the "weltfremdes Genie," the lone individual who left the everyday world in order to search for the secrets of nature. At the same time he stood alone from many others in social matters by his uncompromising support of reason, freedom, peace, and human dignity throughout his life. A later colleague, Abraham Pais (1919–2000), once said, "Einstein was the freest man I have ever known." In his Autobiographical Notes Einstein's comments had the unintended effect of reinforcing the perception of himself as a "weltfremdes Genie." He argued that his internal thoughts originated and played themselves out

classical physics, but Eddington reassured the public that it did not apply outside the limited realm of physical nature. Jeans, to the contrary, offered the public a romantic scientific mythology of a new world fit for human freedom.

The success of the idealistic interpretation of relativity theory in the West encouraged a backlash in the Soviet Union. Here the dominant social ideology was not idealism or individualism but Marxian materialism. Ironically, the myth-makers operating at both political extremes led their followers to react to relativity theory in the same way: rejection of the theory and persecution of its adherents. If relativity was being celebrated as a triumph of idealism and individualism in the West, then it had to be suppressed in Soviet Russia. During the 1920s, Soviet demagogues used Lenin's rejection of Ernst Mach's (1838–1916) "empirio-criticism" to ban relativity theory as "bourgeois idealism." But what of Einstein himself, the legendary hero who brought down fire from the gods? The man who revealed to humankind the hidden secrets of nature? The man who demonstrated the power of transcendent rationality at the height of irrationality to an anxious and confused public caught in the throes of a disintegrating social and intellectual world?

The most famous of
all equations is
formed by the crew
of the air carrier US
Enterprise CV-6,
after 1936

independently of his personal life, "das nur persönlich." "Denn das Wesentliche im Dasein eines Menschen von meiner Art liegt in dem was er denkt und wie er denkt, nicht in dem, was er tut oder erleidet."

Because of this *weltfremder persona*, Einstein's views about the political and social worlds were sometimes considered idealistic, naive, and even simplistic. A hero cannot be a hero in every sphere, it was argued. But often this was said by

Albert Einstein in his study (*Turm-zimmer*) in Haber-landstrasse Berlin, 1927.
A picture of Isaac Newton can be seen on the wall

those who disapproved. For instance, in his opposition to Nazism, many of his closest colleagues were angered that he violated the apolitical position of most academics at that time.

Perhaps because Einstein so exemplified the profound transformation of 20th century physics and was viewed for so long as a legendary hero, the pendulum has now swung to the other extreme regarding Einstein the man. As the result of recent revelations about Einstein's personal faults and his sometimes dysfunctional love life, popularizers of physics – now no longer physicists but usually journalists working for leading newspapers –

continue to exploit public interests and anxieties. This time they are exploiting public fascination with the human faults of a once heroic figure for the familiar aims of personal recognition and financial gain.

Out of this new work of both history and journalism a new mythology of science may emerge. But Einstein as the incomprehensible legendary hero is now gone forever. In its place a deeper understanding of the genuine Einstein is emerging, and with it deeper appreciation of Einstein as one of the greatest physicists and humanitarians of all time.

This text is based upon a paper delivered to a section meeting of the German Physical Society in Augsburg, March 2003

Dieter Hoffmann

"1905 was his Great Year": Interview with Hans Bethe

Hans Bethe talked with Dieter Hoffmann from the Max Planck Institute for the History of Science (Berlin) on 18 September 2004 in Ithaca, NY.

When did you first meet Einstein?

It was 1933, in Princeton, when I was visiting Wigner. He introduced me to Einstein. But the meeting only lasted for two minutes. There was

the special theory is clearly Einstein's most significant achievement – quite apart from the fact that it's used everywhere in physics. Do you mind if I speak English?

Not at all.

I speak German without any accent, but English more easily. The special theory is of utmost im-

Nobel Prize winners in Physics (from left to right): Tsung-Dao Lee, Hans Bethe, Leo James Rainwater, Isador Isaac Rabi, Samuel Chao Chung Ting, Julian Schwinger, William J. McGill, André Frédéric Courmand at Columbia University, New York in 1977

no real exchange. The encounter left no lasting impression whatsoever.

But you'd already heard of Einstein of course?

The first time I heard about Einstein I must have been about twelve years old. Yes, it was in 1918. I was told that Einstein had completely revolutionized physics, but only about ten people in the world understood what he had done. That was wrong of course, because in 1918 his special theory of relativity was both well known and well understood. It's easy to understand! In fact it's one of the easiest works I've ever read. In my opinion

portance to atomic motion, but mainly to nuclear orbits, namely the energy in nuclear physics. And also for the observation of nuclear particles coming out of the nucleus. That, of course, Einstein didn't know in 1905, but in my opinion his three papers of 1905 are by far the most important he ever wrote. I think, the special theory is far more important than general relativity.

Even more important than Einstein's light-quantum hypothesis?

Well, the light-quantum is equally important. But there he had Planck as predecessor, whereas the special relativity had no predecessor.

As a student and young scientist in the 1920s did you regard Einstein not only as a scientist but also as a political person?

For many years, almost up to 1933, purely as a scientist.

Einstein was of course one of the few scientists who said straight away: "The Nazis are criminals." That was my opinion too. Although I wasn't influenced by Einstein, I was in complete agreement with him that the Nazis – judging by their behaviour even before they took over power – were criminals.

In 1933 Einstein declared his resignation from the Prussian Academy of Sciences and the Society of Physics because he no longer agreed with the political conditions in Germany. How did you experience Einstein's conflict with the Academy?

I didn't, because in 1933 I was far too busy with my own concerns. As the Nazis classified half-Jews as Semites, it was clear that my future would be entirely different than planned. Fortunately I had a teacher, Arnold Sommerfeld, who supported his students very selflessly, but then he met with difficulties himself. By the spring of 1933 Sommerfeld had found me a job in England, with Bragg.

Einstein's relationship with Germany was admittedly very complex. As a young man he fled from the Prussian drill and the German school system and went to Switzerland. Then, in 1914, he came to Berlin where he was attacked politically as a Jew and a pacifist in the 20s. In 1933 he finally emigrated and never returned to Germany. Your attitude was quite different, as you returned to Germany already in 1946.

1948!

You revived old contacts, not only with people but also with institutions.

Only with people, really. I visited Heisenberg and Weizsäcker, and then Gerlach in Munich and Madelung in Frankfurt.

You are a member of the Order Pour le Mérite and of German academies. But Einstein refused, in very categorical words, to become a member of the Bavarian and Berlin academies again. After 1945 he always said: "Germany is the country of mass murderers." How do you explain your entirely different attitude, not only to the Third Reich but also to Germany in general?

I have always tried to differentiate between the Third Reich and post-war Germany. I knew many Germans – irrespective of whether they were physicists or non-physicists – who could be both honorable and friends. There were of course many convinced Nazis after 1933 and after 1945 as well! But there were at least as many who stayed clear of it.

What role did Einstein play in American society, especially in the American physicists' community? Did people take notice of him?

As a physicist? No. As a political person? Yes. Hardly anyone took any notice of Einstein's physics during those years. There was a small circle of maybe ten people who were working on it and knew about it.

Hans Bethe (1906–2005) and Norman Ramsey (born 1915)

Could you have imagined being one of Einstein's assistants?

Absolutely not!

Why not?

Because he was working on things that didn't interest me.

Einstein signed the famous letter to Roosevelt which addressed the question of using atomic energy for military purposes.

Einstein didn't write the letter. It was written by Wigner and Szilard. They got Einstein to sign because they thought Roosevelt would take more notice of the issue.

As someone who played an important role in atomic physics and the development of the atomic bomb, how do you see Einstein's attitude to the bomb? In 1945 he had become a critic of atomic armament and had developed various ideas on disarmament and a world government.

Most physicists who came to America from Europe shared Einstein's view that the use of the atomic bomb was not only justified but also the only possible means of ending the war with Japan. In my opinion the use of the bomb saved many human lives. But then, in the late 40s and 50s, atomic armament was overdone and that was the opposite of what Einstein, and we all, wanted. It's a very big difference, whether there are two atomic bombs or 200!

Shortly before he died, one of Einstein's last actions involved the manifesto against atomic armament which he signed at the request of Bertrand Russell. What do you remember about this manifesto?

I gave lectures in the same vein. But of course, Einstein was Einstein! People listened to him. I think I was present at two of his lectures. He put things so simply and so convincingly! That's why Einstein was so very important in the propaganda against increased atomic armament. Einstein was an independent individual, independent from his environment.

In science as well as in his perception of public and political affairs?

Yes. Perhaps just one more little anecdote: Einstein and Bohr were involved in one of their many arguments about the interpretation of the quan-

Hans Bethe, photograph from 1996

tum theory. Rabi was following this discussion and, although he was of the opinion that Bohr was right – and Bohr was of course right! – he still felt that Einstein was much more self-assured and so had actually managed to win this debate.

Do you have an answer to the question that Einstein also asked himself: What is gravitation?

I'd say that Einstein's general theory of relativity is the answer. I don't think it's useful to ask about a graviton. I don't think uniting gravitation with the quantum theory is important. To me they are two completely separate theories. For a long time Einstein thought he could derive the quantum theory from the general theory of relativity. That's impossible of course. The quantum theory is completely independent. There are indeed many physicists who are trying to generalize the modern field theory – using the general theory of relativity as the foundation – but I don't think that's the right path.

Nowadays, if you ask people on the street to name a famous physicist, they're guaranteed to say Einstein. How do you explain Einstein's enormous significance and popularity compared with the other great 20th century physicists who were also responsible for major achievements?

I can give you a real reason for that, namely that Einstein changed the fundamental ideas in physics. And that applies to very few of the others. Planck maybe! But Planck never formulated an explicit and logical presentation. And the first person to actually understand the quantum theory was Einstein. Not Planck! And that was in the three publications of 1905. And that's why I'll say it again: 1905 was his great year.

Hans Bethe died on 6 March 2005. This was the last interview he gave.

Reiner Braun

The Russell-Einstein Manifesto

On Saturday, 9 July 1955, Lord Bertrand Russell presented in Caxton Hall, Westminster what is probably the world's most famous peace manifesto, the "Russell-Einstein Manifesto" against the dangers of nuclear armament.

Lord Russell opened the press conference, headed by Joseph Rotblat, a young physics professor and later Nobel Peace Prize winner in 1995, with the following remarks:

"The accompanying statement, which has been signed by some of the most eminent scientific authorities (nine of eleven signers are Nobel laureates) in different parts of the world, deals with the perils of a nuclear war. It makes it clear that neither side can hope for victory in such a war and that there is a very real danger of the extermination of the human race by dust and rain from radioactive clouds. [...] The only hope for mankind is the avoidance of war. To call for a way of thinking which shall make such avoidance possible is the purpose of this statement."

Lord Russell was the initiator of this manifesto, which would bear his name. Bertrand Russell, philosopher, logician, mathematician, and social critic, was born on 18 May 1872. He studied mathematics and philosophy at Trinity College in Cambridge, where he worked from 1895 to 1901 as a fellow and from 1910 to 1916 as lecturer for logic and mathematics.

Russell was impressed by the certainty of mathematical cognition, sensory experience, and natural science. He believed that this certainty was proof that mathematics was a part of logics. One of the products of this conviction was *Principia Mathematica* (1910–1913), written in collaboration with Alfred North Whitehead. Russell was dismissed from his lectureship at Trinity College because of his political pacifism, and in 1917 he was sentenced to several months in prison for writing an anti-war newspaper article. Between the world wars he traveled to many countries, including China, the Soviet Union, and the United States.

From 1927 to 1934 he ran a progressive school project at Beacon Hill.

In 1938 he was invited to the University of Chicago as visiting professor of philosophy. A professorship offered at the College of the City of New York was blocked because of his liberal stance toward homosexuality.

In 1944 he returned to Trinity College in Great Britain.

Bertrand Russell, around 1950 (photo: Eric Schaal)

His scholarly achievements were honored with an honorary doctorate at the University of Aix-en-Provence and the Order of Merit.

In 1950 he was awarded the Nobel Prize in Literature for his concise scientific prose.

In 1958 Russell initiated the founding of the "Campaign for Nuclear Disarmament" and became its president.

At sixty-eight years of age, he was sentenced to two months in prison for participating in acts of civil disobedience, for "inciting the public against the authority of the state."

In 1963 the Bertrand Russell Foundation was founded. In 1967 he set up an international tribunal on war crimes committed in the Vietnam war, which would also bear his name. Bertrand Russell died in Wales on 2 February 1970.

To launch this manifesto signed by the most prominent scientists of his time, Bertrand Russell first turned to the most famous scientist of the 20th century, Albert Einstein.

On February 11, 1955 he wrote:

"In common with every other thinking person, I am profoundly disquieted by the armaments race in nuclear weapons. You have on various occasions given expression to feelings and opinions with which I am in close agreement. I think that eminent men of science ought to do something dramatic to bring home to the public and governments the disasters that may occur."

This letter and the following correspondence between Einstein and Russell show the motives behind their engagement. It was based on a long, sustained exchange of views between Russell and Einstein, in which they deliberated on the necessity of taking open action against the nuclear arms race.

The development and testing of the hydrogen bomb had raised the perils that a nuclear war posed to the human race to a new dimension. A nuclear war could now mean the destruction of the planet Earth. This development led Bertrand Russell after 1950 to rethink his position toward American nuclear armament, which he had initially favored. The manifold dangers of an ongoing, unchecked nuclear arms race between the West and East, and ignorance of nuclear disarmament proposals, especially in the United States, was a source of growing concern, as was the escalating danger of proliferation.

Not least of all, many scientists were greatly alarmed by the involvement of science in military research and development. The upsurge in the number of scientists who became engaged in the humanist campaign for the survival of the human race was accompanied by strategic, tactical deliberations, such as enlisting independent, neutral countries, India in particular, for an active role in the struggle for disarmament. As for the signatories, they agreed that "there must be no suggestion of seeking advantage for either side or of preferring either side. For this reason it would be a good thing if some are known Communists and others known anti-Communists." (Russell wrote to Einstein in a letter of 11 February 1955.) The political pressures of the Cold War as well as the desire to attain the greatest possible acceptance and support from society were the issues grappled with in the correspondence between

Press statement by Bertrand Russell on the Russell-Einstein Manifesto in Caxton Hall, London, 9 July 1955

London: Bertrand Russell (seated, right) during a demonstration protesting the planned construction of a U.S. submarine base with Polaris missiles in Scotland, 18 February 1961

Bertrand Russell and Albert Einstein. To attain this goal, they agreed to appeal only to prominent scientists as the first signatories of the manifesto. Their correspondence ended with a brief letter from Albert Einstein to Bertrand Russell on 11 April 1955.

"Thank you for your letter of April 5. I am gladly willing to sign your excellent statement. I also agree with your choice of the prospective signers." The signature Alfred Einstein placed under his

letter to Bertrand Russell and under the manifesto were the last signatures of his life. Two days later he fell seriously ill. Albert Einstein died on 18 April 1955.

Joseph Rotblat described the dramatic story of how Russell learned of Einstein's death and his signatures. On 18 April 1955, Russell was flying from a congress in Rome to Paris when the captain notified the passengers of Albert Einstein's death. Russell was gravely dejected, over the death of a prominent scientist but also over the collapse of the peace manifesto, which for him was inseparable from the personage of Einstein. Upon his arrival in Paris, Russell found a letter from London in his hotel – with Albert Einstein's signature under their manifesto.

The Russell-Einstein Manifesto concludes with an urgent appeal to policymakers:

"In view of the fact that in any future world war nuclear weapons will certainly be employed, and that such weapons threaten the continued existence of mankind, we urge the governments of the world to realize, and to acknowledge publicly, that their purposes cannot be furthered by a world war, and we urge them, consequently, to find peaceful means for the settlement of all matters of dispute between them."

Along with Russell and Einstein, eight other scientists signed the manifesto before it was published: Percy W. Bridgman and Hermann J. Muller from the USA, Cecil F. Powell and Joseph Rotblat from

Great Britain, Frédéric Joliot-Curie from France, Leopold Infeld from Poland, and Hideki Yukawa from Japan, and the German Max Born, who was inadvertently forgotten during the announcement at the press conference. Directly after its proclamation, Linus Pauling endorsed the manifesto, increasing the number of signatories to eleven, all of them Nobel Laureates except for Rotblat (at the time) and Infeld.

The echo was resounding: It seemed as if most of the general public had been waiting for this appeal from prominent scientists. Numerous peace actions followed in several countries. Even government officials in the East and West took positive notice of the manifesto. This gave the manifesto immense publicity – even in the United States and England. But in the next few years nothing changed in the armaments policies of the superpowers.

What remained was the appeal for scientists and citizens to take action.

Two years later signatories and other prominent scientists from the East and West founded the "Pugwash Conferences" at the peak of the Cold War. For decades this was the initiative of concerned scientists for nuclear disarmament. In 1995 it was awarded the Nobel Peace Prize for its work.

Since the Russell-Einstein Manifesto of 1955, the engagement of concerned scientists has never diminished. Again and again they have spoken up in various countries and internationally on issues of war and peace – as again in the Albert Einstein Year of 2005.

Horst Kant

German Scientists and the Effects of the Russell-Einstein Manifesto

The relatively extensive attention granted by the international press to the Russell-Einstein Manifesto (see the preceding paper) was presumably due in part to the fact that it was regarded as Albert Einstein's legacy. Yet at that time, with the international peace and anti-atomic war movement enjoying strong gains in popularity, there were certainly more notable activities, some of them even by German scientists. Max Born (1882–1970; Nobel Prize winner in 1954), who had settled in Bad Pyrmont after his British exile, was not only among the first signatories of the Russell-Einstein Manifesto, he also provided Russell with an important impetus for this action in early 1955. In an editorial for No. 1/1955 of the *Physikalischen Blätter* he had exhorted German scientists to face up to their responsibility in the Atomic Age – as many of their English and American colleagues had long since done. In a letter of 1 February 1955 he reported to Otto Hahn (1879–1968; NP 1944), then president of the Max Planck Society, who had discovered atomic nuclear fission with Fritz Strassmann at the Kaiser Wilhelm Institute for Chemistry in Berlin in 1938, about his exchange with Russell; closing with the words: "[...] I believe we should not simply watch idly while the destruction of civilization, so to speak, is being prepared, especially as forces are being used which physics has made available." Hahn answered him in the affirmative on 5 February 1955 and elaborated: "I, too, have been concerned for some time as to whether and what we could do as scientists [...] About ten days ago, before I received your letter, I had already completed the draft for an article [...]." A few days later Hahn found the opportunity to appear before the public with his speech *Cobalt 60 – Gefahr oder Segen für die Menschheit* (Cobalt 60: Danger or Blessing for Humanity) via the regional German radio station *Nordwestdeutscher Rundfunk*. The resonance was great,

Max Born, the intellectual inspiration for Lord Russell and for Otto Hahn (photo: Fritz Eschen)

even abroad, for Hahn, who was a thoroughly dedicated proponent of the peaceful use of nuclear energy, had expressed himself more clearly than others before him about the dangers of its improper utilization. Despite his fears that he might be accused of interfering in politics, and perhaps even in support of the political opposition, Hahn was now willing to take advantage of the positive public reaction to his speech for further activities of this kind. However, these fears kept him from joining initiatives like those already proceeding from the World Federation of Scientific Workers under Frédérick Joliot-Curie (1900–1958; NP 1935) and at that time even from Russell himself, because he was afraid he would be associated as a communist or Soviet sympathizer. In an exchange with Born, Werner Heisenberg (1901–1976; NP 1932) and Carl Friedrich von Weizsäcker (*1912), he had the idea of using the International Meeting of Nobel Laureates, an annual event held in Lindau at Lake Constance since 1951, for this purpose. Born composed a draft for the declaration; the final version was essentially the work of von Weizsäcker. Hahn and Born attempted to win fellow combatants from other countries, and the patron of the Lindau meeting, Count Lennart Bernadotte (1909–2004), suggested to proclaim declaration on his island in Lake Constance, Mainau.

By inviting only Nobel laureates from the natural sciences to sign and thus keeping the manifesto exclusive, Hahn hoped for greater political vigor and, at the same time, a guarantee against communist influences. The decision not to approach the public beforehand was deliberate. The reactions of the Nobel laureates addressed were varied, with quite different reasons being given for both pledges and rejections. While Born felt obligated to sign Russell's proclamation as well, "since he and Einstein would hold it against me if I refused," Hahn declined because of what he saw as "one-sided leftist tendencies."

Hahn succeeded in convincing all sixteen Nobel laureates assembled at Lindau in mid-July 1955 to sign, and the document was delivered to the press as the Mainau Declaration on 15 July (eighteen people had signed, for Arthur Holly Compton and Hideki Yukawa had agreed, but could not make the journey). The manifesto was then sent to all

research to the service of humanity feel obliged to warn emphatically against any abuse of these results."

In the Federal Republic of Germany, which was preparing in the mid-1950s for integration into the NATO alliance as a militarily equal partner, the government's policy was to ignore warnings about atomic weapons and strive for the possession of such weapons. The "Nuclear Physics" working group at the Federal Ministry for Atomic Issues at the time appeared deeply concerned about this at the end of 1956, yet after a meeting with Defense Minister Franz Josef Strauss, its members agreed to forego an appeal to the public for the time being. Comments to the press by Chancellor Konrad Adenauer on 5 April 1957, playing down the danger, prompted several of the atomic scientists to enlighten the public about the real dangers of atomic weapons. Again, it was essentially Hahn and Heisenberg who took the initiative, as well as Weizsäcker who formulated the draft of a corresponding appeal. Its central sentences read:

On 17 April 1957, Otto Hahn (l.), Carl Friedrich von Weizsäcker (r.) and Walter Gerlach follow Konrad Adenauer's summons to the Federal Chancellery to discuss the Göttingen Declaration

Nobel laureates, and one year later a total of 52 signatures had been submitted. The resonance of the Mainau Declaration in the broader public was initially not very great, because Hahn and his fellow combatants had unfortunately neglected to beat the prepare a press campaign – in contrast to Russell. As a result the German scientists were somewhat disappointed to acknowledge that the Mainau Declaration played almost no role at all at the UN conference "Atoms for Peace" held in Geneva in August 1955. Yet at the general meeting of the *Deutsche Physikalische Gesellschaften* (German Physical Societies) in Wiesbaden in late September 1955, both manifestos were referred to explicitly, and a resolution was passed stating that: "The physicists who give the results of their

"We do not feel competent to make concrete proposals for the policies of the great powers. For a small country like the Federal Republic, we believe that it is most conducive for the protection our country and world peace to expressly and voluntarily reject the possession of atomic weapons of any kind. In any case none of the undersigned would be willing to participate in the manufacture, testing or deployment of atomic weapons in any way."

The Göttingen Declaration – supported by eighteen German atomic scientists (no official copy

Mainauer Kundgebung

Wir, die Unterzeichneten, sind Naturforscher aus verschiedenen
Ländern, verschiedener Rasse, verschiedenen Glaubens, verschiedener politischer Überzeugung. Äusserlich verbindet uns nur
der Nobelpreis, den wir haben entgegennehmen dürfen.

Mit Freuden haben wir unser Leben in den Dienst der Wissenschaft gestellt. Sie ist, so glauben wir, ein Weg zu einem
glücklicheren Leben der Menschen. Wir sehen mit Entsetzen,
dass eben diese Wissenschaft der Menschheit Mittel in die
Hand gibt, sich selbst zu zerstören.

Voller kriegerischer Einsatz der heute möglichen Waffen kann
die Erde so sehr radioaktiv verseuchen, dass ganze Völker
vernichtet würden. Dieser Tod kann die Neutralen ebenso
treffen wie die Kriegführenden.

Wenn ein Krieg zwischen den Grossmächten entstünde, wer
könnte garantieren, dass er sich nicht zu einem solchen
tödlichen Kampf entwickelte? So ruft eine Nation, die sich
auf einen totalen Krieg einlässt, ihren eigenen Untergang
herbei und gefährdet die ganze Welt.

Wir leugnen nicht, dass vielleicht heute der Friede gerade
durch die Furcht vor diesen tödlichen Waffen aufrechterhalten
wird. Trotzdem halten wir es für eine Selbsttäuschung, wenn
Regierungen glauben sollten, sie könnten auf lange Zeit gerade durch die Angst vor diesen Waffen den Krieg vermeiden.
Angst und Spannung haben so oft Krieg erzeugt. Ebenso scheint
es uns eine Selbsttäuschung, zu glauben, kleinere Konflikte
könnten weiterhin stets durch die traditionellen Waffen ent-
schieden werden. In äusserster Gefahr wird keine Nation sich
den Gebrauch irgendeiner Waffe versagen, die die wissenschaft-

./.

Original copy of the Mainau Declaration with the signatures of the first signatories (from Otto Hahn's estate)

with signatures exists) – was announced by telephone to the *Deutsche Presseagentur* and several newspapers on the morning of 12 April 1957. This was a Friday, and the tactical consideration for this was apparently that it would not appear in the press until Saturday, and thus have a chance to affect the public before the government could respond. As Chancellor Adenauer reported on 17 April 1957 to Federal President Theodor Heuss about the political influence: "[...] the uprising

of the atomic scientists, which came to me as a complete surprise – Mr. Strauss had told me after a talk with Mr. Hahn that the gentlemen would not take any action –, was very unpleasant, in terms of both domestic and foreign policy. After many hours of discussion, which took place with five of the gentlemen today, the matter appears to have been settled in a manner satisfactory for both parties."

However, the affair was not quite that simple. While the signatories of the Göttingen Declaration held different views about how to proceed, Hahn, Born and Max von Laue (1879–1960; NP 1914), above all, believed that actions against nuclear weapons must be continued. Hahn emphasized that he was acting as an independent scientist in this matter and not as president of the Max Planck Society, however he used the 1957 Convention of the MPG in Lübeck to plead the case of the Göttingen Eighteen once again. In November 1957, at the invitation of the Austrian Cultural Society, he held his acclaimed lecture *Atomenergie für den Frieden oder für den Krieg* (Atomic Energy for Peace or for War) in the Concert Hall in Vienna, and mentioned expliciteley the Declarations of Mainau and of Göttingen.

It can be said that a broader public concern about the dangers of atomic weapons began to take root in the FRG around 1955. The Göttingen Declaration and an accordingly controversial parliamentary

liche Technik erzeugen kann.

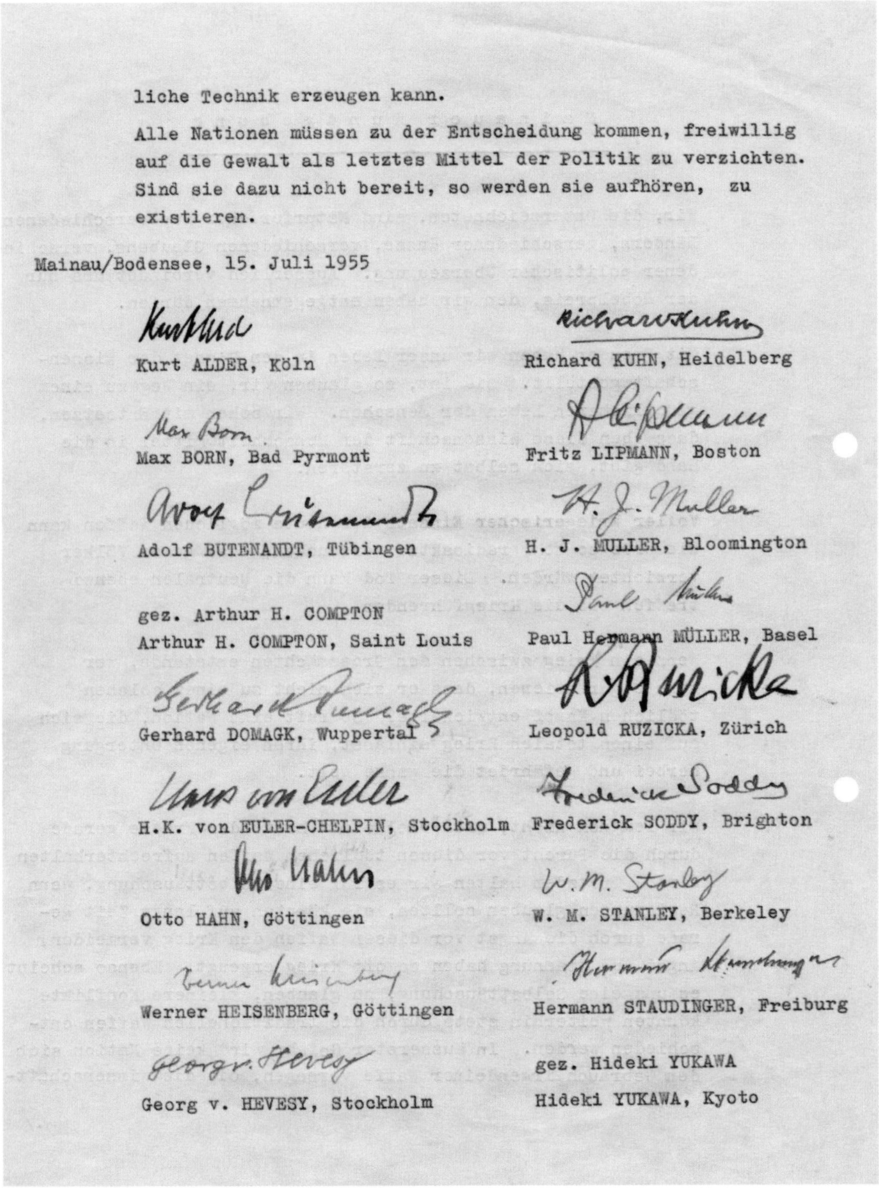

Alle Nationen müssen zu der Entscheidung kommen, freiwillig auf die Gewalt als letztes Mittel der Politik zu verzichten. Sind sie dazu nicht bereit, so werden sie aufhören, zu existieren.

Mainau/Bodensee, 15. Juli 1955

Kurt ALDER, Köln

Max BORN, Bad Pyrmont

Adolf BUTENANDT, Tübingen

gez. Arthur H. COMPTON
Arthur H. COMPTON, Saint Louis

Gerhard DOMAGK, Wuppertal

H.K. von EULER-CHELPIN, Stockholm

Otto HAHN, Göttingen

Werner HEISENBERG, Göttingen

Georg v. HEVESY, Stockholm

Richard KUHN, Heidelberg

Fritz LIPMANN, Boston

H. J. MULLER, Bloomington

Paul Hermann MÜLLER, Basel

Leopold RUZICKA, Zürich

Frederick SODDY, Brighton

W. M. STANLEY, Berkeley

Hermann STAUDINGER, Freiburg

gez. Hideki YUKAWA
Hideki YUKAWA, Kyoto

debate on 10 May 1957 therefore helped to trigger the anti-atomic weapons movement. Yet not until 15 February 1958 did the FRG experience its first protest demonstration in Tübingen against atomic weapons and against equipping the *Bundeswehr* with atomic weapons. This movement brought forth the *Kampf dem Atomtod* (The Fight Against Atomic Death) movement, which received significant support from the SPD (Social Democratic Party), the trade unions, and some circles of the Protestant churches. For the most part, however, the Göttingen Eighteen kept out of this "fight on the street."

This paper does propose to illuminate the diversity of the anti-atomic weapons movement in the Federal Republic of Germany in the mid and late 1950s, but merely highlights the activities of a number of the atomic scientists involved. These activities however, played an important role in the anti-atomic weapons movement into gear. The participating scientists must be taken quite seriously in their honest efforts to prevent the military use of atomic energy. Yet it must not be forgotten that this endeavor had an additional benefit in particular for the number of German atomic scientists. By publicly demonstrating their peaceful intentions, they hoped to achieve a revocation of international sanctions on nuclear research and the use of nuclear energy, which had been imposed on them as a consequence of World War II, thus permitting them to research without restrictions.

Dieter Hoffmann

Einstein's Political File

Albert Einstein was not only an outstanding figure in the field of science during the 20th century. His commitment to human rights and pacifism made him a leading symbolic figure as a scientist of moral standing. These convictions made him many enemies and under authoritarian regimes, such as the German Empire or the Nazi dictatorship, but also in democratic countries such as the Weimar Republic or the United States of America, he was closely scrutinized by the political police and the secret services.

Einstein's political behavior was exceptional in comparison to that of his scientific contempories. His impressions of the First World War and the accompanying enthusiasm led him to take up a decisive position in his ideas about society and social commitment. In the summer of 1914 he refused to join the chorus of chauvinistic and nationalistic approval to which his colleagues thoughtlessly subscribed. In particular he refused to sign the infamous *An die Kulturwelt* (Appeal to the Cultural World), in which Germany's intellectual elite unequivocally condoned German militarism as the defender of German culture: "Without German militarism German culture would long since have been wiped from the surface of the earth [...] The German armed forces and the German people are one". Among the supporters of such pointed statements were the people who had engineered Einstein's move to Berlin and now belonged to his closest circle of acquaintances and colleagues: Max Planck, Walther Nernst and Fritz Haber, even intellectuals who had been more critical towards Wilhelminian Germany, such as Max Liebermann, Gerhart Hauptmann or Max Reinhardt, pledged their "names and honour" to this rabble-rousing testimony of incitement.

It demonstrates not only Einstein's independence in scientific questions when, after the outbreak of the First World War, he opposed the political mainstream and all national chauvinistic tendencies as expressed in the "Appeal to the Cultural

World" and signed the counter-manifesto, the "Appeal to the Europeans," launched by the Berlin physiologist and pacifist Georg Friedrich Nicolai. He called for the power of reason, the fastest possible ending of the war and understanding

between nations. Einstein not only signed this manifesto, he also participated in the founding of the *Bund Neues Vaterland* (Association for a New Fatherland) in autumn 1914. During the war the association developed into an organization of left-wing intellectuals that came under the surveillance of the imperial military and police authorities and was then actually banned in February 1916. During this time Einstein also became a target of police investigations: in March 1916 the Berlin High Command requested the Berlin Academy to provide information about their member, and in January 1918 Einstein's name was included in a police list of prominent pacifists from Berlin and the surrounding area. From 1916 onwards the police went as far as deploying a spy to observe Einstein's activities. During the 1920s, after the spectacular confirmation of the general theory of relativity, Einstein became a prominent public figure, an icon, who now represented more than physics and science. Einstein's international popularity was soon

Einstein at the chair of a "League for Human Rights" event, Berlin 1932. Seated from left to right: Arthur Rosenberg, Albert Einstein, Emil Gumbel

recognized as an asset to German foreign policy and was implemented as such. During his intensive tours of the early 1920s, he increasingly assumed the role of ambassador for intellectual Germany. Although he was not overenthusiastic about this role, he nevertheless carried out the responsibility, more than likely because after the fall of the monarchy he had great hopes for the newly founded Weimar Republic. In the revolutionary period of 1918 he wrote to his mother saying: "I am very happy about the way the situation has developed. At last I'm able to feel comfortable here."

Even though Einstein was seen as an excellent testimonial and a "first rate cultural factor" for the Weimar Republic, he was still viewed as politically suspicious by many people in positions of political responsibility, especially members of the ministerial bureaucracy that to a large extent had been taken over from the Wilhelminian era. Suspicions grew even more during the second half of the 1920s, when the republic became increasingly conservative and the political right gradually strengthened its foothold. At this time, and parallel to these developments, Einstein shifted more to the left. For instance, in the earlier years German ambassadors and diplomatic envoys had compiled comprehensive reports about Einstein's visits abroad, and these went to recipients including the political police. In the second half of the 1920s the authorities became increasingly interested in Einstein's political activities inside Germany itself and recorded these with minute accuracy. The commissar in charge of political surveillance noted in the files that Einstein was still active in the *Liga für Menschenrechte* (League of Human Rights), formerly known as "association for a new fatherland", that he was a committee member of the pro-Soviet *Gesellschaft der Freunde des neuen Russlands* (Society of Friends of the New Russia), and that he was also a trustee on the board of the *Internationale Arbeit-*

erhilfe (International Workers' Aid Association). Police interest in Einstein was further fired by the strengthening of his pacifist views. He spoke out publicly on behalf of conscientious objectors to war service and criticized the military as a

Announcement of Einstein's lecture *What the Worker Must Know about the Theory of Relativity* in the Marxist Workers' School on 26 October 1931

"disgrace to civilization." To him war was nothing less than "plain murder," and military service was a means of "training the body and mind in the art of killing." As a result of this strictly pacifist position he also made repeated categorical statements rejecting all, but especially German, rearmament. This not only intensified the conflict with his conservative adversaries, it also increased the size of the files monitoring his activities.

Despite his sympathies for Berlin's intellectual climate and for his colleagues there, Einstein seriously began weighing up the idea of leaving Berlin and thus Germany at the end of the 1920s, because of the increasingly radicalized political situation. The future of German politics was sealed when the Nazis took over power in January

1933. Einstein was again one of the few prominent German scientists to openly and uncompromisingly criticize the newly established Nazi regime for its persecution and expulsion of Jews and other

Frank Kingdon (journalist), Albert Einstein, Henry Wallace (left-wing presidential candidate) and Paul Robeson, 1948

German citizens categorized as undesirable. In National Socialist Germany he was declared a *persona non grata*, transformed into a target of hatred and excluded from the scientific world. In 1933 Albert Einstein decided not to return to Germany from a research journey to the USA. In American exile he used his fame and reputation to add weight to his support for expelled and persecuted colleagues, friends and relatives. He also considered it his duty to openly denounce the political terror and anti-Semitic campaigns systematically engineered by the Nazi regime. During this period and in the face of Nazi Germany's aggressiveness, he began modifying his radical pacifist views. In a letter written in the summer of 1933 to an English pacifist he said: "I hate the military and violence of all kinds, but I am absolutely convinced that this detested means is now the only effective form of protection." Following this it seems only logical that in 1939 Einstein joined the initiative of the Hungarian physicists Leo Szilard and Eugen Wigner who had also emi-

grated from Germany, and signed a letter to the American President Franklin D. Roosevelt warning of the potential danger of Germany developing an atomic bomb and suggesting that America should begin research in this area. In this way Einstein helped initiate America's atomic bomb project, in which he never actually participated himself.

In Einstein's opinion freedom and human rights were inseparable. Consequently not only did he criticize the conditions in Nazi Germany and the Soviet Union under Stalin's regime, but his critical appraisal of reality in his American home of exile was also based on his experiences in Germany. He appealed that civil liberties be upheld and declared his solidarity with actions of civil disobedience. During the height of the cold war and the so-called McCarthy era, which led to open conflicts and irritation: leftist and liberal circles admired him, while the conservative and rightist spectrum vehemently attacked him and even considered stripping him of his American citizenship and expelling him from the country. This polarization was partially caused by Einstein's insistence on always making his views publicly known, such as when he declared his solidarity with the then governing "Spanish Peoples' Front" during the 1930s and supported the "Friends of the Lincoln Brigade" and other anti-Franco organizations. His uncompromising opposition to Nazism along with his eager support for anti-fascist organizations in the USA, which also included Communist and pro-Soviet groups, was not welcomed by the American establishment. Einstein also met

he indicated to the Polish ambassador that the United States was no longer
a free country and that his activities were carefully scrutinized. He was
a sponsor of a committee to defend the rights of the 12 Communist leaders.
On February 12, 1950, by transcription over NBC network, Einstein advocated
banning all violence among nations to preclude "general annihilation" of
mankind.

CONTACTS AND ASSOCIATES

Einstein's social and professional contacts, since 1938, have
included a number of known members and sympathizers of the Communist Party.
One of his former assistants at Princeton University who was subsequently
denied clearance by the Atomic Energy Commission was recommended favorably by
Einstein. Investigation by the Bureau has shown that his secretary, Helen
Dukas, who resided in Einstein's home has had considerable contact with
individuals known to be Communists, several of whom were suspected as Soviet
agents. The scope of the investigation of Dukas was necessarily limited to
discreet techniques. Information not yet fully developed indicates he may
have had some contact with Emil Klaus Fuchs, who was recently arrested in
England as a Soviet espionage agent.

MISCELLANEOUS

Einstein was one of many distinguished Germans who lent their
influence and prestige to German Communists prior to the rise of Hitler.
In 1940, the Army declined to clear Einstein in connection with the "limited
field of study for which his services were needed" after the Navy had given
its assent. Einstein publicly declared, in 1947, that the only real party
in France with a solid organization and a precise program was the Communist
Party. In May, 1948, he and "10 former Nazi research brain trusters" held
a secret meeting to observe a new beam of light secret weapon which could
be operated from planes to destroy cities, according to the "Arlington Daily,"
Arlington, Virginia, May 21, 1948. The Intelligence Division of the Army
subsequently advised the Bureau that this information could have no foundation
in fact and that no machine could be devised which would be effective outside
the range of a few feet.

- 2 -

Excerpt from an FBI report of 15 April 1950 with comments by FBI boss J. Edgar Hoover

with public opposition for backing the black civil rights movement with his name and statements, and by becoming actively involved in the anti-lynching movement together with the singer Paul Robeson and the ethnologist William du Bois. At the beginning of the 1950s Einstein was once again in the position of the political outsider, and again he became the target of extensive secret service investigations which resulted in a sizable FBI file of some 2,000 pages.

Quite apart from the fact that secret service files throughout the world have always contained large amounts of incorrect or deliberately misleading information as well as trivialities, Einstein's FBI file demonstrates to an unprecedented degree that Einstein was not only actively involved in global politics, propagating the idea of a world government, the banning of nuclear weapons and issuing warnings of an arms race between the super powers. His file also depicts an

CONFIDENTIAL
REPRODUCED AT THE NATIONAL ARCHIVES

85300-B-254
-19-

BIOGRAPHICAL SKETCH

DECLASSIFIED
Authority NND 740058
By M   NARA, Date 6-8-40

DR. ALBERT EINSTEIN,
Princeton University,
Princeton, New Jersey.

Dr. Albert Einstein was born in Ulm-an-der-Donau, Germany, March 14, 1879, the son of Hermann and Pauline (Koch) Einstein. He was educated in Germany and Switzerland. He holds a great many honorary degress from universities all over the world. He was married to Mileva Maree in 1901, and a second time to Elsa Einstein, in 1917. He has one child by his first wife. He is connected with many universities in Europe.

He came to the United States in 1933, and has been located at Princeton University. He is the author of many books on relativity and other matters. His home address is 112 Mercer Street, Princeton, New Jersey.

It is the belief of this office that Professor Einstein is an extreme radical, and that a great deal of material on him can be found in the files of the State Department. This office has knowledge of one incident which was newspaper headlines in 1933 when Einstein and his family left Berlin for the United States.

The American Consul General at Berlin, Mr. George Messersmith, was asked by Mrs. Einstein to visa the passports of herself, Dr. Einstein, and her daughter. She made this request by telephone. Mr. Messersmith told her that Dr. Einstein would have to appear personally at the Consulate and that he would have to swear to an affidavit that he is not a member of any radical organization. Mrs. Einstein then wired prominent Jewish women in New York that Mr. Messersmith was obstructing their coming, and a press campaign was started in American papers demanding that the President recall Mr. Messersmith. The State Department upheld Mr. Messersmith, and Dr. Einstein signed the declaration.

The origin of the case is that in Berlin, even in the political free and easy period of 1923 to 1929, the Einstein home was known as a Communist center and clearing house. Mrs. and Miss Einstein were always prominent at all extreme radical meetings and demonstrations. When the German police tried to bridge some of the extreme Communist activities, the Einstein villa at Wannsee was found to be the hiding place of Moscow envoys, etc. The Berlin Conservative press at the time featured this, but the authorities were hesitant to take any action, as the more radical press immediately accused these reporters as being Anti-Semites.

Incl. 2   CONFIDENTIAL

CONFIDENTIAL
REPRODUCED AT THE NATIONAL ARCHIVES

Dr. Albert Einstein

It is believed that Mr. George Messersmith, at the present time Ambassador to Cuba, will be able to furnish a great many details regarding Dr. Einstein.

In view of his radical background, this office would not recommend the employment of Dr. Einstein on matters of a secret nature, without a very careful investigation, as it seems unlikely that a man of his background could, in such a short time, become a loyal American citizen.

Dr. Einstein, since being ousted from Germany as a Communist, has been sponsoring the principal Communist causes in the United States, has contributed to Communist magazines, and has been an honorary member of the U.S.S.R. Soviet Academy since 1927. The Soviet's enthusiastic birthday greeting to Dr. Einstein appears in the Communist party newspaper, the "Daily Worker", of March 16, 1939.

Einstein who became directly involved in American disputes over domestic policies. For instance, he attacked wrong judgements in the legal system, declared racism America's "worst disease" and called for the upholding of civil rights. He was also among the prominent opponents of Senator Joseph McCarthy and the modern form of "Inquisition" pursued by him and his "House Committee on Un-American Activities." When a New York teacher was suspected of Communist sympathies and was called before one of these committees, Einstein personally intervened with an open letter opposing anti-constitutional snooping into people's political convictions, and he called for civil disobedience: "Every intellectual who is brought before such a committee, must refuse to give evidence. This means being prepared to be jailed and driven to economic ruin, in other words it means sacrificing one's personal interests for the cultural interests of the country." Although American public opinion answered with a "polite refusal" and many of his contemporaries, including intellectuals, refused to take such drastic steps, Einstein's appeal did have an effect and, as he said himself, "it did help to clear the political air a bit."

FBI chief, Edgar Hoover, was almost obsessive in his efforts

FBI report about Albert Einstein and his political environment in Berlin to document Einstein's supposed Communist and subversive convictions, and brand him as the central force behind a network of "red front organizations" – the file listed over seventy such organizations including a dozen civil rights associations with whom Einstein allegedly had some form of contact. But it was undoubtedly for good reason that, in the final instance, nobody dared introduce for the USA. Apart from this, the material that had been gathered was of little use for a sensational spy story or as documentary evidence of subversion, since all of Einstein's activities had taken place in the public eye, and this was contrary to conspiratorial behavior.

As Einstein expert John Stachel says, it was precisely because of this that many of Einstein's con-

Albert Einstein at a television address in February 1950 against the decision of the U.S. government to continue the development of the hydrogen bomb

proceedings against the most famous contemporary scientist and perhaps the most prominent American citizen at that time. In contrast to other victims of the FBI and the Mc Carthy era, Einstein was too popular and, of course, an advertisement temporaries saw, and even admired, him not only as an outstanding physicist, but also as a "symbol of steadfast resistance to the modern inquisition which threatened to destroy civil liberties in this country during the cold war years."

Angelo Baracca

The Dark Side of Einstein's Heritage: The Nuclear Age

Einstein's thought and contributions have had a deep influence on almost every aspect of life in the past century. Not only science and technology, but also art, philosophy, and even common sense and the perception of reality have been deeply affected as a result. It is possible, however, that he would not feel at ease with some of these developments.

Schrödinger and Luis de Broglie, never accepted such a formulation as the final theory and worked actively to point out its internal contradictions. In his criticism, he clashed in particular with Niels Bohr. In 1936 he proposed the so-called "Einstein, Podolsky and Rosen Paradox," which referred to the quantum treatment of coupled atomic systems. This laid the basis for the development of a field of research on the foundations of quantum mechanics. Among the multiply acknowledged aspects of Einstein's heritage, however, one usually seems to be disregarded. Although the relativistic mass-energy relationship $E = mc^2$ even appears on T-shirts, few people are aware of some of the enormous consequences it had for the development of nuclear weapons and war. A further contradiction was between Einstein's pacifist attitude on the

The atomic bomb „Little Boy" – a bomb of this type exploded over Hiroshima in 1945

The world's first nuclear reactor in an underground laboratory of the University of Chicago: The reactor was built from layers of graphite interspersed with uranium. A chain reaction took place on 2 December 1942 as the 57th layer was added. The main breakthrough in the production of atomic bombs had been reached. The decisive moment: the reactor goes critical. Enrico Fermi at the instrument panel, behind him Arthur A. Compton

It sounds paradoxical that although Einstein opened the way to many fields of contemporary physics, the directions they eventually took often diverged from his conceptions. This happened with quantum theory, which he introduced a century ago. It was he who, in 1917, applied a probabilistic treatment of the microscopic interactions between radiation and matter, when he introduced the process of the stimulated emission of radiation, which is the basis of lasers. For him, however, such an approach had a provisional character until the formulation of a complete theory. Yet in the following years the "Orthodox," or "Copenhagen Interpretation" of quantum mechanics adopted a probabilistic framework as a basic ingredient. Einstein, together with Erwin

one hand, and the true revolution that his discoveries brought to military techniques and tools on the other. Einstein could be considered – obviously indirectly – as the "father" of the nuclear age. Much of the influence of the nuclear nightmare, beyond political events, on 20th-century everyday life, anxieties and unconscious uneasiness, still needs to be understood and investigated. Einstein did not contribute directly to nuclear physics, and it seems a twist of fate that he signed the letter to President Roosevelt, proposed by Szilard, pointing out the danger that the Nazis

continues, and in completely new and worrying directions. Einstein would be anything but pleased with the growing connection between "civil" and "military" research and technology. The traditional dual use of nuclear energy was rooted in uranium-enrichment technologies, as

The mushroom cloud of the first American hydrogen bomb, detonated on the Marshall Islands in November 1952

could construct an atomic bomb. He did not play any role in the "Manhattan District," which manufactured the atomic bomb. After the war, Einstein worried about the uncontrolled proliferation of nuclear weapons and the risks of nuclear war, although his death prevented him from seeing the worst phase of the arms race.

Profound hopes of progressive nuclear disarmament arose after the collapse of the Soviet Union. Unfortunately they have been dashed by the establishment of a unipolar world power in which the risks of an effective use of nuclear weapons are more concrete than ever. Nuclear proliferation

well as plutonium production in nuclear reactors. Nowadays, a growing relevance is assumed by tritium, the hydrogen isotope with two neutrons, a volatile radioactive gas that must be continuously produced and which is a fundamental component of modern and efficient nuclear warheads. The latter are boosted by a small quantity of deuterium-tritium mixture, whose nuclear fusion is triggered by the onset of the chain reaction and generates a neutron flux which in turn enhances the performance of the uranium or plutonium fission. The 1996 Comprehensive Test Ban Treaty (although it never entered into force) introduced a moratorium on nuclear tests (apart from those in India and Pakistan in 1998), but accelerated the search for new weapon technologies (super-computer simulations, subcritical tests, new experimental facilities, new nuclear processes). Several countries have the status of virtual, or stand-by proliferators (Japan and Germany are significant, but not unique examples), since they possess both

Obligations for complete nuclear disarmament, whose implementation has been totally disregarded by nuclear weapon states. The next Review Conference, to be held in May 2005, is at serious risk of failure.

The search for new kinds of nuclear arms is proceeding apace in the nuclear weapons laboratories, alongside the building of the anti-missile

The ruined city center of Hiroshima shortly after the explosion of the atomic bomb on 6 August 1945

the materials and know-how needed to build nuclear weapons in a relatively short time. Nowadays it is possible for almost any modern industrial country to design highly efficient and reliable nuclear warheads that work with no need for complete nuclear tests. The current focus on the proliferation dangers from North Korea and Iran is diverting attention from the true risks, which come from the United States, Israel and the other nuclear weapon states. The 2000 Review Conference of the Non-Proliferation Treaty established Thirteen Practical Steps to implement Article VI

shield and space based weapons. There is an ongoing drive towards the miniaturization of nuclear weapons and a related quest for very low yield nuclear explosives, i.e. "fourth generation" nuclear warheads. These would erase the fundamental distinction between conventional and nuclear weapons, allowing the use of the latter as common battlefield arms. Many experts denounce the extreme risks of these trends, and yet the public is kept in the dark.

Such developments are also relevant to another dual-use technology: controlled nuclear fusion.

Although it has been presented for half a century as a way of producing unlimited and cheap energy, such an aim still looks very remote, while military applications seem more immediate. This is particularly true of Internal Confinement Fusion (ICF), in which a small pellet of deuterium-tritium is uniformly bombarded with laser or particle beams, causing its implosion and heating and triggering its ignition. Many countries are building large ICF installations as well as other facilities such as a super-laser of unprecedented power, which may have other military applications.

On the one hand, these investigations clarify the physical properties in very extreme conditions and the process of nuclear fusion, while producing other important fallouts. When the fusion of a deuterium-tritium pellet is achieved, a micro pure-fusion explosion will be produced (with a yield of less than one ton of chemical explosive equivalent) without needing to trigger a nuclear fission explosion as in current two-stage thermonuclear warheads. In order to turn it into a deliverable nuclear weapon, a super-laser or particle accelerator must be miniaturized. Considerable progress

is being made in this direction with the help of new techniques. A few decades ago, nuclear weapon laboratories started the development of nanotechnology, a range of potentially revolutionary engineering techniques for designing microscopic structures in which the materials and their relations are mechanized and controlled atom by atom. The most significant near-term applications of nanotechnology will be in the military domain, to obtain extremely rugged and safe arming and triggering mechanisms, both for conventional and nuclear weapons. Another promising possibility

for triggering nuclear fusion seems to be antiproton beams, which are being researched in high energy physics laboratories.

If Albert Einstein were still alive today, he would probably feel very unhappy about such developments.

Signing of the Treaty on the Non-Proliferation of Nuclear Weapons in Washington: US Foreign Minister Dean Rusk signs in the presence of Lyndon B. Johnson (right); far right Anatoly Dobrynin (Russian Amdassador), next to him Patrick Dean (British Ambassador), 1 July 1968

Karl Fredrik Reutersward: *Anti-Violence*, sculpture of an over-dimensional revolver with a knotted barrel in front of the main United Nations building in New York, 1980 (A gift from the Grand Duchy of Luxembourg 1980)

Danian Hu

Einstein and Relativity in China, 1917–1979

Two Japanese educated Chinese scholars first introduced Einstein's theory of relativity in China in 1917. Three factors preconditioned this introduction. The first is the large number of Chinese students in Japan: thousands of Chinese students went to study in Japan every year between 1905 and 1915. The second is Japanese physicists' early interest in the theory of relativity: their interest

Left:
Zhou Enlai, Minister
President of the
People's Republic
of China from 1949
until his death in
1976

Right:
Hans Driesch,
philosopher and
pacifist, leading
member of the
League for Human
Rights (painting by
Willi Geiger)

began as early as 1907 when most Western physicists did not even know the theory. The third is the May Fourth movement (1917–1921), the intellectual revolution in modern China: it promoted the introduction of Western culture, especially Western sciences.

The theory of relativity and Albert Einstein became widely known in China in the early 1920s. Chinese scholars educated in Japan, Western Europe, and the United States contributed to the quick dissemination of the theory. Bertrand Russell (1872–1970), the British philosopher and mathematician, presented a series of lectures on relativity in Beijing early in 1921; Albert Einstein also planned to visit Peking University at the end of 1922, which was regrettably cancelled. Both events greatly helped the theory's dissemination in China. In the 1920s, the Chinese quickly embraced the theory of relativity, partly because of

the theory's revolutionary nature, which was particularly attractive for left-wing intellectuals such as Zhang Songnian (Zhang Shenfu, 1893–1986) and Zhou Enlai (1898–1976). Zhang was one of the founders of the Chinese communist party in 1921 and Zhou joined the party in 1922 and later became the famous premier of the People's Republic of China. Perhaps a more important reason for the quick and positive Chinese reception of the theory of relativity was the absence in China of a tradition of classical physics. As a result, there was no open Chinese opposition to the theory of relativity before the early 1950s, despite the existing reports on the anti-relativity rally in Berlin in 1920 and on Japanese oppositions to the general theory of relativity. The only critical Chinese essay on relativity published during this time was a translated work by German philosopher Hans Driesch (1867–1941). Furthermore, even the Chinese translator disagreed with Driesch's criticism. The persistent Chinese devotions to the theory of relativity was also evident in the fact that Peiyuan Chou (1902–1993) and Shu Xingbei (1907–1983), two of then only a few Chinese theoretical physicists, carried on their research on general relativity in the late 1930s and 1940s, when most Western physicists had moved away from such study and flocked into the popular research in nuclear physics. The reception and research of the theory of relativity became the starting point of the theoretical physics studies in China.

In the 1920s, Albert Einstein was introduced in China as a scientific hero who revolutionized physical science and our understanding of the universe. Over the next three decades, Einstein and relativity became glorious symbols of modern science and were admired unanimously and persistently in China. The introduction of Einstein's ideas on pacifism and social democracy in the 1930s made him even more popular among the Chinese intellectuals. They not only admired Einstein's scientific achievements, but also appreciated his sincere sympathy and consistent support for the Chinese people when they were under attacks from Japanese militarists and the political persecution of Chiang Kai-shek's authoritarian government.

The communist victory in 1949 and the Korean War (1950–1953) completely changed China's domestic and international environment. Mao Zedong's "leaning to one side (the Soviet Union)" policy in the 1950s helped to break the U.S. led blockade but also subjected the country to the prevalent influence of Stalinism. As a consequence, Soviet-derived political and philosophical criticisms of Einstein and the theory of relativity were introduced and spread in China. Gradually, the Soviet criticism not only tarnished the public image of Einstein, but also induced the Chinese criticism based on dialectical materialism of Marxist philosophy. The criticism had focused on Einstein's philosophical views before 1965 and the critics often cited V. I. Lenin (1870–1924) and mocked Einstein as an "eminent scientist but poor philosopher!" The Chinese criticism in the 1950s and early 1960s prepared far more radical attacks on Einstein and his theory of relativity during the Cultural Revolution (1966–1976), when China was plunged into an unprece-

dented turmoil and when Einstein was not only assaulted in his political and philosophical views but also on his scientific theories.

The Cultural Revolution was nothing less than a tragic calamity and had perhaps the most ruinous

Chiang Kai-shek with his wife, 1927

effects on the Chinese society. Since the movement began by destroying the cultural establishment, Chinese science and scientists were among those that suffered most. During the Cultural Revolution, Einstein and the theory of relativity became primary targets of an organized criticism, which began in Beijing in 1968 and later expanded in Shanghai in the 1970s. The critics consisted of mainly scientists, engineers, philosophers, and journalists of the younger generation, who accused the theory of being based on a groundless hypothesis, advocating reactionary "idealist relativism" and bourgeois viewpoints, and representing "one of the biggest obstacles now blocking the advance of natural sciences." Radical party propagandists such as Chen Boda (1904–1989) and Yao Wenyuan (1931–) sponsored and exploited the criticism for their own political gains. While Chen supported the criticism out of political ambition and cultural prejudice, Yao exploited it to attack his political rival Zhou Enlai and maintain the absolute control of Chinese science by orthodox Marxist ideology. The criticism largely ended in 1976 after both radical leaders fell from power, but not before it did serious damage to Chinese

Mao Zedong (3rd from right) and members of the Politburo of the Chinese Communist Party. Zhou Enlai (minister president, 2nd from left), Deng Xiaoping (party secretary, 2nd from right) and Lin Piao (minister of defense, right), 1966

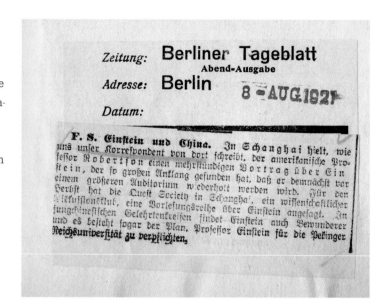

Newspaper notice about an enthusiastically received lecture on Einstein's theory of relativity held by the American Professor Robertson in Shanghai, 1921

science and education. The criticism also induced a "scientific" debate on relativity in a physics journal between 1974 and 1977, which involved many young physicists and demonstrated the deep impact of orthodox Marxist ideology on the scientists: even those who defended the theory of relativity joined the critics to fault Einstein's interpretations based on the teaching of dialectical materialism.

The organized criticism of Einstein and relativity was part of the anti-scientific campaigns during the Cultural Revolution, which encouraged nihilistic attitude toward basic scientific research. As a consequence, scientific research institutes were dismantled and researchers were forced to work in factories or farms. By 1970 basic scientific research, especially theoretical research, had nearly vanished from China. Premier Zhou Enlai endeavored to rescue Chinese science from this dangerous situation in the early 1970s, but attacks by Yao Wenyuan and other radicals prevented improvement of research in the basic sciences before the end of the Cultural Revolution.

The Chinese criticism campaign had some consequences against the critics' expectation. The organized campaign allowed some physicists to resume scientific research and gain access to Western scientific literature; it cleared the way for the

The Cultural Revolution in China: Members of the Red Guard lead a civil servant wearing a "hat of shame" through the streets of Beijing. The prisoner is accused of being a political pickpocket and a member of the anti-revolutionary groups, 1967

Peiyuan Chou, theoretical physicist, President of the Chinese Physical Society in 1979

publication of the Chinese translation of the *Collected Works of Einstein*; and it presented the opportunity for astronomers to turn a meeting of mass criticism into a national scientific conference. The criticism publications ironically constituted a rare publicly accessible source for Chinese youths to learn about contemporary scientific discoveries in the early 1970s, which, despite their distortion of the truth, inevitably aroused curiosity about modern physics and motivated many to seek more information.

As soon as the radical leaders fell from power at the end of 1976, some Chinese physicists began to plan a grand ceremony for the centennial of Einstein's birthday in 1979 in order to rehabilitate the great physicist publicly. The plan was eventually approved by Deng Xiaoping (1904–1997) and other top Chinese leaders. To demonstrate their sincerity and seriousness, Chinese organizers chose to convene the celebration ahead of other major ceremonies around the world. On 20 February 1979, more than one thousand Chinese scientists assembled in Beijing to honor Albert Einstein. Peiyuan Chou, now the chairman of the Chinese Association of Science and Technology and the president of the Chinese Physical Society, gave the keynote speech, in which he offered a comprehensive reevaluation of Einstein and an official condemnation of the criticism campaign during the Cultural Revolution. Chou praised Einstein's life-long scientific achievements, refuted the malicious charges against him, and, most significantly, extolled his social democratic ideas. Albert Einstein has since been not only restored as a great scientific hero, but also esteemed as a champion of social democracy and justice.

Diana Kormos-Buchwald

The Einstein Papers Project 1955–2005

The publication of the *Collected Papers of Albert Einstein* has been a long-standing editorial project. So far, nine volumes of Einstein's collected writings and correspondence have been published since 1987 (with one double volume), covering Einstein's life and work from 1879 to 1921.

Albert Einstein (1879–1955) lived in Germany, Switzerland, and eventually the United States, through six major epochs of modern history: the

Albert Einstein at the blackboard with a formula about the Milky Way, Pasadena 1931

Wilhelmine Empire, World War I, the Weimar Republic, Fascism, World War II, and the post-war period. He had been the prime mover of the transformation of modern physics, starting with the publication of his revolutionary papers on the photoelectric effect and on the special theory of relativity in what we call "the miraculous year 1905," and became the most important scientist after Isaac Newton. Einstein never wished for a memorial, and insisted that his ashes be scattered at an undisclosed location. The Einstein Papers Project is our intellectual memorial to his life and work.

Although its first volume appeared more than thirty years after Einstein's death in Princeton, N.J., in 1955, plans for such an edition were initiated early. In his last will and testament, Einstein bequeathed his entire literary estate, with no stipulations or strings attached, to the Hebrew University in Jerusalem, an institution which he helped establish and on whose behalf he had begun working as early as 1919. Upon his death, the task of organizing his written legacy was entrusted to the executors of his literary estate, his long-time assistant Miss Helen Dukas and his friend Otto Nathan, an economist living in New York. Both, Dukas and Nathan, decided that a publication plan should be initiated with dispatch, yet the process ultimately turned out to be arduous and at times painful. Not only did various complications arise in the process of collecting material, but also Einstein's various family branches disagreed with the executors regarding the status of physical and intellectual ownership to material which was variously held in private hands, libraries and archives, as well as in Einstein's former home and office.

Other aspects of the envisioned project needed to be clarified and implemented, such as the possible location, funding sources, editorial policies, and management. Eventually, in the early 1980s, John Stachel, professor of physics at Boston University, officially began the work of planning for what was already a highly challenging enterprise. By that time, Helen Dukas and Otto Nathan, with assistance and advice from a number of historians of science, had organized Einstein's manuscripts and letters at the Institute for Advanced Study in Princeton, where Einstein had worked since 1933. Dukas and Nathan corresponded with hundreds of individuals, institutions, libraries, former friends, colleagues, and family members over a period of three decades. By the mid 1980s, they had collected more than 40,000 items, either in the original or in photocopy, from the U.S. and Europe, and had established an impressive card-catalogue which Dukas had typed up. Between 1979 and 1981, with assistance from Princeton University staff, the catalogue was entered into a computerized database. The materials themselves

were photographed on microfilm, several hard-copies were produced, while the original Einstein Archive, as it was known by then, was crated and shipped to the newly established Albert Einstein Archives at the Jewish National & University Library at the Hebrew University in Jerusalem, where it has resided ever since. Helen Dukas died shortly thereafter, having devoted almost six decades to Einstein and his legacy. A biography of this remarkable woman is still to be written! The editorial project was established at Boston

University, in a building housing nearby the Faculty of Philosophy. Princeton University Press, whose director Herb Bailey had been an early and enthusiastic supporter of the project, took on the responsibility of publishing the large format bound books, as well as a paperback companion translation project. The editors decided that all of Einstein's scientific writings, both published and unpublished, as well as drafts, notebooks, scientific and personal correspondence would appear in chronological order. Unpublished materials would be faithfully transcribed, no silent corrections of typographical or other errors would be applied, and punctuation and style would be reproduced, while errors of fact or calculation would be explained in the annotation. Published items would appear in facsimile, while comparisons to drafts or versions would be examined and detailed in the footnotes. An Introduction and various Editorial Headnotes analyzed major themes in Einstein's life and work. It was to be a project that combined the most rigorous American and European editorial standards, whereby historical events, places and names, as well as scientific developments would form the core of the detailed scientific-historical annotation.

The first volume of Einstein's papers was the most challenging to produce. Little material documenting his childhood and youth was extant, and, most distressingly, manuscripts of his early work, among them one of his most remarkable papers on the special theory of relativity, published in 1905, when Einstein was only twenty-six years old, were not available. Nevertheless, the editors embarked on an ultimately highly successful hunt for unknown letters and documents, so that by 1987 the deeply personal and revealing letters between Albert Einstein and his classmate, sweetheart, and first wife Mileva Marić could be included in a volume that also contained copies of his school certificates, his final leaving examinations, his university notebooks, and other highly illumi-

Otto Nathan in conversation with Albert Einstein © The Hebrew University of Jerusalem, Albert Einstein Archives

nating and never before examined materials. This achievement of the first editorial team at Boston University, consisting of John Stachel as senior editor, together with David Cassidy (now at Hofstra

Journal of the American trip in 1930/31. Shown here is the first page, with Einstein's entry about the departure from Berlin, Zoologischer Garten station, 30 November 1930

University), Robert Schulmann (now residing in Bethesda), and Jürgen Renn (now at the Max Planck Institute for the History of Science in Berlin), was followed by a series of important analyses of Einstein's scientific development. Among them are the volumes documenting his most revolutionary scientific contributions between 1905 and 1916, included in Volumes 2–4, and most significantly, Einstein's strenuous path towards his greatest achievement, the general theory of relativity. Over the years, the editorial team, consisting of historians and philosophers of science whose specialty was the complex

physical, mathematical, and conceptual foundations of modern physics, included Martin J. Klein (Yale), A. J. Kox (now at the University of Amsterdam), Michel Janssen (now at the University of Minnesota), and many other associated and contributing editors, graduate students and editorial staff. Ein-stein scholarship has benefited immensely from new insights gained over the past eighteen years. With almost every volume, new biographies for the general public are published, and our received view of Einstein deepens and changes. The four first volumes covering Einstein's writings were followed by the first volume devoted exclusively to correspondence. Since then, the published volumes have been divided into a *Writings* series and a *Correspondence* series, a division that was dictated by the rising number of letters after 1915. Thus Volume 5 covered Einstein's correspondence from 1902 to 1914, a significant time period, while the latest Volume 9, published in 2004, only covers a period of sixteen months.

This dramatic increase in Einstein's correspondence was due in large part to his rising fame shortly after the end of World War I, his many additional administrative responsibilities, his status as Germany's pre-eminent scientist, and his gradual engagement in social, political, and humanitarian causes. It was in the spring of 1919, only several months after the end of the war, the dissolution of the German empire, and before the Versailles Treaty had even been concluded, that two British astronomical expeditions set sail for the northeastern coast of Brazil and the northwestern coast of Africa to observe a solar eclipse on May 29. The expedition's major goal was to test one of the three fundamental observational predictions of the theory of general relativity, namely, that light from distant stars will be deflected when it passes the strong gravitational field in the vicinity of the sun. The official report on the results of the expeditions was published in England in the fall. As a

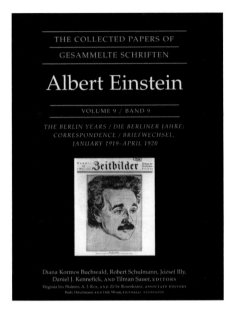

Jacket of volume 9 of the *Collected Papers of Albert Einstein*, published in late 2004

consequence, Einstein, who had eagerly awaited the results of the expedition, became known world-wide. Besieged by reporters, Einstein started appearing on the covers of popular magazines and newspapers. He became a celebrity in Berlin and elsewhere, was asked for popular renditions of his most recent work, and invited to lecture not only at home but also abroad. As the only German prominent scientist identified with an opposition to World War I and among the few German intellectuals who had not signed the *Manifesto of the 93* that had defended German's war aims, Einstein thus became a spokesman for international reconciliation among scientists and intellectuals and increasingly a public figure rather than a private savant.

The project moved in the summer of 2000 from Boston to the California Institute of Technology in Pasadena when I was appointed the new general editor. Einstein had been a visiting scientist at Caltech during three winter terms in 1931, 1932, and 1933. He had come here to engage with the astronomers and physicists who, during the late 1920s and early 1930s, were making some of the most revolutionary advances in our understanding of the structure of the universe. He met extensively with Edwin Hubble, who had shown that, contrary to Einstein's initial assumptions, we live in a dynamical and expanding universe of a much larger scale than Einstein had known when he developed his theory of generalized relativity in the first two decades of the century. The project, now housed in its own rather charming villa on the

campus of Caltech, consists of a staff of about five full-time equivalent editors with varying areas of expertise and experience on the project (Jozsef Illy, Daniel Kennefick, Tilman Sauer, Virginia Iris Holmes, A. J. Kox, Ze'ev Rosenkranz and Jeroen van Dongen), supplemented by gifted undergraduate students, an editorial and technical staff, visitors, and a number of researchers in Europe, the US, and Israel who contribute material from local libraries and archives. Over the past two years, our infrastructure has changed significantly, especially in that we have moved to a web-based database consisting of more than 60,000 items (or more than half a million photocopied pages in our collection). We work from copies, in close cooperation in particular with the Albert Einstein Archives in Jerusalem, with whom we have launched, in a highly intense and successful collaboration, a new

website containing high-quality images of 900 digitized scientific and non-scientific manuscripts in Einstein's hand, as well as a finding aid and database. The website, launched in May 2003 on the occasion of a symposium celebrating twenty-five years of editorial work on the Collected Papers, held at the Museum of Natural History in

Homepage of the Online Albert Einstein Archives' website, launched in 2003

New York on the occasion of its Einstein exhibition, was accessed during its first few weeks by several hundred thousand users. (www.alberteinstein.info) The challenges of selectivity are one of the most significant current editorial challenges. While Einstein's writings for the period 1922–1925 would probably fit into no more than one volume of Writings, his complete correspondence of incoming and outgoing letters for the same period would probably necessitate several Correspondence Volumes. A decision was thus made to Calendar administrative items pertaining to Einstein's activities as member of the Prussian Academy of Sciences and as director of the Kaiser-Wilhelm-Institute of Physics, a calendar which reached eighty pages in Volume 9. We are now publishing approximately 50% of his correspondence and are debating ways of reducing the number even further without prejudicing certain aspects of Einstein's work and life, presenting a fair representation of the broad incoming correspondence from publishers, students, colleagues, and public figures without imposing the historian's judgment too severely onto extant material.

What emerges however is that over the next few volumes much new, unexplored material will continue to come to light, providing an ever expanding and deepening understanding of Einstein's life and work.

Abhay Ashtekar

Einstein's Heritage for the 21th Century

Einstein's geometrization of physics ranks as one of the deepest ideas in science. Like a shining beacon, it will continue to guide fundamental physics in the 21th century. Politically, this century began with a burst of violence that has polarized the world. The rhetoric of "you are either with us or against us" has added further instability. Einstein's staunch pacifism and his uncompromising stand against the arrogant use of military might will provide the much needed wise philosophical core during this difficult century.

Olivier Darrigol

The Virtues of Conscious Brooding

In a letter to Besso written shortly before his death, Einstein despaired over the "fifty years of conscious brooding [*die ganzen 50 Jahre bewußter Grübelei*]" that failed to bring him closer to a solution of the quantum enigma. All through his life he refused intellectual complacency, and condemned the premature closing of human endeavors. In contrast, some contemporary intellectuals pronounce the end of physics, the inevitable disunity of knowledge, and the reduction of science to forms of social life. These dark prophets need not have the last word. Einstein has shown us a better way: free thinking bridled by human sympathy and inspired by a true love of nature.

Hans-Olaf Henkel

Einstein's work reminds us of the fact that the strength of the human mind and his creativity can only blossom when it is not hampered by bureaucracy. His departure from Germany and the exodus of other Jewish scientists at that time was a severe blow to Germany's scientific community from which the country is still suffering. Even today we must attend with great care to the needs of the national and international scientific elite in our country.

Michel Janssen

Einstein was a revolutionary, I think, more in his scientific methodology than in his scientific imagination. He had a remarkable talent for sniffing out the salient features of large bodies of experimental data and theoretical ideas and of wrestling conclusions from them with relentless logic. Unfazed when those conclusions clashed violently with common sense or received wisdom, he had the intellectual courage to let the chips fall where they may. Proceeding in this manner, he showed how the clash between the two postulates of his special theory of relativity produced a new space-time structure. Similarly, he showed that some general and well-established experimental and theoretical ideas about matter and radiation inescapably led to the conclusion that a satisfactory theory of light must combine aspects of a wave and a particle theory. Unfortunately, after the success of general relativity, Einstein abandoned this methodology and began to rely on his own 19th century intuitions about the physical world instead. He came to believe that his own mathematical aesthetics provided the most reliable guide to physical truth and rewrote the history of his own thinking accordingly. The excesses of modern string theory are a sad reminder that the older Einstein's pronouncements on methodology have had a much greater impact on modern theoretical physics than the methodology that actually brought the young Einstein the astonishing breakthroughs for which he is still celebrated today.

Stefan Kaufmann

Einstein is more than a physicist to me. He is a researcher who took his contract with society seriously. He knew that those who had acquired specific knowledge also had the responsibility to distinguish what could be done from would should be done.

Wilhelm Krull

At the beginning of the 21th century, a multi-faceted genius like Albert Einstein still poses many challenges, not only to scientists, but also to research funders.
Looking back at the major achievements made by him in 1905, when he was still working at the Swiss Patent Office, we must ask ourselves: to what extent would we have been able to recognize his genius, appreciate his unconventional way of thinking, and ultimately provide him with the necessary resources so early in his career? Responding positively to these challenges is not an easy task but it is worth every effort in order to make sure that some of our best, often quite vulnerable minds can have trust in the ability of our institutions to engage in truly intellecutal risk taking.

Adolf Muschg

One of the reasons I admire Erich Mendelsohn's Einstein tower is that its constructors cheated on the engineering: concrete technology was not far enough along after World War I to cast the new shape desired in one mold. This is more or less the way my understanding of Einstein's theoretical edifice works: I feel like I have to help out with some bricks. I dream of someday misunderstanding Einstein as productively as Goethe misunderstood Newton. Unfortunately, even this involves all kinds of genius.

Susan Neiman

In celebrating Einstein we must avoid the cliches of the saint, the clown, or the sad fool. Einstein was not only a world-historical genius, but a savvy, sophisticated observer who knew exactly how to use his influence for the moral and political causes he supported. The positions he took were brave, subversive, and nearly always right, and his active engagement should serve as a model for contemporary intellectuals.

Joseph Rotblat

A World Without War: A Tribute to Einstein's Quest for World Peace

Einstein's fame and unique status in the world are mainly due to his scientific discoveries. Much less is known about his political activities: his ant-war-campaigns and his advocacy of a world government. Yet, next to science, these matters were nearest to him, he devoted to them much time until the very end of his life.

In many papers on a world without war, I have posed two pertinent questions related to Einstein's peace activities: is world without war desirable? And, is it feasible? The first question is surely rhetorical. After many millions of lives lost in the two World Wars of the last century, and in the many wars since, a world without war is assuredly more desirable. And is has been made all the more desirable by the events that have occurred since the end of the Second World War; not only is a war-free world desirable, it is now necessary. It is essential, if humankind is to survive.

In this "Einstein Year" when we honor the great discoveries he made at the start of his life as a scientist, we should also remember his efforts at the end of his life, to create a world without war.

Sam Schweber

We should remember the context between 1900 and 1925 during which Einstein was most productive, and in particular, the role of the communities he was part of, and the interactions he took part in. Science is a communal activity. The understanding of the physical world revealed by Einstein's amazing works is indicative of the capabilities of the human mind. Such productions and the consequent understanding can only occur in places where "pure" research is encouraged and supported, and which value the knowledge which is produced.

Lee Smolin

For me personally, Einstein is the master who showed us the kind of moral clarity necessary to contribute to science on the highest level. He did so much, not because he was smarter, but because he cared more deeply than anyone else that our understanding of nature be logically coherent and complete. At the end of his career, as at the beginning, he stood alone because the purity of his standards for clarity of understanding set him apart from most professional scientists. At the same time, he was the most opportunistic of purists, as he knew not just how to dream, but exactly what to do to force science to take a step forward.

Thus, as we celebrate his discoveries, we must beware of the kind of hypocrisy which denies the fact that it is still true that those most like him have the hardest time making careers in the modern academy.

It is obvious, but it must be said, that if we could tolerate more like him, science would progress faster. But few academics have the kind of reckless courage he had to put his demand for a clear understanding ahead of all considerations of career success. His story is a lesson for us professors to learn to give preference to those students and colleagues whose critical judgements make us most uncomfortable, and to make room for those loners like Einstein whose passionate independence makes them the least likely colleagues and team players.

Ernst-Ludwig Winnacker

Einstein Stands as a Symbol for:

– the unity of science.

– the power of scientific truth and scientific excellence.

– the fact that science thrives best in glass houses, i. e. under conditions which are as transparent as possible.

– the significance of basic research.

– peace and democracy since Einstein's work and life-long efforts have demonstrated that creativity is not and must not be restricted by gender, nationality, skin color, religion etc.

Abhay Ashtekar
Physicist, Pennsylvania State University

Angelo Baracca
Historian of physics, Università di Firenze

Volker Barth
Historian, Paris

Bruno Bertotti
Physicist, Università di Pavia

Hans Bethe
Physicist, Nobel Prize winner, Ithaca, NY
(† 6 March 2005)

Fabio Bevilacqua
Historian of science, Università degli Studi di Pavia

Richard H. Beyler
Historian of science, Portland State University,OR

Charlotte Bigg
Historian of science, Swiss Federal Institute of
Technology, Zurich

Katja Bödeker
Psychologist, Max Planck Institute for
the History of Science, Berlin

Gerhard Börner
Physicist, Max Planck Institute for Astrophysics,
Garching

Stefano Bordoni
Historian of science, Rimini

Reiner Braun
Journalist, Berlin

Horst Bredekamp
Art historian, Humboldt University, Berlin

Erika Britzke
Einstein researcher, Caputh / Wilhelmshorst

Diana Kormos-Buchwald
Historian, Director of the *Einstein Papers Project*,
Pasadena CA

Jochen Büttner
Historian of science, Max Planck Institute for
the History of Science, Berlin

Lea Cardinali
Historian of science, Università di Pavia

David Cassidy
Historian of science, Hofstra University, Hempstead NY

Jordi Cat
Historian of science, Indiana University, Bloomington IN

Olivier Darrigol
Historian of science, Centre National de la
Recherche Scientifique, Paris

Christian Dirks
Historian, New Synagogue Berlin – Centrum Judaicum
Foundation

Tevian Dray
Mathematician, Oregon State University, Corvallis OR

Michael Eckert
Historian of science, Deutsches Museum, Munich

Jürgen Ehlers
Physicist, former Director of the Max Planck Institute
for Gravitational Physics, Golm – Potsdam

Yehuda Elkana
Historian of science, President and Rector of the
Central European University, Budapest

Lidia Falomo
Historian of science, Università degli Studi di Pavia

Tibor Frank
Historian, Eötvös Loránd University, Budapest

Lucio Fregonese
Historian of science, Università di Pavia

Peter Galison
Historian of science, Harvard University, Cambridge MA

Carla Garbarino
Historian of science, Università degli Studi di Pavia

Clayton Gearhart
Physicist, St. John's University, Collegeville MN

Enrico Antonio Giannetto
Historian of science, Università degli Studi di Pavia

Domenico Giulini
Physicist, University of Freiburg

Hubert Goenner
Physicist, University Gottingen

Karl Wolfgang Graff
Engineer and historian of science, Ludwigsburg

Hanoch Gutfreund
Physicist, former President of the
Hebrew University, Jerusalem

Gerhard Hartl
Historian of astronomy, Deutsches Museum, Munich

Günther Hasinger
Physicist, Max Planck Institute for Extraterrestrial
Physics, Garching

Hans-Olaf Henkel
President of the Leibniz Association, Berlin

Dieter B. Herrmann
Historian of science, Archenhold Planetarium, Berlin

Roger Highfield
Science journalist, *The Daily Telegraph*, London

Dieter Hoffmann
Historian of science, Max Planck Institute for the
History of Science, Berlin

Gerald Holton
Historian of science, Harvard University, Cambridge MA

Danian Hu
Historian, The City College of New York

Michel Janssen
Historian of science, previous co-editor of the *Collected
Papers of Albert Einstein*, University of Minnesota,
Minneapolis MN

Axel Jessner
Physicist, Max Planck Institute for Radioastronomy,
Radio-Observatory Effelsberg, Bad Münstereifel

Thomas Jung
Radiobiologist, Federal Office for Radiation Protection,
Oberschleissheim

David Kaiser
Historian of science, Massachusetts Institute
of Technology, Cambridge MA

Horst Kant
Historian of science, Max Planck Institute for
the History of Science, Berlin

Shaul Katzir
Physicist, Bar-Ilan University, Ramat Gan, Israel

Stefan Kaufmann
Biologist, Max Planck Institute for Infection Biology,
Berlin

Daniel Kennefick
Physicist, California Institute of Technology,
Pasadena CA

Andreas Kleinert
Historian of science, Martin-Luther-University,
Halle-Wittenberg

Eberhard Knobloch
Historian of science, Technical University, Berlin

Anne J. Kox
Historian of physics, University of Amsterdam

Helge Kragh
Historian of science, University of Arhus

Wilhelm Krull
Secretary General of the Volkswagen Foundation,
Hannover

Wolf-Dieter Mechler
Historian, Historisches Museum am Hohen Ufer,
Hannover

Karl von Meyenn
Historian of science, Neuburg a.D.

Adolf Muschg
Author, President of the Akademie der Künste, Berlin

Falk Müller
Historian of science, Johann-Wolfgang-Goethe
University, Frankfurt am Main

Susan Neiman
Philosopher, Director of the Einstein Forum, Potsdam

Thomas de Padova
Physicist and science journalist, *Der Tagesspiegel*,
Berlin

Barbara Picht
Historian, Berlin

Ulf von Rauchhaupt
Physicist and science journalist, *Frankfurter Allgemeine
Sonntagszeitung*, Frankfurt am Main

Jürgen Renn
Historian of science, Director at the Max Planck
Institute for the History of Science, Berlin

Ze'ev Rosenkranz
Historian, *Einstein Papers Project*, California Institute
of Technology, Pasadena CA

Joseph Rotblat
Physicist and winner of the Nobel Peace prize, London

Don Salisbury
Physicist, Austin College, Sherman TX

Tilman Sauer
Historian of science, *Einstein Papers Project*,
California Institute of Technology, Pasadena CA

Britta Scheideler
Historian, University of Osnabrück

Matthias Schemmel
Historian of science, Max Planck Institute for the
History of Science, Berlin

Gregor Schiemann
Philosopher of science, University of Wuppertal,
Wuppertal

Erhard Scholz
Historian of science, University of Wuppertal,
Wuppertal

Gerhard Schröder
Federal Chancellor, Berlin

Volkmar Schüller
Historian of science, Max Planck Institute for the
History of Science, Berlin

Robert Schulmann
Historian of science, Bethesda MD

Michael Schüring
Historian of science, Max Planck Institute for the
History of Science, Berlin

Bernard Schutz
Physicist, Director at the Max Planck Institute for
Gravitational Physics, Golm – Potsdam

Silvan Schweber
Physicist and historian of science, Brandeis University,
Waltham MA

Christian Sichau
Historian of science, Deutsches Museum, Munich

Stefan Siemer
Historian, Deutsches Museum, Munich

Circe Mary Silva da Silva
Historian of science, Universitário de Goiabeiras,
Vitória-ES, Brasilien

Lee Smolin
Physicist, Perimeter Institute for Theoretical Physics,
Waterloo Ontario, Canada

John Stachel
Physicist, Boston University MA

Kenji Sugimoto
Historian of science, Kinki University Higashi – Osaka,
Japan

Alfredo Tiomno Tolmasquim
Historian of science, Director of the Museu de
Astronomia e Ciencias Afins, Rio de Janeiro

Wolfgang Trageser
Archivist, Archive of the Johann-Wolfgang-Goethe
University, Frankfurt am Main

Henning Vierck
Historian of science, Comenius-Garten, Berlin

C.V. Vishveshwara
Historian of science, Jawaharlal Nehru Planetarium,
Bangalore, Indien

Klaus A. Vogel
Historian of science and Captain,
Eddigehausen / On Sea

Renate Wahsner
Philosopher, Max Planck Institute for the
History of Science, Berlin

Mark Walker
Historian, Union College, Schenectady NY

Scott Walter
Historian of science, Université Nancy

Milena Wazeck
Historian of science, Max Planck Institute for the
History of Science, Berlin

Hans Wilderotter
Art historian, University of Applied Studies, Berlin

Ernst-Ludwig Winnacker
Biochemist, President of the Deutsche Forschungs-
gemeinschaft, Bonn

Erdmut Wizisla
Germanist, Head of the Bertolt-Brecht-Archive, Berlin

Barbara Wolff
Archivist, Albert Einstein Archives,
Hebrew University of Jerusalem

Gereon Wolters
Philosopher, University of Constance

Jörg Zaun,
Physicist and historian of science, Berlin

Anton Zeilinger
Physicist, University of Vienna

Einstein und die Folgen. Heidelberg: Spektrum der Wissenschaft Verlagsgesellschaft, 2005.

Innovative Structures in Basic Research. Ringberg Symposium October 2000. München: Max-Planck-Gesellschaft, 2002.

Aczel, Amir D.: *God's Equation. Einstein, Relativity, and the Expanding Universe.* New York: Four Walls Eight Windows, 1999.

Aczel, Amir D.: *The Riddle of the Compass. The Invention that Changed the World.* New York et al.: Hartcourt, 2001.

Albrecht, Ulrich, Ulrike Beisiegel, Reiner Braun and Werner Buckel (Eds.): *Der Griff nach dem atomaren Feuer.* Frankfurt am Main et al.: Lang, 1996.

Almeida, Theophilo Nolasco: *Einstein vs. Michelson.* Rio de Janeiro, 1930.

Ashtekar, Abhay, Robert S. Cohen, Don Howard, Jürgen Renn, Sahotra Sarkar and Abner Shimony (Eds.): *Revisiting the Foundations of Relativistic Physics - Festschrift in Honour of John Stachel.* Dordrecht: Kluwer, 2003.

Baracca, Angelo, Stefano Ruffo and Arturo Russo: *Scienza e industria 1848-1914.* Bari: Laterza, 1979.

Barkan, Diana Kormos: *Walther Nernst and the Transition to Modern Physical Science.* Cambridge: Cambridge University Press, 1999.

Barthes, Roland: *Mythologies.* Paris: Éd. du Seuil, 1970.

Bartusiak, Marcia: *Einstein's Unfinished Symphony. Listening to the Sounds of Space-Time.* Washington: Henry, 2000.

Baumgarth, Christa: *Geschichte des Futurismus.* Reinbek bei Hamburg: Rowohlt, 1966.

Beller, Mara: *Quantum Dialogue. The Making of a Revolution.* Chicago et al.: University of Chicago Press, 1999.

Beller, Mara, Robert S. Cohen and Jürgen Renn (Eds.): *Einstein in Context.* Cambridge: Cambridge University Press, 1993.

Bergia, Silvio: *Einstein. Das neue Weltbild der Physik.* Heidelberg: Spektrum der Wissenschaft Verlagsgesellschaft, 2002.

Bernal, John Desmond: *Science in History (4 Vols.).* Middlesex et al.: Penguin Books, 1969.

Bernal, John Desmond: *The Social Function of Science.* Cambridge et al.: MIT Press, 1967.

Bernstein, Aaron: *Naturwissenschaftliche Volksbücher (21 Vols.).* Berlin: Duncker, 1897.

Bevilacqua, Fabio and Lucio Fregonese (Eds.): *Nuova Voltiana. Studies on Volta and His Times (5 Vols.).* Milan: Università degli studi di Pavia, 2000-2004.

Beyerchen, Alan D.: *Scientists Under Hitler. Politics and the Physics Community in the Third Reich.* New Haven et al.: Yale University Press, 1977.

Bigg, Charlotte: *Behind the Lines. Spectroscopic Enterprises in Early Twentieth Century Europe.* Cambridge: University of Cambridge, 2001.

Börner, Gerhard: *The Early Universe - Facts and Fiction.* Berlin et al.: Springer, 2004.

Bordoni, Stefano: *Una indagine sullo stato dell'etere - Un ragazzo tedesco interroga la fisica di fine Ottocento.* La Goliardica Pavese: Università degli Studi di Pavia, 2005.

Born, Max: *Einstein's Theory of Relativity.* New York: Dover Publications, 1962.

Bosch, Hans-Stephan and Alexander Bradshaw: Kernfusion als Energiequelle der Zukunft. *Physikalische Blätter*, 57 (2001), p. 55-60.

Braun, Reiner and David Krieger: *Einstein - Peace Now! Vision and Ideas.* Weinheim et al.: Wiley-VCH, 2005.

Bredekamp, Horst: *The Lure of Antiquity and the Cult of the Machine. The Kunstkammer and the Evolution of Nature, Art and Technology.* Princeton: Wiener, 1995.

Bredekamp, Horst, Jochen Brüning and Cornelia Weber: *Theater der Natur und Kunst. Wunderkammern des Wissens.* Berlin: Henschel, 2000.

Brian, Denis: *Einstein. A Life.* New York et al.: Wiley, 1996.

Brian, Denis: *The Unexpected Einstein - The Real Man Behind the Icon.* Weinheim: Wiley-VCH, 2005.

Brush, Stephen G.: A History of Random Processes. I. Brownian Movement from Brown to Perrin. *Archive for the History of Exact Sciences,* 5 (1968), p. 1-36.

Brush, Stephen G.: The Kind of Motion We Call Heat. A History of the Kinetic Theory of Gases in the 19th Century. (2 Vols.). Amsterdam: North Holland, 1986.

Brush, Steven G. and C. W. Francis Everitt: Maxwell, Osborne Reynolds and the Radiometer. *Historical Studies in the Physical Sciences,* 1 (1969), p. 105-125.

Bucky, Peter R.: *The Private Albert Einstein.* Kansas City: Andrews and McMeel, 1992.

Bührke, Thomas: *Albert Einstein.* München: Deutscher Taschenbuchverlag, 2004.

Büttner, Jochen, Jürgen Renn and Matthias Schemmel: Exploring the Limits of Classical Physics. Planck, Einstein, and the Structure of a Scientific Revolution. *Studies in History and Philosophy of Modern Physics,* 34 (2003), p. 37-59.

Byrne, Patrick: *The Significance of Einstein's Use of History of Science. Dialectica,* 34 (1980), p. 263-276.

Calaprice, Alice (Ed.): *Dear Professor Einstein. Albert Einstein's Letters to and from Children.* New York: Prometheus Books, 2002.

Carroll, Sean: The Cosmological Constant. *Living Reviews in Relativity,* 4 (2001), p. 1–77. http://relativity.livingreviews.org/Articles/lrr-2001-1/index.html.

Cassidy, David C.: *Einstein and Our World.* Revised ed. New York: Humanity Books, 2004.

Cassidy, David C.: *Uncertainty. The Life and Science of Werner Heisenberg.* New York: Freeman, 1992.

Cassirer, Ernst: *Determinismus und Indeterminismus in der modernen Physik. Historische und systematische Studien zum Kausalproblem.* Gesammelte Werke, Bd. 19. Text und Anmerkungen von Claus Rosenkranz. Hamburg: Meiner, 2004.

Castagnetti, Giuseppe, Peter Damerow, Werner Heinrich, Jürgen Renn and Tilman Sauer: *Wissenschaft zwischen Grundlagenkrise und Politik.* Einstein in Berlin. Berlin: Max-Planck-Institut für Bildungsforschung. Forschungsbereich Entwicklung und Sozialisation. Arbeitsstelle Albert Einstein, 1994.

Castagnetti, Giuseppe and Hubert Goenner: *Einstein and the Kaiser Wilhelm Institute for Physics (1917–1922). Institutional Aims and Scientific Results.* Preprint 261. Berlin: Max Planck Institute for the History of Science, 2004.

Clark, Ronald W.: *Einstein. The Life and Times.* London: Sceptre, 1996.

Corry, Leo et al.: Belated Decision in the Hilbert-Einstein Priority Dispute. *Science,* 278 (1997) 5341, p. 1270–1273.

Corry, Leo: *David Hilbert and the Axiomatization of Physics (1898–1918).* Dordrecht: Kluwer, 2004.

Costa, Manuel Amoroso: *Introduçao à teoria da relatividade.* 2. ed. Rio de Janeiro: Ed. UFRJ, 1995.

Damerow, Peter, Gideon Freudenthal, Peter McLaughlin and Jürgen Renn: *Exploring the Limits of Preclassical Mechanics.* 2. ed. New York: Springer, 2004.

Darrigol, Olivier: *Electrodynamics from Ampère to Einstein.* Oxford: Oxford University Press, 2000.

Dijksterhuis, Eduard J.: *The Mechanization of the World Picture. Pythagoras to Newton.* Princeton, NJ: Princeton Univ. Press, 1986.

Duhem, Pierre Maurice Marie: *Medieval Cosmology. Theories of Infinity, Place, Time, Void and the Plurality of Worlds.* Chicago: University of Chicago Press, 1985.

Eckert, Michael: *Die Atomphysiker. Eine Geschichte der theoretischen Physik am Beispiel der Sommerfeldschule.* Braunschweig: Vieweg, 1993.

Eckert, Michael and Karl Märker (Eds.): *Arnold Sommerfeld. Wissenschaftlicher Briefwechsel.* Berlin et al.: Verlag für Geschichte der Naturwissenschaften und Technik, 2000 and 2004.

Ehlers, Jürgen and Gerhard Börner: *Gravitation.* Heidelberg et al.: Spektrum, 1996.

Einstein, Albert: *Akademie-Vorträge. Die Wiederabdrucke der Akademie-Vorträge Albert Einsteins.* Berlin: Akademie Verlag, 1979.

Einstein, Albert: *Albert Einstein, the Human Side. New Glimpses from His Archives.* Ed. by Helen Dukas and Banesh Hoffmann. Princeton: Princeton University Press, 1979.

Einstein, Albert: *Albert Einsteins Relativitätstheorie. Die grundlegenden Arbeiten zur Relativitätstheorie.* Ed. by Karl von Meyenn. Braunschweig: Vieweg, 1990.

Einstein, Albert: *The Collected Papers of Albert Einstein,* Vol. 1–9. Princeton: Princeton University Press, 1987–2004.

Einstein, Albert: *Einstein on Peace.* Ed. by Otto Nathan and Heinz Norden. New York: Avenel Books, 1981.

Einstein, Albert: *Einstein's Annalen Papers. The Complete Collection 1901–1922.* Ed. by Jürgen Renn. Berlin: Wiley-VCH, 2005.

Einstein, Albert: *Einstein's Miraculous Year. Five Papers that Changed the Face of Physics.* Ed. by John Stachel. Princeton et al.: Princeton University Press, 2005.

Einstein, Albert: *Letters to Solovine.* New York: Philosophical Library, 1987.

Einstein, Albert: *The Meaning of Relativity.* Expanded ed. Princeton: Princeton University Press, 2005.

Einstein, Albert: *Out of my Later Years. The Scientist, Philosopher and Man Portrayed Through his Own Words.* New York et al.: Wings Books, 1993.

Einstein, Albert: *The Quotable Einstein.* Ed. by Alice Calaprice. Princeton: Princeton University Press, 1996.

Einstein, Albert: *Relativity. The Special and the General Theory; a Popular Exposition.* London et al.: Routledge, 2002.

Einstein, Albert: *The World as I see it.* Ed. by Carl Seelig. New York: Citadel Press Book, 1999.

Einstein, Albert: Zur einheitlichen Feldtheorie. *Sitzungsberichte der Preussischen Akademie der Wissenschaften,* (1929), I, p. 2–7.

Einstein, Albert and Michele Besso: Le manuscrit Einstein-Besso. *De la relativité restreinte à la relativité générale.* Paris: Scriptura, 2003.

Einstein, Albert and Max Born: *The Born-Einstein Letters. Friendship, Politics, and Physics in Uncertain Times ; Correspondence Between Albert Einstein and Max and Hedwig Born from 1916 to 1955.* New York: Macmillan, 2005.

Einstein, Albert and Wanders J. de Haas: Experimenteller Nachweis der Ampèreschen Molekularströme. *Verhandlungen der Deutschen Physikalischen Gesellschaft,* 17 (1915), p. 152-170.

Einstein, Albert and Leopold Infeld: *The Evolution of Physics. The Growth of Ideas from Early Concepts to Relativity and Quanta.* Cambridge: Cambridge University Press, 1971.

Einstein, Albert and Mileva Marić: *The Love Letters.* Ed. by Jürgen Renn and Robert Schulmann. Princeton: Princeton University Press, 1992.

Eisenstaedt, Jean: *Einstein et la relativité générale. Les chemins de l'espace-temps.* Paris: CNRS Editions, 2002.

Elkana, Yehuda: *Anthropologie der Erkenntnis. Die Entwicklung des Wissens als episches Theater einer listigen Vernunft.* Frankfurt am Main: Suhrkamp, 1986.

Filk, Thomas and Domenico Giulini: *Am Anfang war die Ewigkeit. Auf der Suche nach dem Ursprung der Zeit.* München: Beck, 2004.

Fischbeck, Hans-Jürgen and Regine Kollek (Eds.): *Fortschritt wohin? Wissenschaft in der Verantwortung – Politik in der Herausforderung.* Münster: Agenda-Verlag, 1994.

Fischer, Ernst Peter: *Aristoteles, Einstein & Co. Eine kleine Geschichte der Wissenschaft in Porträts.* München: Piper, 1996.

Fischer, Ernst Peter and Klaus Wiegandt (Eds.): *Mensch und Kosmos. Unser Bild des Universums.* Frankfurt am Main: Fischer Taschenbuch Verlag, 2004.

Flückinger, Max: *Albert Einstein in Bern. Das Ringen um ein neues Weltbild ; eine dokumentarische Darstellung über den Aufstieg eines Genies.* Bern: Haupt, 1974.

Fölsing, Albrecht: *Albert Einstein. A Biography.* New York: Viking, 1997.

Frank, Tibor: *Ever Ready to Go. The Multiple Exiles of Leo Szilard.* Preprint 275. Berlin: Max Planck Institute for the History of Science, 2004.

Freudenthal, Gideon: *Atom and Individual in the Age of Newton. On the Genesis of the Mechanistic World View.* Dordrecht et al.: Reidel, 1986.

Fritzsch, Harald: *An Equation that Changed the World. Newton, Einstein, and the Theory of Relativity.* Chicago et al.: University of Chicago Press, 1994.

Galison, Peter: *Einstein's Clocks and Poincaré's Maps.* New York: Norton, 2003.

Galison, Peter: *How Experiments End.* Chicago et al. University of Chicago Press, 1987.

Galison, Peter: *Image and Logic. The Material Culture of Microphysics.* Chicago: University of Chicago Press, 1997.

Galluzzi, Paolo: *Renaissance Engineers from Brunelleschi to Leonardo da Vinci.* Florence: Giunti, 1996.

Gearhart, Clayton A.: Planck, the Quantum, and the Historians. *Physics in Perspective,* 4 (2002), p. 170–215.

Giedion, Sigfried: *Space, Time, Architecture. The Growth of a New Tradition.* Cambridge: Harvard University Press, 1982.

Gillispie, Charles Coulston (Ed.): *Dictionary of Scientific Biography (18 Vols.).* New York: Scribner, 1981.

Giulini, Domenico: Das Problem der Trägheit. *Philosophia Naturalis,* 39 (2002), p. 343–374.

Giulini, Domenico: *Special Relativity. A First Encounter, 100 Years since Einstein.* New York: Oxford University Press, 2005.

Goenner, Hubert: *Einführung in die Kosmologie.* Heidelberg et al.: Spektrum Akademischer Verlag, 1994.

Goenner, Hubert: *Einstein in Berlin. 1914–1933.* München: Beck, 2005.

Goenner, Hubert: *Einsteins Relativitätstheorien. Raum, Zeit, Masse, Gravitation.* 4. ed. München: Beck, 2005.

Graham, Loren R.: The Reception of Einstein's Ideas. Two Examples from Contrasting Political Cultures. In: Holton, Gerald and Yehuda Elkana (Eds.): *Albert Einstein. Historical and Cultural Perspectives.* Princeton: Princeton University Press, 1982, p. 107–136.

Greene, Brian: *The Fabric of the Cosmos. Space, Time, and the Texture of Reality.* New York: Knopf, 2004.

Greither, Aloys and Armin Zweite (Eds.): *Josef Scharl 1896–1954.* München: Prestel, 1982.

Gruber, Howard and Katja Bödeker (Eds.): *Creativity, Psychology, and the History of Science.* Dordrecht: Springer, 2005.

Grundmann, Siegfried: *Einsteins Akte. Wissenschaft und Politik – Einsteins Berliner Zeit.* 2. ed. Berlin et al.: Springer, 2004.

Grüning, Michael (Ed.): *Ein Haus für Albert Einstein. Erinnerungen, Briefe, Dokumente.* Berlin: Verlag der Nation, 1990.

Guth, Alan H.: *The Inflationary Universe. The Quest for a New Theory of Cosmic Origins.* Reading, Mass. et al.: Addison-Wesley Publ., 1997.

Habermas, Jürgen: *The Future of Human Nature.* Cambridge: Polity, 2003.

Habermas, Jürgen: *Zeitdiagnosen. Zwölf Essays 1980–2001.* Frankfurt am Main: Suhrkamp, 2003.

Hartl, Gerhard: *Planeten, Sterne, Weltinseln. Astronomie im Deutschen Museum.* München: Deutsches Museum, 1993.

Hasinger, Günther: *The X-Ray Background. Echo of Black Hole Formation?* Preprint series AIP 99/01. Potsdam: Astrophysikalisches Institut, 1999.

Hawking, Stephen: *A Brief History of Time.* New York: Bantam Books, 2005.

Heilbron, John L.: *The Dilemmas of an Upright Man. Max Planck as Spokesman for German Science*. Berkeley et al.: University of California Press, 1986.

Heisenberg, Werner: *Der Teil und das Ganze. Gespräche im Umkreis der Atomphysik*. München: Piper, 1969.

Held, Carsten: *Die Bohr-Einstein-Debatte. Quantenmechanik und physikalische Wirklichkeit*. Paderborn et al.: Schöningh, 1998.

Henderson, Linda Dalrymple: *The Fourth Dimension and Non-Euclidean Geometry in Modern Art*. Princeton: Princeton University Press, 1983.

Hentschel, Klaus: *The Einstein Tower. An Intertexture of Dynamic Construction, Relativity Theory, and Astronomy*. Stanford: Stanford University Press, 1997.

Hentschel, Klaus: *Interpretationen und Fehlinterpretationen der speziellen und allgemeinen Relativitätstheorie durch Zeitgenossen Albert Einsteins*. Basel et al.: Birkhäuser, 1990.

Hermann, Armin: *Einstein. Der Weltweise und sein Jahrhundert; eine Biographie*. München: Piper, 2004.

Herrmann, Dieter B.: *Astronomiegeschichte. Ausgewählte Beiträge zur Entwicklung der Himmelskunde*. Berlin et al.: Paetec, Verl. für Bildungsmedien, 2004.

Hertz, Heinrich: *Die Constitution der Materie*. Ed. by Albrecht Fölsing. Berlin et al.: Springer, 1999.

Highfield, Roger and Paul Carter: *The Private Lives of Albert Einstein*. New York: St. Martin's Press, 1994.

Hoffmann, Banesh: *Relativity and its Roots*. Mineola, NY: Dover Publications, 1999.

Hoffmann, Dieter: *Einsteins Berlin – Auf den Spuren eines Genies*. Weinheim: Wiley-VCH, 2005.

Hoffmann, Dieter, Hubert Laitko and Staffan Müller-Wille (Eds.): *Lexikon der bedeutenden Naturwissenschaftler (3 Vols.)*. Heidelberg et al.: Spektrum Akademischer Verlag, 2003–2004.

Hoffmann, Dieter and Robert Schulmann: *Albert Einstein (1879–1955)*. Teetz: Hentrich & Hentrich, 2005.

Holton, Gerald: *Einstein, History, and Other Passions*. Woodbury, NY: American Institute of Physics, 1995.

Holton, Gerald: *Science and Anti-Science*. Cambridge et al.: Harvard University Press, 1993.

Holton, Gerald: *Thematic Origins of Scientific Thought. Kepler to Einstein*. Cambridge: Harvard University Press, 1988.

Holton, Gerald and Yehuda Elkana (Eds.): *Albert Einstein. Historical and Cultural Perspectives*. Princeton: Princeton University Press, 1982.

Howard, Don and John Stachel (Eds.): *Einstein Studies (11 Vols.)*. Basel et al.: Birkhäuser, 2000.

Howard, Don and John Stachel (Eds.): *Einstein. The Formative Years, 1879–1909*. Basel et al.: Birkhäuser, 2000.

Hu, Danian: *China and Albert Einstein. The Reception of the Physicist and His Theory in China, 1917–1979*. Cambridge: Harvard University Press, 2005.

Humboldt, Alexander von: *Kosmos. Entwurf einer physischen Weltbeschreibung*. Ed. by Ottmar Ette and Oliver Lubrich. Frankfurt am Main: Eichborn, 2004.

Infeld, Eryk (Ed.): *Leopold Infeld. His Life and Scientific Work*. Warszawa: Polish Scientific Publishers, 1978.

Infeld, Leopold: *Leben mit Einstein. Kontur einer Erinnerung*. Wien et al.: Europa-Verl., 1969.

Israel, Hans, Erich Ruckhaber and Rudolf Weinmann (Eds.): *100 Autoren gegen Einstein*. Leipzig: Voigtländer, 1931.

Janssen, Michel: *A Comparison Between Lorentz's Ether Theory and Special Relativity in the Light of the Experiments of Trouton and Noble*. Ann Arbor: UMI, 1995.

Janssen, Michel and Christoph Lehner (Eds.): *Cambridge Companion to Einstein*. New York: Cambridge University Press, in press.

Janssen, Michel and Jürgen Renn: *Untying the Knot. How Einstein Found His Way Back to Field Equations Discarded in the Zurich Notebook*. Preprint 264. Berlin: Max Planck Institute for the History of Science, 2004.

Jerome, Fred: *The Einstein File. J. Edgar Hoover's Secret War Against the World's Most Famous Scientist*. New York: St. Martin's Press, 2002.

Jorda, Stefan (Ed.): *Albert Einstein und das Wunderjahr 1905. Sonderausgabe von Physik Journal 4 (2005) 3*. Weinheim: Wiley-VCH, 2005.

Jungk, Robert: *The Big Machine*. London: Deutsch, 1969.

Jungnickel, Christa and Russell MacCormmach: *Intellectual Mastery of Nature. Theoretical Physics from Ohm to Einstein (2 Vols.)*. Chicago: University of Chicago Press, 1986.

Kant, Horst: *Werner Heisenberg and the German Uranium Project. Otto Hahn and the Declarations of Mainau and Göttingen*. Preprint 203. Berlin: Max Planck Institute for the History of Science, 2002.

Karlsch, Rainer: *Hitlers Bombe. Die geheime Geschichte der deutschen Kernwaffenversuche*. München: Deutsche Verlags-Anstalt, 2005.

Katzir, Shaul: *A History of Piezoelectricity. The First Two Decades*. Tel Aviv: Tel Aviv University, 2001.

Kirsten, Christa and Hans-Jürgen Treder (Eds.): *Albert Einstein in Berlin 1913-1933 (2 Vols.)*. Berlin: Akademie Verlag, 1979.

Klafter, Joseph et al.: Beyond Brownian Motion. *Physics Today*, 49 (1996), p. 33-39.

Klein, Ursula: *Verbindung und Affinität. Die Grundlegung der neuzeitlichen Chemie an der Wende vom 17. zum 18. Jahrhundert*. Basel et al.: Birkhäuser, 1994.

Kleinert, Andreas: Nationalistische und antisemitische Ressentiments von Wissenschaftlern gegen Einstein. In: Nelkowski, Horst et al. (Eds.): *Einstein Symposion Berlin, aus Anlass der 100. Wiederkehr seines Geburtstages 25. bis 30. März 1979*. Berlin et al.: Springer, 1979, p. 501-516.

Kocka, Jürgen (Ed.): *Die Königlich Preußische Akademie der Wissenschaften zu Berlin im Kaiserreich*. Berlin: Akademie Verlag, 1999.

Kragh, Helge: *Cosmology and Controversy. The Historical Development of Two Theories of the Universe*. Princeton: Princeton University Press, 1996.

Kuhn, Thomas S.: *The Copernican Revolution. Planetary Astronomy in the Development of Western Thought*. Cambridge et al.: Harvard University Press, 2002.

Kuhn, Thomas S.: *The Structure of Scientific Revolutions*. Chicago et al.: University of Chicago Press, 2004.

Laporte, Paul M.: Cubism and Relativity. With a Letter from Einstein. *Art Journal,* 25 (1966), p. 246-249.

Lefèvre, Wolfgang, Jürgen Renn and Urs Schoepflin (Eds.): *The Power of Images in Early Modern Science*. Basel et al.: Birkhäuser, 2003.

Levenson, Thomas: *Einstein in Berlin*. New York et al.: Bantam Books, 2003.

Lissitzky, El: K. und Pangeometrie. In: Einstein, Carl and Paul Westheim (Eds.): *Europa-Almanach*. Potsdam 1925, p. 103-113.

Lissitzky-Küppers, Sophie (Ed.): *El Lissitzky. Life, Letters, Texts*. London et al.: Thames and Hudson, 1992.

Lorentz, Dominique: *Affaires atomiques*. Paris: Les Arènes, 2001.

Lorimer, Duncan and Michael Kramer: *Handbook of Pulsar Astronomy*. New York: Cambridge University Press, 2005.

Lux-Steiner et al.: Strom von der Sonne. *Physikalische Blätter*, 57 (2001), p. 47-53.

Lyotard, Jean-François: *Die Logik, die wir brauchen. Nietzsche und die Sophisten*. Ed. by Patrick Baum and Günter Seubold. Bonn: DenkMal-Verl., 2004.

Mach, Ernst: *Knowledge and Error. Sketches on the Psychology of Enquiry*. Translation from the 5th edition, 1926. Dordrecht: Reidel, 1976.

Mach, Ernst: *Die Mechanik in ihrer Entwicklung. Historisch-kritisch dargestellt*. Ed. by Renate Wahsner and Horst-Heino von Borzeskowski. Berlin: Akademie Verlag, 1988.

Maeterlinck, Maurice: *Geheimnisse des Weltalls*. Berlin et al.: Deutsche Verlags-Anstalt, 1930.

Markl, Hubert: *Schöner neuer Mensch?* München et al.: Piper, 2002.

Miller, Arthur I.: *Albert Einstein's Special Theory of Relativity. Emergence (1905) and Early Interpretation (1905-1911)*. New York et al.: Springer, 1998.

Miller, Arthur I.: *Einstein, Picasso. Space, Time, and the Beauty that Causes Havoc*. New York: Basic Books, 2002.

Misner, Charles W., Kip S. Thorne and John A. Wheeler: *Gravitation*. New York: Freeman, 1973.

Mittelstraß, Jürgen (Ed.): *Enzyklopädie Philosophie und Wissenschaftstheorie (4 Vols.)*. Stuttgart et al.: Metzler, 2004.

Montesinos, José and Carlos Solís (Eds.): *Largo Campo di Filosofare. Eurosymposium Galileo 2001*. La Orotava: Fundación Canaria Orotava de Historia de la Ciencia, 2001.

Moreira, Ildeu C. and Antonio A. Videira (Eds.): *Einstein e o Brasil*. Rio de Janeiro: Ed. UFRJ, 1995.

Müller, Falk: *Gasentladungsforschung im 19. Jahrhundert*. Berlin et al.: Verlag für Geschichte der Naturwissenschaften und der Technik, 2004.

Müller, Ulrich: *Raum, Bewegung und Zeit im Werk von Walter Gropius und Ludwig Mies van der Rohe*. Berlin: Akademie Verlag, 2004.

Neffe, Jürgen: *Einstein. Eine Biographie*. Reinbek bei Hamburg: Rowohlt, 2005.

Neumann, Thomas (Ed.): *Albert Einstein*. Berlin: Elefanten Press, 1989.

Newton, Isaac: *Die mathematischen Prinzipien der Physik*. Ed. by Volkmar Schüller. Berlin: de Gruyter, 1999.

Nickel, Jens Uwe: Mikrolaser als Photonen-Billards. Wie Chaos ans Licht kommt. *Physikalische Blätter,* 54 (1998), p. 927-930.

Nietzsche, Friedrich: *Vom Nutzen und Nachteil der Historie für das Leben*. Ed. by Michael Landmann. Zürich: Diogenes, 1984.

North, John David: *The Measure of the Universe. A History of Modern Cosmology*. New York: Dover Publications, 1990.

Nye, Mary Jo: *Molecular Reality. A Perspective on the Scientific Work of Jean Perrin*. London et al.: Macdonald, 1972.

Olschki, Leonardo: *Geschichte der neusprachlichen wissenschaftlichen Literatur (3 Vols.)*. Vaduz: Kraus Reprint, 1965.

Overbye, Dennis: *Einstein in Love. A Scientific Romance.*
New York: Viking Penguin, 2000.

Overduin, James Martin and P. S. Wesson: Dark Matter and Background Light. *Physics Reports,* 402 (2004), p. 267–406.

Padova, Thomas de: *Am Anfang war kein Mond. 40 Science-Stories, wie unser Sonnensystem entstand und das Leben auf die Erde kam.* Stuttgart: Klett-Cotta, 2004.

Pais, Abraham: *"Subtle is the Lord …". The Science and the Life of Albert Einstein.* Oxford et al.: Oxford University Press, 1983.

Pauli, Wolfgang: *Theory of Relativity.* New York et al.: Dover, 1981.

Penrose, Roger: *The Emperors New Mind. Concerning Computers, Minds, and the Laws of Physics.* New York: Penguin Books, 1991.

Perlick, Volker: Gravitational Lensing from a Spacetime Perspective. *Living Reviews in Relativity,* 7 (2004), p. 9. http://www.livingreviews.org/lrr-2004-9.

Perlmutter, Saul, Greg Aldering and G. Goldhaber et al.: Measurement of W and L from 42 High-Redshift Supernovae. *Astrophysical Journal,* 517 (1999), p. 565–586.

Piaget, Jean: *Le développement de la notion de temps chez l'enfant.* 3. éd. Paris: Presses Universitaires de France, 1982.

Piaget, Jean: *Introduction à l'épistémologie génétique (2 Vols.).* 2. éd. Paris: Presses Universitaires de France, 1973–1974.

Pössel, Markus: *Das Einstein-Fenster. Eine Reise in die Raumzeit.* Hamburg: Hoffmann und Campe, 2005.

Randow, Gero von (Ed.): *Jetzt kommt die Wissenschaft. Von Wahrheiten, Irrtümern und kuriosen Erfindungen.* Frankfurt am Main: FAZ-Institut für Management-, Markt- und Medieninformationen, 2003.

Renn, Jürgen: *Auf den Schultern von Riesen und Zwergen. Einsteins unvollendete Revolution.* Berlin: Wiley-VCH, 2005.

Renn, Jürgen (Ed.): *Galileo in Context.* Cambridge: Cambridge University Press, 2001.

Renn, Jürgen (Ed.): *Genesis of General Relativity (4 Vols.).* Dordrecht: Springer, in press.

Renn, Jürgen and Tilman Sauer: Eclipses of the Stars. Mandl, Einstein, and the Early History of Gravitational Lensing. In: Ashtekar, Abhay et al. (Eds.): *Revisiting the Foundations of Relativistic Physics.* Dordrecht: Kluwer, 2003, p. 69–92.

Renn, Jürgen and Tilman Sauer: *Einsteins Züricher Notizbuch. Die Entdeckung der Feldgleichungen der Gravitation im Jahre 1912.* Preprint 28. Berlin: Max Planck Institute for the History of Science, 1995.

Renn, Jürgen and John Stachel: *Hilbert's Foundation of Physics. From a Theory of Everything to a Constituent of General Relativity.* Preprint 118. Berlin: Max Planck Institute for the History of Science, 1999.

Renn, Jürgen and Henning Vierck: *Künstler, Wisssenschaftler, Kinder und das Nichts.* Ein Werkstattbericht. Berlin: Comenius Garten, 2004.

Rhodes, Richard: *The Making of the Atomic Bomb.* London et al.: Penguin Books, 1988.

Rosenberger, Ferdinand: *Die Geschichte der Physik in Grundzügen (3 Vols.).* Braunschweig: Vieweg, 1882–1890.

Rosenkranz, Ze'ev: *The Einstein Scrapbook.* Baltimore: John Hopkins University Press, 2002.

Salmon, Merrilee H., John Earman, Clark Glymour, James G. Lennox, Peter Machamer, J. A. MacGuire, John D. Norton, Wesley C. Salmon and Kenneth F. Schaffner (Eds.): *Introduction to the Philosophy of Science.* Indianapolis et al.: Hackett, 1999.

Sayen, Jamie: *Einstein in America. The Scientists' Conscience in the Age of Hitler and Hiroshima.* New York: Crown Publications, 1985.

Scheideler, Britta: The Scientist as Moral Authority. Albert Einstein Between Elitism and Democracy, 1914–1933. *Historical Studies in the Physical and Biological Sciences,* 32 (2002), p. 319–346.

Schemmel, Matthias: An Astronomical Road to General Relativity. The Continuity Between Classical and Relativistic Cosmology in the Work of Karl Schwarzschild. In: Renn, Jürgen (Ed.): *The Genesis of General Relativity.* Dordrecht: Springer, in press.

Schiemann, Gregor: *Wahrheitsgewissheitsverlust. Hermann von Helmholtz' Mechanismus im Aufbruch der Moderne.* Darmstadt: Wissenschaftliche Buchgesellschaft, 1979.

Schilpp, Paul Arthur (Ed.): *Albert Einstein. Philosopher – Scientist.* New York: Tudorg, 1970.

Scholz, Erhard: *Herman Weyls Raum-Zeit-Materie and a General Introduction to his Scientific Work.* Basel et al.: Birkhäuser, 2000.

Scholz, Erhard: *An Outline of Weyl Geometric Models in Cosmology.* Preprint. Wuppertal, 2004. http://arXiv.org/astro-ph/0403446.

Schönbeck, Charlotte: *Albert Einstein und Philipp Lenard. Antipoden im Spannungsfeld von Physik und Zeitgeschichte.* Berlin et al.: Springer, 2000.

Schüring, Michael: Der Vorgänger. Carl Neubergs Verhältnis zu Adolf Butenandt. In: Schieder, Wolfgang and Achim Trunk (Eds.): *Adolf Butenandt und die Kaiser-Wilhelm-Gesellschaft. Wissenschaft, Industrie und Politik im Dritten Reich.* Göttingen: Wallstein, 2004, p. 369–403.

Schüring, Michael: Ein Dilemma der Kontinuität. Das Selbst-verständnis der Max-Planck-Gesellschaft und der Umgang mit Emigranten in den 50er Jahren. In: VomBruch, Rüdiger and Brigitte Kaderas (Eds.): *Wissenschaften und Wissenschaftspolitik. Bestandsaufnahmen zu Formationen, Brüchen und Kontinuitäten im Deutschland des 20. Jahrhunderts.* Stuttgart: Steiner, 2002, p. 453–464.

Schutz, Bernard F.: *Gravity from the Ground Up.* Cambridge: Cambridge University Press, 2003.

Schwartz, Joseph and Michael McGuinness: *Einstein for Beginners.* Barton et al.: Icon Books, 1992.

Seelig, Carl: *Albert Einstein. A Documentary Biography.* London: Staples Press, 1956.

Smolin, Lee: *The Life of the Cosmos.* London: Weidenfeld & Nicolson, 1997.

Sommerfeld, Arnold: *Wissenschaftlicher Briefwechsel (2 Vols.).* Ed. by Michael Eckert and Karl Märker. Berlin: Verlag für Geschichte der Naturwissenschaften und Technik, 2000–2004.

Stachel, John: *Einstein from "B" to "Z".* Boston et al.: Birkhäuser, 2002.

Stern, Fritz: *Einstein's German World.* Princeton: Princeton University Press, 1999.

Sugimoto, Kenji (Ed.): *Albert Einstein. A Photographic Biography.* New York: Schocken Books, 1989.

Tauber, Gerald E.: Einstein and Zionism. In: French, Anthony P. (Ed.): *Einstein, a Centenary Volume.* Cambridge: Harvard University Press, 1979, p. 199–207.

Teichmann, Jürgen: *Wandel des Weltbildes. Astronomie, Physik und Meßtechnik in der Kulturgeschichte.* Stuttgart: Teubner, 1999.

Thorne, Kip S.: *Black Holes and Time Warps. Einstein's Outrageous Legacy.* New York et al.: Norton, 1994.

Tolmasquim, Alfredo: *Einstein. O viajante da relatividade na América do Sul.* Rio de Janeiro: Vieira & Lent, 2003.

Toth, Imre: Gott und Geometrie. Eine viktorianische Kontroverse. In: Henrich, Dieter (Ed.): *Evolutionstheorie und ihre Evolution. Vortragsreihe der Universität Regensburg zum 100. Todestag von Charles Darwin.* Regensburg: Mittelbayerische Druckerei- und Verlagsgesellschaft, 1982, p. 141–204.

Tropp, Eduard A., Viktor Frenkel and Artur D. Chernin: *Alexander A. Friedmann. The Man Who Made the Universe Expand.* Cambridge: Cambridge University Press, 1993.

Vierhaus, Rudolf and Bernhard Vom Brocke (Eds.): *Forschung im Spannungsfeld von Politik und Gesellschaft. Geschichte und Struktur der Kaiser-Wilhelm/Max-Planck-Gesellschaft.* Stuttgart: Deutsche Verlags-Anstalt, 1990.

Vogel, Klaus A.: Cosmography. In: Daston, Lorraine and Katherine Park (Eds.): *The Cambridge History of Early Modern Science.* Cambridge: Cambridge University Press, in press.

Vom Brocke, Bernhard and Hubert Laitko (Eds.): *Die Kaiser-Wilhelm-, Max-Planck-Gesellschaft und ihre Institute. Studien zu ihrer Geschichte.* Berlin: de Gruyter, 1996.

Walker, Mark: *German National Socialism and the Quest for Nuclear Power. 1939–1949.* Cambridge et al.: Cambridge University Press, 1989.

Walker, Mark: *Nazi Science. Myth, Truth, and the German Atomic Bomb.* New York: Plenum Press, 1995.

Walther, Herbert: Photonen unter Kontrolle. *Physik Journal,* 4 (2005), p. 45-51.

Wambsganss, Joachim: Gravitational Lensing in Astronomy. *Living Reviews in Relativity,* 1 (1998). http://livingreviews.org/lrr-1998-12.

Wazeck, Milena: Einstein in the Daily Press. A Glimpse into the Gehrcke Papers. In: Eisenstaedt, Jean and A. J. Kox (Eds.): *Universe of General Relativity.* New York: Birkhäuser, 2005, p. 339-357.

Weart, Spencer R.: *Nuclear Fear. A History of Images.* Cambridge: Harvard University Press, 1988.

Weinberg, Steven: *Dreams of a Final Theory.* New York: Vintage Books, 1994.

Wilderotter, Hans (Ed.): *Ein Turm für Albert Einstein. Potsdam, das Licht und die Erforschung des Himmels.* Potsdam: Haus der Brandenburgisch-Preußischen Geschichte, 2005.

Will, Clifford M.: *Was Einstein Right? – Putting General Relativity to the Test.* New York: Basic Books, 1986.

Wohlwill, Emil: *Galilei und sein Kampf für die Copernicanische Lehre (2 Vols.).* Hamburg: Voss, 1909–1926.

Wolters, Gereon: *Mach I, Mach II, Einstein und die Relativitäts-theorie. Eine Fälschung und ihre Folgen.* Berlin et al.: de Gruyter, 1987.

Yam, Philip: Everyday Einstein. *Scientific American,* 9 (2004), p. 50–55.

Zeilinger, Anton: *Einsteins Schleier. Die neue Welt der Quanten-physik.* München: Beck, 2003.

Zuelzer, Wolf: *The Nicolai Case. A Biography.* Detroit: Wayne State University Press, 1982.

http://www.einsteinausstellung.de/
Exhibition "Albert Einstein: Chief Engineer of the
Universe"

http://www.einsteinjahr.de/
Einstein Year 2005

http://www.physics2005.org/
World Year of Physics 2005

http://www.mpiwg-berlin.mpg.de
Max Planck Institute for the History of Science

http://www.aei-potsdam.mpg.de/
Max Planck Institute for Gravitational Physics
(Albert Einstein Institute)

http://www.mpq.mpg.de
Max-Planck-Institute of Quantum Optics

http://www.mpa-garching.mpg.de
Max Planck Institute for Astrophysics

www.mpe.mpg.de/institute.html
Max Planck Institute for Extraterrestrial Physics

http://www.mpia-hd.mpg.de/Public/index_en.html
Max Planck Institute for Astronomy

http://public.web.cern.ch/Public/Welcome.html
CERN

http://www.imss.fi.it/index.html
Institute and Museum of the History of Science
Florence

http://www.deutsches-museum.de/
Deutsches Museum München

http://www.mensch-einstein.de/
Einstein-Portal des Rundfunks Berlin-Brandenburg
RBB

http://www.zdf.de/ZDFde/in-
halt/17/0,1872,2254097,00.html
Einstein-Portal des Zweiten Deutschen Fernsehens

http://www.3sat.de/einstein/
Einstein-Portal von 3sat

http://www.dw-world.de/einsteinjahr
Einstein-Portal der Deutschen Welle
http://www3.bbaw.de/bibliothek/digital/index.html
History of *Königlich Preußische Akademie der
Wissenschaften*, Sources

http://echo.mpiwg-berlin.mpg.de
"European Cultural Heritage Online" (ECHO), Sources

http://archimedes2.mpiwg-berlin.mpg.de/
archimedes_templates
History of Mechanics, Sources

http://www.alberteinstein.info/
"Einstein Archives Online"

http://www.einstein.caltech.edu/
"Collected Papers of Albert Einstein"

http://www.living-einstein.de/
"Living Einstein", Einstein Sources

http://einstein-annalen.mpiwg-berlin.mpg.de/home
"Annalen der Physik": Articles of Albert Einstein

http://www.lrz-muenchen.de/~Sommerfeld/
Arnold Sommerfeld (1868-1951): Scientific
Correspondence

http://www.mpiwg-
berlin.mpg.de/KWG/publications.htm#Ergebnisse
Preprints of the Research program "History of the
Kaiser Wilhelm Society in the National Socialist Era"

http://foia.fbi.gov/foiaindex/einstein.htm
Einstein's FBI Files (Freedom of Information Privacy
Act)

http://www.mpiwg-berlin.mpg.de/PREPRINT.HTM
Max Planck Institute for the History of Science:
Preprints

http://www.livingreviews.org/
"Living Reviews", Open Access for Physical Journals

http://arxiv.org/archive/gr-qc
"General Relativity and Quantum Cosmology e-Prints"

http://www-groups.dcs.st-and.ac.uk/~history/
History of Mathematics Archive

http://www.aip.org/history/
Center for History of Physics

http://ppp.unipv.it/
"Pavia Project Physics"

http://hyperphysics.phy-astr.gsu.edu/hbase/hph.html
General Physical Knowledge

http://www.colorado.edu/physics/2000/index.pl
General Physical Knowledge

http://www.einstein-online.info/
Einstein's Theory of Relativity
http://www.colorado.edu/physics/2000/index.pl
"Physics 2000", Interactive Demonstrations of Physics

http://shatters.net/celestia/
"Celestia", Space Simulations

http://www.desy.de/html/home/
Deutsches Elektronen-Synchroton

http://www.zarm.uni-bremen.de/
ZARM, Center of Applied Space Technology and
Microgravity

http://coolcosmos.ipac.caltech.edu/index.html
Infrared-Astronomy

http://www.nrao.edu/whatisra/index.shtml
Radio Astronomy

http://map.gsfc.nasa.gov/m_uni.html
Cosmology

http://archive.ncsa.uiuc.edu/Cyberia/NumRel/
EinsteinLegacy.html
Einstein's Legacy

http://carina.astro.cf.ac.uk/groups/relativity/
research/part1.html
Gravitational waves

http://cosmology.berkeley.edu/Education/BHfaq.html
Black Holes

http://www.damtp.cam.ac.uk/user/gr/public/
Black Holes, Quantum Gravity

http://einstein.stanford.edu/
Gravity Probe B

http://www.ipp.mpg.de/ippcms/de/pr/fusion21/
Nuclear Fusion

http://www.desy.de/html/forschung/
forschungsueberblick.html#elementar
Elementary Particle Physics

http://www.solarserver.de/wissen/photovoltaik.html
Photovoltaik

http://www.iap.uni-bonn.de/P2K/bec/
Bose Einstein Condensation

http://www.ethikrat.org/
Nationaler Ethikrat

http://www.zim.mpg.de/openaccess-berlin/
berlindeclaration.html
Berlin Declaration on Open Access to Knowledge in
the Sciences and Humanities

http://www.fas.org
Federation of American Scientists (US)

http://www.sgr.org.uk
Scientists for Global Responsibility (UK)

http://www.inesap.org
INESAP (International Network of Engineers and
Scientists Against Proliferation, Darmstadt University
of Technology, Germany), Information Bulletins

Sky and Telescope: 177 (graphic: Joe Bergeron)

Springer Verlag, Heidelberg: 347

Spiegel Verlag, Hamburg: 108

Stadtarchiv München, München: 129, 130 a., 132 l.

Staatsbibliothek zu Berlin, Preußischer Kulturbesitz, Berlin: 77

Stiftung Archiv der Akademie der Künste, Bertolt-Brecht-Archiv, Berlin: 350–353

Stiftung Archiv der Akademie der Künste, Archivabteilung: Baukunst, Konrad-Wachsmann-Archiv, Berlin: 272, 273 l., 274 r. a., 277, 278

SV-Bilderdienst, DIZ, München: 228, 242 l., 243, 383

Taudin, Laurent, illustration: 28, 29

Technische Universität Berlin, Berlin: 113 (Petzoldt-Archive), 132 r.

Tsinghua University, Beijing: 447 r.

ullstein bild, Berlin: 31 b., 35, 189, 190, 192, 253, 349 l. and r., 377, 402, 410, 430, 431, 434, 440–443, 444 l., 447 l.

United Press International, Washington DC: 436

VG Bild-Kunst, Bonn 2005: 256 (Zentrum Paul Klee, Bern), 259 (Städtische Kunsthalle Mannheim)

Weidle-Verlag, Bonn: 16, 322, 331, 338, 348, 426

Zionist Archives, Jerusalem: 335

Private Lenders

Besso, Laurent: 142

Britzke, Erika: 273 r., 275 r., 276, 279

Cardinali, Lea: 142

Fiegel, Susanne: 340 r., 341, 342, 343 r.

Herrmann, Dieter B.: 234–237

Murken, Jan: 18, 20

Sugimoto, Kenji: 285–287, 289

Wörner, Martin: 146, 149 l., 150 l., 151 r.

Zaun, Hans-Otto: 47

Sources

Auerbach, Felix / Rothe, Rudolf (Eds.): *Taschenbuch für Mathematiker und Physiker.* Leipzig/Berlin: Teubner, 1911, vol. 2: 164.

Comenius, Johann Amos: *Orbis sensualium pictus*, Nürnberg: Endter, 1658: 34 r.

Costa, Manoel Amoroso: *Introdução a teoria da relatividade*, Rio de Janeiro: Editoria UFRJ, 1995: 294.

Crookes, William: On Repulsion resulting from radiation, Part III. In: *Philosophical Transactions of the Royal Society of London* 166 (1876): 50, 51.

Crookes, William: On radient matter, part I & II. In: *Nature* (London) 20 (1879): 50.

Einstein, Albert / Infeld, Leopold: *The Evolution of Physics*, Cambridge: University Press, 1938: 56.

Ganot, Adolphe: *Traité élémentaire de physique expérimentale et appliqué et de météorologie*, Paris: l'auteur-éditeur, 1870: 57.

Hellmann, Ullrich: *Künstliche Kälte – die Geschichte der Kühlung im Haushalt*, Gießen: Anabas-Verlag, 1990: 238.

Huygens, Christiaan: De motu corporum (1659). In: *Oeuvres complètes*, vol. 16 (1888–1916): 175.

Koralle: *Magazin für alle Freunde von Natur und Technik*, Berlin: Ullstein, 1926: 193.

Lummer, Otto: *Die Lehre von der strahlenden Energie (Optik)*, Braunschweig: Vieweg, 1909 (Müller-Pouillet's Lehrbuch der Physik und Meteorologie; vol. 2, 3. book): 116, 117.

Maxwell, James Clerk: *The scientific papers of James Clerk Maxwell*, ed. by W.D. Niven. Cambridge: University Press, 1890, Vol. 1: 125 l.

O Estado, Sao Paulo, 14. Mai 1925: S. 297.

O Jornal, Rio de Janeiro, 16. Mai 1925: S. 296.

Perrin, Jean: Mouvement brownien et réalité moléculaire. In: *Annales de chimie et de physique*, ser. 8, vol. 18 (1909): 120.

Pfarr, Bernd: *Komische Kunst*, Zürich: Kein & Aber, 2003: 392.

Poincaré, Henri: *Oeuvres*, Bd. 7: Tome VII, Paris: Gauthier-Villars, 1952: 163.

Sommerfeld, Arnold: Zur Relativitätstheorie, II: Vierdimensionale Vektoranalysis. In: *Annalen der Physik* (Leipzig) 33 (1910): 164.

Stark, Johannes / Kleinert, Andreas (Eds.): *Erinnerungen eines deutschen Naturforschers*. Mannheim: Bionomica-Verlag, 1987: 226.

Straaten, Evert van (Ed.): *Theo van Doesburg: painter and architect*, Den Haag: SDU Publ., 1988: 257.

Vallentin, Antonina: *Le drame d'Albert Einstein*, Paris: Plon, 1954: 247.

Wörner, Martin: *Die Welt an einem Ort: illustrierte Geschichte der Weltausstellungen*, Berlin: Reimer, 2000: 146, 149 l., 150 l., 151 r.